Manufacturing Productivity in China

Manufacturing Productivity in China

Li Zheng

Simin Huang

Zhihai Zhang

CRC Press
Taylor & Francis Group
Boca Raton   London   New York

CRC Press is an imprint of the
Taylor & Francis Group, an **informa** business

CRC Press
Taylor & Francis Group
6000 Broken Sound Parkway NW, Suite 300
Boca Raton, FL 33487-2742

Printed and bound by CPI UK on sustainably sourced paper
Version Date: 20151030

International Standard Book Number-13: 978-1-4665-9542-2 (Hardback)

Visit the Taylor & Francis Web site at
http://www.taylorandfrancis.com

and the CRC Press Web site at
http://www.crcpress.com

Contents

Preface

This book is concerned with Made in China and tries to explore the stories behind successful manufacturing practices in China. In the past 30 years, the value added to manufacturing industries in China has grown from US$123 billion in 1980 to US$29 trillion in 2013. China has been the number one manufacturing country for several years now. Numerous researchers attribute its success to competitiveness in terms of cost (low labor cost and low material cost), but from my observation, that is not the only reason. There have also been tremendous efforts that improved productivity through manufacturing innovation. We, as professors of industrial and manufacturing engineering, have spent 12 years observing hundreds of factories in China, from small private companies to large state-owned companies and joint ventures; some of these factories have been investigated for about 10 years. We tried to sum up the reasons for the success of Made in China, but it is very difficult or too simplified to summarize its success in just a few paragraphs. So, instead, in this book, we came up with a series of typical cases to show the real picture of the Chinese factory. It is our wish that, through *Manufacturing Productivity in China*, readers can better understand Made in China.

The Chinese higher education system pays abundant attention to manufacturing engineering education. In fact, manufacturing engineering is one of the biggest academic disciplines in the country. Industrial engineering (IE) was introduced in China in 1993. To date, there are more than 200 IE programs and 1000 manufacturing engineering programs in the country. Thousands of industrial and manufacturing engineering students graduate from universities each year and move on to work in manufacturing industries; these individuals contribute their wisdom to improve and innovate manufacturing processes and systems in a broad field. Thus, *Manufacturing Productivity in China* should be of great value to all industrial and manufacturing specialists, factory managers, as well as investors and businessmen who are involved in manufacturing industries.

The 12 cases described in this book have been written by the foremost authorities from Tsinghua University and from industry; these cases cover a broad section of the manufacturing industry. Each chapter is heavily tilted toward different manufacturing applications, and a significant number of figures and tables are utilized to facilitate the usability of the material.

We give a special thanks to Professor Salvendy, who is the founding chair of the IE program at Tsinghua University, Beijing, China. He served as the department head for 10 years at Tsinghua University. Without his leadership, we cannot imagine that we would be continuously working on manufacturing improvement and innovation projects and finishing the book after more than 10 years of observation and investigation. We had the privilege of working on this project with a helpful editor, who significantly facilitated our editorial work. We also acknowledge the invaluable contribution of an assistant.

Editors

Li Zheng earned his BS and PhD degrees from Tsinghua University, Beijing, China, in 1986 and 1991, respectively. He is currently a Chang Jiang Scholar, chair professor in the Department of Industrial Engineering, Tsinghua University. He was a visiting professor at the Georgia Institute of Technology from 1994 to 1996. He has authored/coauthored more than 300 publications and 10 book chapters. His research interests include production system analysis and information-driven manufacturing. He has won several important awards, such as the National Science and Technology Progress Award in 2005, the National Invention Award in 1990, and the Excellence Young Faculty Award from the Ministry of Education (MOE) of the People's Republic of China in 2000. Dr. Zheng is a senior member of the Institute of Industrial Engineers (IIE) and the founding chair of the IIE, China chapter.

Simin Huang is a professor and associate head of the Department of Industrial Engineering, Tsinghua University, Beijing, China. He earned his PhD in operations research at the State University of New York at Buffalo, New York, USA. His current research interests include supply chain risk management, production scheduling, and network design. He has served as associate editor of the *IIE Transactions* since 2005 and the *Asia-Pacific Journal of Operational Research* since 2014. He has also served as an editorial board member of the *Industrial Engineering Journal* (Chinese) since 2011. His work has been published in the *Annals of Operation Research, Computer and Industrial Engineering, Computers & Operations Research, European Journal of Operational Research, IEEE Transactions on Engineering Management, IIE Transactions, Interfaces, International Journal of Production Research, Medical Decision Making, Naval Research Logistics, Omega, Transportation Research Part B, Transportation Research Part E,* among others.

 **Zhihai Zhang** is an associate professor and director of the Institute of Engineering Systems in the Industrial Engineering Department, Tsinghua University, Beijing, China. Dr. Zhang was a visiting professor at the Institute of Production Systems at Aachen University, Aachen, Germany, in 2003. He was also a visiting professor in the Department of Industrial Engineering and operations research at the University of California, Berkeley, USA, in 2008. He earned his BS and PhD degrees in mechanical engineering from Tsinghua University in 1997 and 2002, respectively. His current research interests focus on resource allocation optimization, supply chain and logistics management, production planning and scheduling, and large-scale optimization. He has published numerous articles in many journals such as *Computer and Industrial Engineering*, the *International Journal of Advanced Manufacturing Technology*, and the *Asia-Pacific Journal of Operational Research*. He has also participated in many government and industrial projects.

Contributors

Linning Cai
Department of Industrial
    Engineering
Tsinghua University
Beijing, China

Ye Cheng
Department of Industrial
    Engineering
Tsinghua University
Beijing, China

Qin Gao
Department of Industrial
    Engineering
Tsinghua University
Beijing, China

Xiguang Hu
China South Industries Group
    Corporation
Beijing, China

Simin Huang
Department of Industrial
    Engineering
Tsinghua University
Beijing, China

Yixiao Huang
Department of Industrial
    Engineering
Tsinghua University
Beijing, China

Binfeng Li
Department of Industrial
    Engineering
Tsinghua University
Beijing, China

Yin Liu
Beijing SpaceCrafts of CAST
Beijing, China

Li Mo
China South Industries Group
    Corporation
Beijing, China

Patrick Rau
Department of Industrial
    Engineering
Tsinghua University
Beijing, China

Chenjie Wang
Department of Industrial
    Engineering
Tsinghua University
Beijing, China

Liang Wen
Department of Management
Mechanical Engineering College
Beijing, China

Su Wu
Department of Industrial
  Engineering
Tsinghua University
Beijing, China

Mingxing Xie
Department of Industrial
  Engineering
Tsinghua University
Beijing, China

Zhihai Zhang
Department of Industrial
  Engineering
Tsinghua University
Beijing, China

Lei Zhao
Department of Industrial
  Engineering
Tsinghua University
Beijing, China

Zuozhuo Zhao
Department of Industrial
  Engineering
Tsinghua University
Beijing, China

Li Zheng
Department of Industrial
  Engineering
Tsinghua University
Beijing, China

Jiarao Zhu
Anhui Liugong Crane Co., Ltd
Anhui, China

chapter one

Made in China

Li Zheng

Contents

As many economic miracles of "Made in China" have historical characteristics, Made in China can be analyzed historically. This chapter will review the milestones of Made in China since the founding of the People's Republic of China.

1.1 Factories under planned economy (1949–1977)

In 1949, when the People's Republic of China was established, the planned economy, which was a highly centralized system, entailed that all enterprises are to be run by the state. As set out in the first state-owned Enterprise Work Regulations (Draft), a state-owned enterprise is an economic organization under a wholly owned socialist system; each enterprise shall produce goods as per the unified state plan, and the state shall allocate goods and determine profits and taxes, which were paid by enterprises in accordance with the state regulation. The government, and not the enterprise, was directly responsible for enterprise competency, state government–involved enterprise operation, and management improvement. This was carried out via multiple ways in a top–down level, and enterprise had no power of autonomy.

In 1950, the Central People's Government reformed the structure of state-owned enterprises to enhance enterprise management. On one hand, it established a factory management committee consisting of representatives elected by workers in identical number with that of the administrative staff; executed factory management; and realized the democracy of

factory management. On the other hand, it reformed production management. This included standardizing product family and managing product
development stage; performing work study and measurement; making
the production plan and establishing a statistical system; reinforcing the
process standard and enhancing quality control; liquidating the assets,
verifying of funds, initially establishing the financial management system, and trying cost accounting; implementing the eight-level skill wage
system and incentive system; improving production organization and
labor protection; conducting the rationalization campaign and production
record-breaking campaign for the purpose of improving quality, increasing output, reducing costs, eliminating accidents, etc.

During this period, the reform focused on the several areas.

1. Scientization

   Standardized work is establishing, controlling, and managing technological and management standards in an enterprise.
   Standardization forms a solid base for technology development
   and operation management, which promote the rationalization,
   standardization, and scientization of production, technology, and
   business activities. Until the 1970s, the state standards, departmen-
   tal standards, and enterprise standards have been gradually established in different categories for mechanical and electrical products.
   In addition, the management standard of the mechanical and electrical industry was continuously developed and adapted to the
   mechanical industry.

2. Work and time study standard data

   This involves various technical–economic standard data that
   relate to labor, capital, energy, and material consumptions. In early
   1958, a book (*Operation Rationalization*) was written for training of
   work and time study. Until the 1970s, the management system of
   technical–economic standards of work and time had been well
   developed as a scientific base for planning production, checking the
   execution of production plan, making economic accounting, and
   implementing incentive systems according to work, gradually ele-
   vating the level of scientific management.

3. Job training

   Job training means the most fundamental career education considered necessary for performing one's job responsibilities, including
   ideological discipline education, law-related education, occupational
   ethics education, technical and operation education, and management skills education. Since the 1960s, most enterprises in the
   mechanical industry had opened the organization or designated the
   personnel for on-the-job training and improved the business competence and technological level of workers as per the requirement

of job responsibilities and production operation and development, according to governmental direction and rules.

4. Regulations

Regulation means the authoritative standard or code of conduct that workers should strictly follow for the normal operation of production business activity, including regulations in relation with schedule, production, technology, quality, supply, labor, safety, finance, and economic accounting. Until the 1970s, the enterprise gradually established and developed the regulations, thus making clear the division of labor, responsibility, and coordination in production operation activity. A good production operation was therefore created to promote the continuous development of the mechanical industry as per the established regulations.

5. Socialist labor campaign

Labor campaign is one of the important media of relying on the masses, tapping the initiative and talents of the workers, vigorously developing the production, and managing the enterprise well. It is also a component of enterprise management and the evident feature of socialist enterprise management. There are several different forms of the socialist labor campaign, whose objectives include productivity improvement and cost reduction. The rationalization suggestion campaign, which is another type of socialist labor campaign, solves the key problems of production. The technological improvement campaign focuses on new technology, whereas the promotion collaboration campaign is used to spread best practice.

6. Production planning management

For the purpose of balancing production, production planning management means establishing the powerful production management system; building and developing the hierarchical production management system and responsibility system, which is the three-level production plan structure (i.e., factory, workshop, work section [unit]); preparing well the production technology, work study, and the technological and economic standards of production. This also entails carefully dispatching and controlling the production as planned. The plan attainment rate of parts or components should be 80% or higher; for final products, the plan attainment rate should be 100%.

1.2   Factories after the reform and opening up (1978–2000)

Enterprises gradually grew to be the market entity with the gradual development of the socialist market economy system and the stepwise implementation of state-owned enterprise reform including expanding the autonomy of enterprise operations. Eventually, enterprise began to take the initiative to solve the problems of enterprise management by improving management. In particular, since China joined the World Trade Organization, enterprise had gradually been engaged in global economics; thus, enterprise had gradually and involuntarily enhanced its core competitiveness and adopted modern management.

Earlier in the reform and opening up, the advanced management practices of foreign enterprises that local enterprises learned or drew from were primarily prompted by specific management-related problems, including problems relating to production, quality, finance, supply, and sales. However, this only partially solved the problems of enterprise management—as vividly described by the saying "cure the symptoms, not the disease." As the reform and opening up took deeper root, enterprise—while trying to adapt to the intensified market competition at that time—came to realize that partial management optimization was not enough if it intended to scale up and sustain its growth. Instead, it was essential to fully improve enterprise management—from inception, idea, regulation, system, organization, human resources (HRs), culture to business process; as well as fully and systematically integrate, restructure, and transform its mechanism, method, and instrument. It also involved the government, which encouraged enterprises to learn Western management theories and methods and incorporate traditional Chinese culture and managerial practice, and then to gradually establish the modern management system and finally to fully modernize the enterprise management.

The reform and opening up also led to a material change in the national economic system, as primarily indicated in the gradual transition from planned economy to market economy. The external environment facing enterprise also changed substantially—where the past focus on output value, yield, and planned national task shifted to quality, article, delivery, cost reduction, and market demand. Under this new setup, the role of enterprise changed from that of a national workshop to a self-financing entity. Although it left many enterprises in a dilemma of transformation, the enterprise and the operator were invigorated in the long run.

In 1982, the 3-year national streamlining campaign was launched in a planned and stepwise manner to promote the transformation of state-owned enterprises. This campaign aimed to resolve problems relating to incompetent management, overstaffing, redundancy, and nonstrict labor discipline, as well as low product quality, large waste, and bad economic profit. The streamlining campaign stressed improving the economic profit and governing the enterprise. The state policy asserts that all state-owned enterprises should establish "three constructions," build "four systems," and meet "six good requirements" through the streamlining campaign.

To be exact, these "three constructions" are as follows: (1) Gradually establish the rational management system. The factory manager should be politically conscious, an expert in production technology and business management, honest and upright, willing to face challenges, close to the masses, energetic, and able to establish the production and business system; the factory manager should also establish and develop workers' congress and gather their requirements. (2) Through the streamlining campaign, gradually build a competent team of workers. Workers are the masters of enterprise. Their quality level has a direct effect on the enterprise's performance such as production capacity, inventory level, and procurement cost. Enterprise should fully establish an HR management system; conduct staff training; educate the workers on professional ethics; transfer knowledge and skills to the production and management teams; enhance the political, cultural, and technological level of the staff members; introduce the code of conduct; reinforce labor discipline; and establish a strict reward system. (3) Through the streamlining campaign, gradually establish a set of scientific and civilized management system. Scientific management means that enterprise should comply with the management of natural and economic laws, whereas civilized management implies that the enterprise should be managed to reflect modern material civilization and socialist spiritual civilization.

During the streamlining campaign, the enterprise should gradually establish four systems. (1) Fully establish the production planning system. The enterprise should conduct market research and correctly identify the enterprise objective and make business decisions. The production plan should be driven by market demand, and should strictly comply with the state technical and economic indexes. Make an aggregated production and business plan, as well as a detailed plan that breaks down the enterprise's tasks into each department and workstation for the whole process, from new product development to sales and service; all activities shall be strictly executed based on the plan. (2) Fully establish the Quality Management System (QMS). Implement the policy of "Quality First," and ensure that its continued implementation and development are based on quality and variety, respectively. Plan quality upgrades and improvement programs based on customer needs. Closely combine management, technology, and statistical methods to achieve goals. Set up a comprehensive quality assurance system to ensure full management from research and development (R&D) to sales. (3) Fully establish the economic accounting system. Use economic measures to manage the production activities of the enterprise; conduct economic accounting in all respects, processes, and employees; rationally use materials and labor; estimate labor consumption and production results based on market prices; calculate expenses and enterprise revenues. In addition, try to ensure that the output exceeds the input, and do your utmost to improve the economic profitability, as well as increase the state, enterprise, and individual incomes. (4) Fully establish the HR management system. Combine and systematize the training, evaluation, appointment, promotion, and awarding of staff members in a planned manner; motivate the entrepreneurial spirit of staff members; continuously improve the quality of staff members; and stimulate initiative and creativity among staff members. Through "three constructions," build "six good enterprises"—a good mix of "State, Enterprise, and Individual," good product quality, good economic benefit, good labor discipline, good civilized production, and good political work.

On July 4, 1986, the State Council set out in Decision of Several Problems to Strengthen the Industrial Enterprise Management that enterprises should fully launch the campaign for "management strengthening, upgrading, and quality improvement." Specifically targeted at "on-site process technique," enterprises optimize on-site management, vigorously improve management, and conduct enterprise upgrading. The optimization of on-site management is intended to serve the customers and the market, which is the start point and end point of production management, and assures the overall optimization of enterprise management, enterprise upgrading, and management modernization. The optimization of on-site management is to impose comprehensive on-site management, by using modern management concepts, methods, and strategies to rationally allocate production factors; effectively plan, organize, and control the total production process; improve the operational efficiency of the production site; assure the implementation of a balanced, safe, and civilized production process; and attain the objective of quality and productivity. (1) 5S (Sort, Set in Order, Shine, Standardize, and Sustain) management. Change the "dirty, messy, and bad" condition of the production site; ensure that the personnel and material flow operate in an orderly and smooth manner; keep the site environment clean; and keep the civilized production in order. (2) In compliance with the process requirement of a product, further adjust the work routing and process arrangement; improve the process; precisely organize the production flow as per the process requirement in order to improve the product quality and output stability. (3) To achieve a rationalized and efficient production organization system, continuously optimize the scientific allocation of production factors, optimize the labor arrangement, rationalize the division of labor, reduce all kinds of labor inefficiencies and time wastage, and maintain a flexible production plan and high labor efficiency. (4) Develop regulations, technical standards, work standards, time standards, inspection detection, statistical ledgers, or other related basic regulations; systematize, standardize, and precisely evaluate these basic regulations; keep the information flow running in a timely, accurate, and smooth manner. (5) Develop the management assurance system; coordinate the management system in respect of quality, process, consumption, production, equipment, finance, and safety; impose an effective control on the total process from input to output via an advanced management method; and improve the operation efficiency of on-site management. (6) Conduct team-building activities for the work center and promote democratic management within the work center; encourage rationalized suggestion, technological innovation, and "Double Increasing, Double Saving" as daily Plan, Do, Check, Action activities; motivate workers; vigorously mobilize the creativity of the workers; and develop a team of staff members with ideological awareness, solid skills, and good discipline.

Since the reform and opening up, further developments of enterprise management had been vigorously geared following the streamlining campaign, enterprise upgrading, and management modernization, transformation of enterprise business mechanism, optimization of on-site management, and overall optimization of enterprise management. These campaigns gave rise to many new experiences related to enterprise management, in particular, emerging enterprises with high level of management, such as "Lean Production" in FAW Group, "People-oriented Management" in Jier Machine-tool Factory, "Quality Loss Compensation System" in Hanchuan Machine-tool factory, "Product Maintenance Assurance System" in Yuchai Machinery, "Backflush Costing" in Shandong Sida Industrial and Trading, and MAPII Management Information System implemented in Kelon Electrical. To further push for the development of industrial enterprise, the "Double Standardization" campaign was launched nationwide in 1997; it featured the basic work standardization qualifier campaign ("Standardization") and model enterprise of advanced management ("Model Enterprise").

The following is a list of additional requirements for standardization qualifier.

1. Work Standard

    Product standard with a formal approval procedure is a must before launching production, and product standard should fit user needs and include the following: raw materials, semifinished products, purchased parts, packaging, storage and transport of main parts, production process and tooling, safety, health and environment, inspection, and testing. Management processes and work standards, such as production, technology, quality, equipment, material, energy, business operation, and finance, are also needed.

2. Metering

    Grade 3 or above metering qualified enterprises should be well equipped and must include rational measurement inspection spots, network maps, and measurement instruments. The equipment rate of measurement instrument is 100% with the applicable accuracy; the weekly inspection rate and pass rate of standard measurement instrument are 100%.

3. Work and time standard

Establish the standard usage systems for labor, material, energy consumption, cost, expense, and equipment. These standards should cover more than 90% of the products, and the execution rate should be set at 100%. The enterprise should implement centralized management by specified departments, maintain and improve standards, and ensure good implementation.

4. Information management

Establish an information archiving system; this should ensure that gathering, collection, analysis, transfer, systemization, and standardization of data related to products, processes, and works can be easily and immediately applied to decision making during production operations. Various information including the original data, ledger, statistical reports, and dispatch order are to be maintained completely and updated regularly. Enterprise should provide a development plan and implementation for information and archiving work and gradually set up computer-aided management of information.

5. Basic regulations

Enterprises should establish and develop basic regulations, including those for the enterprise leadership system. Under the factory director (manager) responsibility system, "3 rules" are worked out. In the modern enterprise system, the standard Shareholders General Meeting, Board of Directors, Board of Supervisors, or other enterprise leadership systems are laid out. All enterprises should establish rules on business decisions, planning, production, technology, quality, material, finance, equipment, power, labor, sales and service, ideological political work, public welfare or other management regulations, and operation instructions. Under the economic responsibility system and assets management responsibility system, clarify the duties and responsibilities of various positions; implement the enterprise policies and objectives into each department, workshop, and unit; impose evaluations; and calculate the workers' salaries according to the work result. Regulations should be enforced strictly and maintained with feasible measures and evaluation methods, which can be amended and supplemented as needed from time to time.

6. Worker education

According to the needs of enterprise development, enterprises should provide a long-term training plan and annual implementation plan. Safety training and job skills are a must for all staff members; new workers should be given on-the-job training, and operators assigned to special tasks/jobs must have the necessary certificates and/or training for such tasks.

7. Team building

Enterprises should establish an economic responsibility system for work units and, based on that, form the management system of such work units. This includes enhancing cultural and professional skills, adopting democratic management, conducting technological innovation and technological reforms, and fully motivating the workers as a team.

8. On-site management

"Process breakthrough" has five components: build and develop the unified and effective process management system; with a series of methods to motivate the process planners, develop the complete process document and regulations; precisely execute "3 As Per" (as per drawing, process, standard) and "3 Fixed" (fixed number of workers, fixed number of machines, fixed profession); create the order of civilized production; build and perform the standard work procedure and eliminate quality-related accidents caused by inadvertent process management. On-site management should be in compliance with six basic requirements: settle the on-site environment, implement the 5S system, improve the "dirty, messy, and bad" condition of the production site, ensure that personnel and material flow are operating in an orderly and smooth manner, keep the site environment tidy, establish the order of civilized production, continuously optimize labor management, rationalize the division of labor, reduce all kinds of labor inefficiencies and time wastage, and maintain a flexible production direction and high labor efficiency.

To make the enterprise more adaptive in a fiercely competitive, market-driven economy, it must exert more efforts to ensure management development and continuous improvements, in addition to the factors as just described, and establish a modern management system. For this purpose, the enterprise should find a way to comply with basic qualification standards and subject itself to reviews and pass certification programs with model enterprise of management advancements. If it wants to achieve/maintain high comprehensive indexes of profitability and main technical and economic indexes, while ensuring excellent product quality and an exemplary safety record (no serious safety violations or default), it also needs to meet the following 10 criteria:

1. The enterprise should launch standard restructuring in accordance with the system of modern enterprise, and establish, develop, and begin to effectively implement the corporate governance structure.
2. Its leadership should emphasize, vigorously promote, and implement management, technology, and system innovations.

3. It should implement strategic management and work out medium- and long-term development strategies; it should also scientifically and effectively make material decisions, as well as adjust and optimize the internal organization structure, technological structure, product structure, and personnel structure.
4. It should understand the trend of world enterprise management, and draw, absorb, and innovate its management based on the situation. Its unique experience and innovation of enterprise management may have an important value for other enterprises in its particular industrial sector.
5. It should open a technical center and continuously drive technological innovation; it should maintain an R&D team to improve its core technology and products with emphasis on intellectual property rights and market competitiveness.
6. It should establish its own QMS as well as an Environment System. It should strive to gain ISO 9000 QMS certification, and be well prepared to qualify for ISO 14000 EMS (Environment Management System) certification.
7. It should actively conduct and implement information-based management; optimize, adjust, or restructure the organizational structure and business process via modern information technology and business management models; and enhance corporate competitiveness with good achievements.
8. It should harness modern information technology to enable the transfer and sharing of global information resources, and actively apply new business services and trade terms such as e-commerce according to the product and market reality.
9. It should emphasize the development and management of HRs, establish a scientific and strict performance appraisal system, with the developed personnel training system and regulations, and create a scientific and effective incentive and restraint mechanism.
10. It should emphasize the cultural establishment, create a corporate culture with its own in-house features, reinforce the structure of staff members, and create a civilized enterprise with good public image and strong unification values.

Until 2005, the qualifiers of basic management standards were 2040, those of model enterprises of management advance were 207, and those of modern management enterprises were 54. Moreover, a group of advanced models and managerial experience had been identified and chosen, which promoted and pushed for deepening reforms, management reinforcement, mechanism transformation, and enhancement of different enterprises.

1.3   Emergence of industrial engineering and innovation of Made in China (1990–2014)

In 1986, in order to improve product quality, the government reinforced process management, hardened process discipline, improved manufacturing processes, and called for total quality management. Thus, some enterprises took on a new look by the reinforcement of workshop management. In 1989, the Chinese Society of Mechanical Engineering (CMES) held a symposium on industrial engineering (IE). The event marked the first time that IE was introduced to China. In this symposium, Mr. Wang Renkang, the former deputy chief engineer of Science and Technology Division of the Ministry of Electrical and Mechanical Industry mentioned that competent authority realized that process work should not be only applied to production preparation, but also on the basic work of the enterprise, based on his 3 years of experience in the field of reinforcing manufacturing process. Many medium and large enterprises were required to deepen and strengthen their manufacturing processes, suggesting the need for "planning systematically and integrated implementation," when it was high time for Chinese enterprises to adopt IE processes. In this symposium, seven experts who have visited the United States, Canada, and Japan introduced various technologies and their applications and development in foreign countries, and discussed the issues concerning the definition of IE, industrial influence, education, and training. The Shanghai Mechanical Engineering Society noted that they had the IE handbook written by Professor Gaverial Salvendy translated in 1987. Huayi Electrical Appliance, under the Shanghai Electrical Appliance Group, held a training session, in which workers quickly mastered various methods such as motion and time study. Based on the improved processes, it was observed that production efficiency could be enhanced by tens or hundreds of times. According to the feedback from workers, IE technology was feasible and operable, production output was increased, and workload was surprisingly reduced. Based on the scientific work and time study, staff from the Labor Division of the Mechanical and Electrical Department vigorously shared their IE knowledge in Shanghai Jinlin Radio Factory, Shanghai Radio No. 27 Factory, Chengdu Hongguang Electronic Tube Factory, and Shanghai Instrument Factory; this endeavor earned good economic returns and was greatly popular among workers. In the final session of this symposium, it was agreed that the Ministry of Education would start to offer IE subjects/courses to develop local talents. The CMES organized an IE commission to organize more academic activities and promote IE. It was the first time that such a meeting had been held in China, and after this notable event the academic field of IE was launched and was consequently applied to various enterprises nationwide.

In 1990, the CMES established the Industrial Engineering Institute (renamed as Industrial Engineering Subsociety 3 years later). In October 1992, an IE higher education symposium was convened in Xi'an Jiaotong University, Xi'an, China. During the discussion, it was agreed that IE courses could be offered in universities, which should then be substantially developed. In 1993, with the approval of the Ministry of Education, Tianjin University, Xi'an Jiaotong University, and Chongqing University were among the first to offer IE undergraduate programs, whereas other universities including Tsinghua University were ready to offer postgraduate courses. In 1995, Shenyang Aircraft Corporation organized the fourth national IE conference, marking the first time that an enterprise organized such an event. In this period, many universities opened their own respective IE programs, such as Beihang University, Northeastern University, Zhengzhou Institute of Aeronautical Industry, and Nanjing Aeronautical Institute. In addition, typical IE applications from Shenyang Aircraft Corporation marked the beginning of enterprise application and innovation (Shenyang Aircraft Corporate was the often-cited example to illustrate this trend). At the same time, IE applications and achievements in various enterprises such as Motorola, FAW-Volkswagen, and Shanghai Volkswagen captured nationwide attention. Thus, in the developed Pearl River Delta, companies such as Guangdong-based Kelon Group, Midea Group, and Konka Group began to cooperate with the IE Institute and made remarkable attainments. Since 2000, with the rapid development of the Chinese manufacturing industry, a growing number of enterprises have applied IE into their processes; correspondingly, an increasing number of schools have also opened their IE programs to develop talents for enterprise. (As of 2010, more than 200 universities have included IE programs in their curriculum.) Famous Chinese universities such as Tsinghua University, Peking University, Shanghai Jiaotong University, and Nanjing University are offering IE programs to develop talents and allow students to further their studies in this field, whereas large enterprises have also established their own IE departments to promote IE applications. FAW, Foxconn, and other large business enterprises were hailed for their IE-related achievements; a few consulting companies with an annual revenue of more than 100 million yuan have also joined the fray.

Tsinghua University opened IE as a second degree in 1989, and began to offer IE postgraduate courses starting in 1993. The institution began to enroll IE undergraduates in 1997. The IE Department was eventually established, and Professor Salvendy (from the IE Department of Purdue University) was appointed as department chair and chair professor, and Mr. Zheng Li was named the executive chair in 2001. Over a period of 10 years, Tsinghua University's IE Department has opened and maintained an IE program and educational system in precise alignment with world-class universities by referring to their IE education standards, conducting quality IE studies, and engaging in pushing for IE applications nationwide. Since 2005, more than 100,000 people have attended classes and undergone various training courses in the IE auditorium. Enterprise IE was also popularized as reflected in all kinds of enterprise services in the form of student internships, dissertations, R&D projects, enterprise consultations, and joint research centers. For more than a decade, IE has been utilized and has benefited large multinational corporations such as Intel, GM, Siemens, as well as Chinese Small and Medium Enterprises.

This book has selected several typical cases to illustrate the innovation of Chinese manufacturing and enhancement of production efficiency, and to demonstrate the secret of Made in China from one angle. Chapter 2 ("New rules for running production lines in shoemaking industry"), which Professor Cheng Ye wrote on the basis of his personal experience, describes the partial work of the first joint research center established by the IE Department (Tsinghua University) and enterprise. Under the context of the 2008 Global Financial Crisis, the enterprise needs to find a way to survive the challenges facing the original equipment manufacturers sector, which is currently grappling with the surging labor cost. Cooperating for 6 years, more than 30 faculty members and students were engaged in work to increase the production efficiency of labor-intensive enterprises. Chapter 3 ("Culture: Management style comparison") reports on a consulting project entrusted by a transnational corporation, which responds to the basic problem: How is the management of their Chinese factory? To this end, Professor Gao Qin has selected five transnational enterprises from different countries and one Chinese enterprise to investigate, contrast, and identify the influence of different cultures on factory management. I was so impressed by the amazed directors when the Tsinghua team reported the results of their study to this large transnational enterprise (particularly their Project Summary). Chapter 4 focuses on the design of a production line. The bulk of Chinese production lines are imported, and enterprise is used to construct a new production line by choosing the template through an overseas investigation. This case study shows how to consider the actual environment of the enterprise in the design of production lines, including factors such as material flow and worker's operation, and to finally design a good production line. Chapters 5 ("Optimizing selection of cracking material in petrochemical enterprise") and 6 ("Optimizing oil production") examine the process industry, and concentrate on the effect of operations research on the optimization of decision and show that enterprise can earn substantial economic returns through the optimization of decision via the tools of operations research. Chapter 7 ("Ways of Enterprise T to improve laptop computer quality") focuses on how enterprise improves the quality of its electronic products. This work is sourced from the dissertation by a part-time IE master, who was advised by Professor Wu Su. The chapter describes how to effectively increase the first unpacking pass rate by means of an IE tool. Chapter 8 ("China's implementation of manufacturing management reform") describes a mechanical enterprise engaged in many lines of business, including automobiles in large-volume production and heavy machinery in small- or middle-volume production, and how to enhance production competitiveness in such an enterprise. For this purpose, the enterprise–university collaboration calls for the production management system—China South Production System—in order to enhance production competitiveness. China South is among the early implementers of this production system, which contains four social factors and eight technical factors. Chapter 9 ("Restructure of supply chain") is the case study of material flow and supply chain, in which Professor Cai Linning and the enterprise worked out the material flow network. Chapter 10 ("Indirect labor improvement project at Source Photonics [Chengdu]") stems largely from a site visit made by Professor Salvendy, who suggested applying an IE method with good potential for substantial improvements as prompted by the unevenly distributed workload. To this end, Professor Li Binfeng with four undergraduate students carried out a 2-month field work, which improved personnel utilization with a terrific effect. Chapter 11 ("Groundwork for industrial engineering inside enterprise"), is written by my student, Zhu Jiarao, who has been engaged in manufacturing after graduation. He was the general manager of a factory under Liugong Group when he wrote it. He had the experience of implementing IE with good effect in an enterprise where no one knew about IE. This chapter can be used as a good reference for Chinese enterprise production management because of its inspiring benefit. Chapter 12 ("Implementation of cell manufacturing mode of aerospace enterprise") was written by another student, Liu Yi, who was primarily engaged in IE-related work because the enterprise lacked IE personnel, although his academic major was in manufacturing engineering. In this chapter, he explains how to restructure the production system by using cell production. The last chapter ("User-centered design"), which is written by Professor Rao Peilun, is the only one related to product design.

In 12 chapters, three cases were chosen from transnational enterprises, five were from state-owned large-sized enterprises, and four were from private enterprises; 10 cases were chosen from discrete manufacturing (three in electrical appliances, two in the auto industry, two in heavy machinery, one in aerospace, and two in light industry and daily chemicals), and two cases from the process industry, if classified by industry; five cases were concentrated on the efficiency of workshops, one on the improvement of manufacturing quality, two on the improvement of material flow and supply chain, and three on how to construct a system to promote lean production; one case was concentrated on product design. These cases were sourced from project summaries contributed by the faculty of Tsinghua University, whereas some were produced by its graduates based on work-related experience. These are first-hand experiences.

Bibliography

Decoding World Class Production Management: Method, Case and Trend, Li Zheng, Li Mo, Tsinghua University press, 2014, ISBN: 978-7-302-34836-8.

Growing Footprint: 60 Year of Enterprise Management in Chinese Mechanical Industry, Edited by Zumei Sun, Tianhu Song et al., China Machine Press, 2009, ISBN: 978-7-111-28774-2.

Manufacturing Industry Development of China: In a Global Perspective, Linyan Sun, Tsinghua University press, 2008, ISBN: 978-7-302-18150-7.

chapter two

New rules for running production lines in shoemaking industry

Ye Cheng and Li Zheng

Contents

In 2004–2006, "Made in China" was thriving at its peak, with global orders flooding into China. China's cheap labor costs stimulated the development of export-oriented and labor-intensive enterprises, and many entrepreneurs who had been counting on this trend to continue were caught unprepared when the 2008 Global Financial Crisis broke out.

However, some farsighted entrepreneurs began to feel uneasy as early as 2006 (when things were still booming). Some were bothered by several nagging questions. Will Chinese workers continue to work diligently generation after generation for foreigners? Will the next generation of workers who grew up in a more well-off environment be willing to see themselves as "working slaves" for the rest of their lives?

On July 19, 2008, the former premier, Wen Jiabao, paid a visit to HuaJian Group. The premier agreed with Mr. Zhang Huarong's (the president and chairman of Huajian Group) operating ideology of "development, transformation, and upgrading." Premier Wen commented that "Development is the absolute principle, upgrading is the core, and transformation is the restructuring. Three simple words, as extracted from the experience, reflect the milestone of enterprise, which can covert pressure and difficulty into power and opportunity respectively." It was evident that both the statesman and the entrepreneur were both sensitive to the changing world and figured out the countermove (Figure 2.1).

Figure 2.1 Premier Wen's visit to HuaJian Group.

HuaJian Group is one of China's largest original equipment manufacturer manufacturers of middle- and high-class women's leather shoes (Figures 2.2 and 2.3). It is also known for its flagship status in the shoemaking industry and for its notable overall strengths, which include 47 production lines, more 24,000 employees, annual production capacity of more than 20 million pairs of shoes, and loyal customers with globally famous brands (e.g., Clarks,

Figure 2.2 Huabao Plant in Dongguan, HuaJian Group.

Figure 2.3 About 10,000 workers convened in Ganzhou Plant, HuaJian Group.

Coach, Guess, Nine West, Easy Spirit, Calvin Klein, Nina, Marc Fisher; 15 of the company's customers were among the world's top 50 companies specializing in middle- and high-class women's footwear). The crippling global financial crisis of 2008 witnessed thousands of small and medium enterprise shoemakers struggling to avoid bankruptcy; the size of orders from foreign customers drastically decreased to thousands or hundreds per order (compared with tens of thousands or hundreds of thousands before). This became more evident in the winter of 2009, which should have been a busy season for manufacturing boots, when 70–80% of orders were for less than 800 pairs. In previous years, several production lines were usually loaded day and night for a few weeks for a single order; but now production lines are typically "changed over" for another product model in every 1 or 2 days. As a consequence, manufacturing became vulnerable to substantial variations, and this, in turn, destabilized throughput and quality.

"Good luck would never come in pairs, while misfortunes never come singly," as the old saying goes.

The source of headaches for the majority of shoemakers was usually attributed to the new employees of production lines. With the reinforcement of the social welfare system and rising minimum salary levels, migrant workers would now prefer to find jobs in areas near their hometowns rather than flood into Shenzhen or Dongguan in the Pearl Delta. For this reason, it became more increasingly difficult for labor-intensive enterprises to recruit workers; this problem evolved into "labor shortage" year after year. After each Spring festival, manufacturers resorted to tricks to attract job seekers. For this purpose, they almost set no requirements for new employees, and manual workers were always welcomed regardless of age, gender, educational background, or skills. Gradually, this resulted in an urgent shortage of skillful workers, inactivity, low production efficiency, and declining quality. Foremen found that certain management actions no longer worked as they did before, especially the "carrot-and-stick" policy. These days, it is the employees who select their employer rather than the other way around. A new generation of employees would prefer to work for an employer that provides air-conditioned free housing and well-equipped computers in the Internet bar. Certainly, a high base salary should be offered, perhaps piece-rate wage or even extra work allowance would be attractive. In this type of setup, however, some bosses began to wonder as to who is the real boss.

In August 2008, the HuaJian Group management led by President Zhang paid a visit to the Department of Industrial Engineering, Tsinghua University, to solicit advice on the governance of production lines. An agreement was soon made between the two parties, which suggested solving the problems of production lines by harnessing industrial engineering expertise. In October 2008, a delegation of faculty led by department head Professor Salvendy headed for the shoemaking plants of HuaJian Group.

They studied its production organization, facility layout, planning and scheduling, personnel management, quality control, and material supply, understood the enterprise basics, and tailored the systematic optimization solution (Figure 2.4).

In April 2009, the Tsinghua–HuaJian Research Institute of Industrial Engineering Applications was officially founded. This marked the beginning of a collaborative work where 20 faculty members and several graduate students from the Department of Industrial Engineering, Tsinghua University, were stationed in HuaJian Group, and launched a 3-year study of industrial engineering applications (Figure 2.5).

Figure 2.4 Professor Salvendy investigates shoemaking line for industrial engineering.

Figure 2.5 Experimental line of industrial engineering applications.

When we utilized the methods and tools of industrial engineering to diagnose the status of the company's shoemaking production, we were struck with the following problems:

1. How to select a new employee who is fit for manual skill operations?
2. How to group employees with varying skill levels into production lines appropriately?
3. How to respond to small orders requiring multiple models?
4. How to respond to uncertainties during production?
5. How to determine the strengths of different production lines?
6. How to rationalize the structure and scale of production organization?

We did so because it was difficult to find the solution for production problems from the classic knowledge system of industrial engineering expertise.

For this purpose, we worked with HuaJian management, technicians, and foremen to invent a set of effective methods in 3-year trials against complex, dynamic, and uncertain problems in relation to production management. These methods will be further discussed in the following sections.

2.1 Essential operational skill test for new employees

People are the lifeblood of an enterprise. Operators in production lines can determine the majority of production performance, especially in labor-intensive enterprises such as shoemaking.

Since 2009, many young job hoppers have been increasingly lured by the promise of easy and quick job seeking. In shoemaking, an operator who works for more than half a year can teach newcomers, and an operator who works more than 2 months can be treated as a skillful worker. Foremen in the production lines were troubled by a shortage of operators, especially skillful workers or masters. For instance, one production line can be started regardless of an insufficient number of on-duty operators. In some operations requiring special skills, senior employees with good skills can fill in for various positions as needed. By doing so, the production line can continue to operate; however, this also disturbs the production takt, which should have been strict enough, and amasses work-in-process (WIP) shoes stuck in each workstation. If the production order size is as small as 100 pairs, for example, the last pair is finished in the first workstation, when the first pair may not go to the last workstation. This is why two or three models may exist within the same line, which will result in low efficiency and uncontrolled quality.

Through a long course of observation, the faculty and students of Tsinghua University found that a large number of the employees did not posses the necessary level of skills required for the operations they were handling. To make matters worse, a few of them were violent or rude, and one worker could not catch up with the production pace. In short, they were not eligible workers for the production line, who not only created little value but also caused chaos. One who wants to do well does not necessarily translate to a worker who can perform well, if he or she is not assigned to a job that fits his/her strengths and weaknesses (e.g., hiring a robust man to sew buttons or weave tassels and braids, or assigning a petite woman to handle the heavy shoe last or push a heavily loaded cart).

As a matter of fact, a shoemaking plant has a set of skills test methods for various positions. However, these methods are useless for newcomers who have no experience in shoemaking, because the new recruits are unable to handle any operation.

When we reviewed the literature on skill testing of common manual operations, we identified one characteristic of methods in use, where one specialized expert is supposed to use a unit of testing instrument or tools to test the employees one by one. Unfortunately, this efficiency tool cannot be applied to a shoemaking plant that recruits hundreds of new employees in a single day.

To address this concern, we strove for developing a test method, which resembles a classroom examination. Tens or hundreds of examinees are allowed to take the test simultaneously. A special expert is assigned to promptly check the examination test sheets. After several experiments, a test method was developed, widely applied, and proved to be effective. We referred to this test method as "8 min plus 20 cents." The evaluation result of new employees was found to be precisely aligned with their skill performance in production lines.

This test method is carried out in four pieces of A4 papers imprinted with various figures. The cost incurred from copying each paper is 5 cents, so the cost of the test medium per examinee is only 0.20 RMB.

During the test, each examinee was required to draw certain figures using a red marker, and to ensure that they aligned with the lines as much as possible. As the human hand has limited control accuracy, the drawn lines would deviate from the original ones by 1 mm or more. It was discovered that someone stopped and continued the drawing.

Concerning the characteristics of manual operations, we summarized most work contents in shoemaking into four essential manual operation styles: (1) positioning operation at a target location of small range; (2) straight moving operation; (3) curve moving operation; (4) closed-curve operation in large range.

In contrast with the four essential operations as mentioned above, we designed four test papers, which were printed with figures: (1) small circles of

1 mm diameter, which are stamped sparsely to densely; (2) inclined straight-line segments, which are stamped sparsely to densely, in short to long length; (3) S-shaped curve segments, which are also stamped sparsely to densely, in short to long length; (4) several groups of big closed circles, which are concentric or nonconcentric circles, with different diameters. These figures were provided to each employee, who was then instructed to draw using a red marker. The allowed time for each paper was set at 2 min. In this way, each batch of tests taken by several tens or hundreds of new employees only took 8 min. Several typical papers of new employees are exhibited in Figure 2.6.

One employee carefully drew a small circle in 1 mm diameter by using red marker, which could more or less cover the printed small black circles, which deviated a little; however, this employee drew it slowly and failed to finish, as shown in Figure 2.6 (1A). Another employee drew it quickly, but several small red circles were irregularly shaped, which significantly deviated from the printed small black circles by 0.5 mm or more, as shown in Figure 2.6 (1B). Another new employee drew it precisely and quickly, which indicated a better and precise control of marker point and tiny actions, as shown in Figure 2.6 (1C).

Similarly, one employee drew red straight lines basically along the printed black ones, but these lines waved a little; sometimes it could not be drawn in one stroke, or stopped and continued, as shown in Figure 2.6 (2A). Most straight lines were well drawn, but the last line was deviated totally upward, as shown in Figure 2.6 (2B). A quick hand drew lines randomly, which basically disregarded the original black lines, so that most red lines were left at one side away from black lines, as shown in Figure 2.6 (2C).

The "S" curve was a challenge for newcomers, whose manual operations were forced to change in transition from convex curve to concave

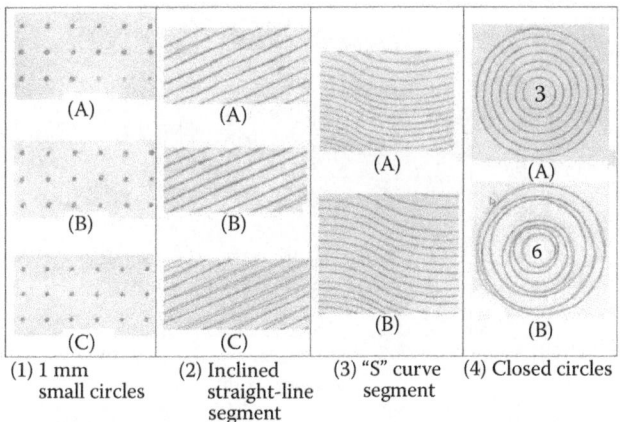

| (1) 1 mm small circles | (2) Inclined straight-line segment | (3) "S" curve segment | (4) Closed circles |

Figure 2.6 Answer sheets from a basic operation skills test for new employees.

curve; some hesitated, slowed down, or vibrated in this moment, as shown in Figure 2.6 (3A). However, some adapted quickly to this change and went smoothly through the convex–concave transition, as shown in Figure 2.6 (3B).

When drawing closed big circles, the sight and gesture of tested employees needed to rotate 360°, which challenged the precise movement of hands at all directions. Before closing the circle, the eyes should stare at both the marker point and the starting point of the circle, so that the circle could be closed smoothly, as shown in Figure 2.6 (3A). When drawing nonconcentric circles, the space between circles was varied, which may hint the tested employee to attract or repel it, as shown in Figure 2.6 (3B).

When the examinee has quickly finished the examination, the industrial engineer may judge the quality and speed of several figures drawn by the employee, and score the quality and efficiency, respectively, according to the defined evaluation criteria. By doing so, the instinctive manual operation skills of new employees can be evaluated in these two dimensions.

Figure 2.7 shows the diagram of efficiency and quality level distribution of tested essential skills for a group of new employees. A few employees posted excellent scores; these people finished the examination quickly and performed well. These "good hands" are what we need, so it is necessary to train their skills at each key operation as soon as possible; we need to transform them to be the "cornerstones" in production lines, try hard to nurture these talents, and gradually develop them as key persons and foremen. The examinees with good quality but had low efficiency can be called "precise hands." They are among those who we hope to recruit, as they can be assigned with fine and precise manual duties. Some of the new

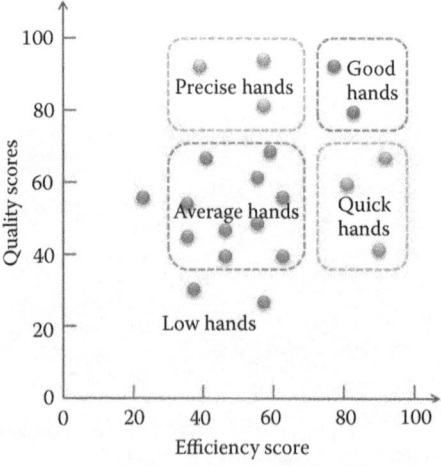

Figure 2.7 Efficiency and quality level distribution of essential skills of new employees.

employees with unreliable quality, but with quick or agile movements can be called "quick hands"; they are suitable for operating machines or can be assigned to operations with efficiency bottleneck. The majority of examinees demonstrated an average level of quality and efficiency—they are considered "average" employees. They can be assigned to some insignificant common tasks, which can exert little negative influence on the overall efficiency and quality level of production lines. However, there are others who may have a detrimental effect on quality and efficiency, who will contribute little, will cause quality problems frequently, or will become the bottleneck of efficiency, and consequently drag the entire team, unless that they can be assigned to somewhat less important operations. Those "low hands" should not be recruited.

The test method for new employees turned out to be a success in HuaJian Group. The outcomes of the essential skill test were precisely aligned with the live performance of new employees, according to the feedback from foremen of production lines. In this way, foremen are able to be well prepared for developing potential talents before the probation period.

The Human Resources (HR) department also benefited from this method. Before this method was applied, complaints such as "what cheap guys!" were often heard from production foremen. With this essential skills test, HR can now screen out those new employees who are not suitable for shoemaking operations, which leads to reduced costs and less trouble for the company.

2.2   Adaptive production line balancing

As described in the preceding section, people greatly differ in operating skills, not only for ordinary operations. The differences in skills among people are still quite remarkable even after they have undergone specialized training for various professions.

The relative production efficiency of one employee can be expressed in terms of the efficiency coefficient of his/her profession. In one profession, the ratio of one employee's working efficiency to the average efficiency of other employees of the same profession is the coefficient of profession efficiency.

For example, the standard time of one stitching operation is 100 s. If Operator A can finish this operation within 80 s, the coefficient of this profession can be calculated as follows:

$$\text{Standard throughput} = 3600 \text{ s}/100 \text{ s} = 36 \text{ pairs/h}$$

$$\text{Operator A's throughput} = 3600 \text{ s}/80 \text{ s} = 45 \text{ pairs/h}$$

$$\text{Operator A's coefficient of stitching efficiency} = 45/36 = 125\%$$

It can be inferred that Operator A's efficiency is 25% higher than the average efficiency of common stitching operators. Another simple way to calculate this is: 100 s/80 s = 125%.

Similarly, a stitching newcomer may take 120 s for the same stitching. Thus, this person's coefficient of profession efficiency is: 100 s/120 s = 83.3%, which is lower than the average rate of common stitching operators.

According to the traditional production line balancing method, it is assumed that each employee's working efficiency is equal to that of others, when assigning tasks to each operation in production lines. In other words, anyone could be assigned to either operation to do the required work. Moreover, work time is irrelevant to the operator. This assumption is more or less correct in mature industries of developed countries, because standardized employers always train new employees to be eligible enough before they are assigned to their designated post. For example, a Japanese automaker sets a training station besides the production line, where new employees get trained and go live until they can handle their tasks precisely and control time within seconds.

However, the domestic labor-intensive industries of China, including clothes, shoes, hats, suitcases, bags, toys, simple electronic products, are totally dependent on low-cost labor. Employers habitually try their best to reduce the requirements for employees' knowledge and skill levels in order to be competitive, by keeping labor costs as low as possible. However, they usually end up hiring operators with different levels of knowledge and skills, which hinders the standard arrangement of production.

Theoretically, production line balancing is an essential method to arrange the operational tasks for transfer line production. However, it cannot be implemented in the shoemaking lines of HuaJian Group, for two reasons: (1) the execution time of each task differs from operator to operator; (2) in theory, rational assignment may be inappropriate if there is a lack of employees with the necessary skills.

This phenomenon can be supported by a short example.

Generally, a vamp stitching line requires 25 to 30 operators, and production takt time is about 2 to 4 min. The vamp stitching production of a model involves 100 to 200 operations, of which several operations require the use industrial sewing machines, trimming machines, setting machines, or other equipment, require special skills, or are related to complex situations, to name just a few.

To expound on the new method of production line balancing, we present a hypothetical vamp stitching line that is highly simplified. In this setup, 10 operators are needed to perform 15 operations, as shown in Table 2.1.

The total time consumed by 15 operations is 1085 s, and each workstation takes an average time of 108.5 s, when these tasks are assigned to 10 workstations. In consideration of uncertain times and unbalanced

Table 2.1 Operational tasks of vamp stitching line
(fictional and simplified)

Operational task no.	Profession type	Standard time (s)
1	Hand work	105
2	Stitching	110
3	Stitching	90
4	Hand work	40
5	Hand work	130
6	Folding	110
7	Stitching	80
8	Hand work	30
9	Hand work	60
10	Hand work	80
11	Hand work	40
12	Folding	100
13	Hand work	50
14	Hand work	40
15	Hand work	20

tasks, production takt (cycle) time is set as 130 s, where the corresponding throughput per hour is 27.7 pairs.

A feasible assignment can be figured out in accordance with the traditional production line balancing method, as shown in Figure 2.2; however, this assignment is only theoretically feasible.

First, each employee cannot handle all operations in this production line. There is a limited number of employees who have the necessary specialized skills to perform difficult tasks, and it is not necessarily true that eligible operators are sufficient to meet this requirement. For example, Operation 2 is stitching, whereas Operation 6 is folding—both tasks are so complex that only expert or skillful operators can handle them. Only those common hand works can be handled by all employees. As a matter of fact, this production line lacks technical experts, so the foreman has to borrow workers from another other line when he assigns tasks to each operation. If these workers are not available, the production line is in "paralysis": very few technical experts shift from here to there among key workstations, which causes the production line to stop intermittently. In order to avoid downtime, the production foreman usually takes action: WIP buffer stocks, some 3–5 or 8–10 pairs, are placed in each workstation. The total production cycle will be extended if the WIP stocks are accumulated in production lines, which may lead to delayed delivery of products (Table 2.2).

Table 2.2 Operational task assignment of normal
production line balancing

Workstation no.	Operational task no.	Standard time (s)
Workstation 1	1	105
Workstation 2	2	110
Workstation 3	3, 4	130
Workstation 4	5	130
Workstation 5	6	110
Workstation 6	7	80
Workstation 7	8, 9	90
Workstation 8	10, 11	120
Workstation 9	12	100
Workstation 10	13, 14, 15	110

Second, operating skills differ substantially among employees. Each operator can perform only limited types of jobs. The actual work hours also vary if different people are assigned with the same task. Operational task No. 5 is a common hand work, for instance. Any employee may handle it, but the difference is that a skilled worker may spend 100–110 s to finish this operation compared to 150–160 s for a newcomer (this is also where error happens frequently). Once error occurs, rework may take longer than the corresponding normal operational times. This explains why a newcomer always piles up shoes when the downstream workstation has no alternative but to wait until the upstream workstation closes. A newcomer may become the bottleneck of the entire production line and restrict the throughput rate of the whole production, if he or she is assigned with a heavy load like this workstation.

Each employee of this production line can handle their assigned tasks with the coefficients of operational efficiency as shown in Table 2.3. It is evident that stitching is the most difficult operation, and only three operators can handle it. Betty has the highest efficiency rate in stitching and is considered the "master hand" of this line for stitching operations. She is also a little quicker when performing ordinary hand work compared with others (110%). Moreover, Betty can also do folding but is a little slower than other folding operators (80%). In contrast, Frieda, who is accredited as the "second master hand," is even slower especially in folding (60%). Andrew is considered the true "folding master hand," so he is responsible for the longest folding operation. When performing common hand work, people differed considerably. The senior employee, Chris, is able to perform common hand work more quickly than technical work (130%). "Green hand" Diana, who has been working in the shoes plant for only

Table 2.3 Coefficient of profession efficiency
of production line employees

Name of employees	Coefficient of profession efficiency (%)		
	Stitching	Folding	Hand work
Andrew		120	100
Betty	130	80	110
Chris			130
Diana			60
Eric	100		80
Frieda	110	60	100
George		100	100
Helen			110
Jack			90
Kathy			100

1 week, did rather poorly—her efficiency rate for basic hand work is only 60%. The foreman claims that Diana was chosen just because he only has two available hands with some potential to become an expert.

Production foremen can gradually figure out the production line scheme by considering the available manpower and assignment of operational tasks.

The first step is to schedule the stitching operations. Operation 2 (consisting of the stitching job alone) takes 110 s. Operations 3 and 4 may be combined (130 s in total), where a stitching worker can manage two operations, although Operation 4 is a hand work. Similarly, Operation 7 (stitching, 80 s) and Operation 8 (common hand work, 30 s) can be assigned to a stitching operator. The foreman is relieved at the thought of assigning three stitch operators to work on this model. "If the next model is boots, another one or two stitching operators may be needed, when I have no alternative but assign myself to this stitching job, work as two, and run to and fro," added the foreman. The "first master hand" Betty comes to his mind, who should be assigned to "Operation 3 stitching, 90 s + Operation 4 common hand work + 40 s"), which is considered a difficult and heavy load. Two operations combined took 130 s before, but he will just spend about 100 s. The "second master hand" Frieda is naturally assigned to "Operation 7 stitching, 80 s + Operation 8 common hand work, 30 s," which is less loaded when compared with Betty's workload. The remaining stitching operation "Operation 2 stitching, 110 s" is assigned to no one else but Eric.

The second step is to schedule folding operations, which is comparatively easy. Two operations are covered, which are: Operation 6, 110 s; and Operation 12, 100 s. Naturally, "folding expert" Andrew should be assigned to the longer Operation 6. The shorter Operation 12 is left to George, the last of those who can fold.

Common manual operations with more operations are discussed next. Actually, two common hand work operations, Operation 4 and Operation 8, have been scheduled into stitching, as described above. In consideration of two scheduled folding operations, the remaining common hand operations can be divided into several chunks:

- Chunk 1: Operation 1 (105 s)
- Chunk 2: Operation 5 (130 s)
- Chunk 3: Operation 9 (60 s), Operation 10 (80 s), Operation 11 (40 s)
- Chunk 4: Operation 13 (50 s), Operation 14 (40 s), Operation 15 (20 s)

Among these chunks, Chunk 4 has many operations, but one operator can manage them in the total standard time of 110 s. Chunks 1 and 2 should be assigned to two operators.

Chunk 3 is a little problematic. If Operation 9 (60 s) and Operation 10 (80 s) are combined into a single workstation, the total standard time is 140 s, which is longer than the preset takt of 130 s. This result is evidently undesirable. However, if Operation 10 (80 s) and Operation 11 (40 s) are combined, the total standard time is 120 s, which is acceptable. In this way, a small operation—Operation 9 (60 s)—is left behind.

An idea occurred to the foreman: Operation 9 only takes 60 s. Why not assign it to the "green hand" Diana? Her coefficient of efficiency is 60%, which means he would spend 100 s on a task that only takes others 60 s to complete. However, this does not matter because the upper limit is 130 s!

The next step comes easily, in that the remaining common hand works are assorted in sequence, according to the length of work hours:

1. Chunk 2: Operation 5 (130 s)
2. Chunk 3 left: Operation 10, Operation 11 (120 s in total)
3. Chunk 4: Operation 13, Operation 14, Operation 15 (110 s in total)
4. Chunk 1: Operation 1 (105 s)

Available employees can be sequenced according to the coefficient of profession efficiency of common hand works:

1. Chris: 130%
2. Helen: 110%
3. Kathy: 100%
4. Jack: 90%

Following the principle of assigning master hands to difficult or heavy workloads, Chris, Helen, Kathy, and Jack are assigned to Chunk 2

(Operation 5), Chunk 3 two left operations (Operations 10, 11), Chunk 4 (Operations 13, 14, 15), and Chunk 1 (Operation 1), respectively.

Until now, all operational tasks have been assigned well, as shown in Table 2.4, wherein we calculated the actual execution time of each operation against the assigned operator, and fill in the last row.

Based on the said table, the foreman was surprised to find that the total actual work hours of 10 workstations is 1046 s, which is 40 s less than the total standard work hours (1085 s). He finally realized that he had saved time by assigning master hands with difficult and heavy workloads.

To our huge surprise, there is no 120 s of actual work hours in either workstation! Immediately, the foreman decided to shorten the production takt to 120 s, which means that the throughput per hour will be increased to 30 pairs, up 8.3% from the theoretical line balance (throughput per hour: 27.7 pairs).

By doing so, the embarrassing "Team of Low Efficiency" is upgraded to "Dream Team" by harnessing these tricky combinations.

The faculty members and students from the Department of Industrial Engineering, Tsinghua University, stayed and communicated with foremen and employees in production lines for half a year, who later demonstrated the rationality of the foremen's "tailored action." They made and examined the profession efficiency rates based on standard time; discussed in depth and studied the balancing method of production assignments; standardized such method; developed the production balancing program package tailored for professions and coefficients of efficiency; and promoted such application in each vamp workshop and plant, which was recognized and accepted by the production site. This method is termed as the "Adaptive Balancing Method of Production Lines." Under this method, the gist of the matter lies in assigning operations and adapting

Table 2.4 Production line balancing with available profession and coefficient of efficiency

Workstation no.	Operational task no.	Employee	Actual work hours (s)
Workstation 1	1	Jack	117
Workstation 2	2	Eric	110
Workstation 3	3, 4	Betty	106
Workstation 4	5	Chris	100
Workstation 5	6	Andrew	92
Workstation 6	7, 8	Frieda	103
Workstation 7	9	Diana	100
Workstation 8	10, 11	Helen	109
Workstation 9	12	George	100
Workstation 10	13, 14, 15	Kathy	110

to the actual operating capabilities of available employees; therefore, the balanced result can be practical, feasible, and effective.

2.3   Quick changeover of production lines

As described in the beginning of this chapter, since the aftermath of the 2008 World Financial Crisis, the international market of shoemaking has been rapidly changed, as indicated in the drastic downsizing of orders from customers. To cope with this, the production lines had to be frequently changed over in every 2 or 3 days, or every day, as required by circumstances.

According to the traditional changeover mode of shoemaking production line, the technician has to instruct the operators (one by one) on key points of operations, quality requirements, and special notes about new models, from the first workstation to the last workstation. The first pair of new model should go offline after proceeding on this instruction, which is also termed as "run-through the whole line." In other words, each operation along the production line is processing the new model. "Instruction time" at each workstation can be determined by the complexity of the new product, which is generally 3- to 5-fold of the normal production takt time; some special workstations may take much longer "instruction time" depending on the complexity or difficulty of the operation. One phenomenon may frequently happen to a vamp stitching line with 25–30 workstations; when the technician instructs at the fifth to seventh workstations, the last pair of the old model may flow to the end of the line, if it runs at the original production takt, which signals the production closure of the old model. At this moment, operators assigned to the last 3/4 of the line have no choice but to wait in the downtime. In the early days with low labor costs, this was commonplace and no one felt strange about it. Moreover, when employees worked on an old model for 10 or more days or even 1 or 2 months, 2 or 3 h of rest can be treated as a human-based concern. In contrast, when labor cost is very high and line changeover occurs too frequently, such 2- or 3-h daily downtime may prove too lavish to labor resources. For employees on a piece-rate salary scheme, the downtime may impose a big loss to the employees, because it means no throughput, and thus zero earnings.

In this regard, how do we reduce the production changeover time of a new product model? This question has been posed to plant managers and faculty members of the Department of Industrial Engineering, Tsinghua University.

We tried to draw on the Single Minute Exchange of Dies (SMED) as contained in the Japanese Lean Production System, which means exchange the production setups or tools within minutes. It did not take us long to realize the limited usage of this method. We can prepare the sewing machine, needles, trimming machine, setting die, marking template or other machines, tools, and auxiliary materials for the next model. With the aid of the maintenance staff, operators can quickly exchange the setups or tools for an old model with new ones, after the last piece of old model shoes goes through one operation. It only takes 1 or 2 min at each workstation, whereas in theory the operator can start the manufacturing of new products. However, most shoemaking operations are highly dependent on operating proficiency. Take, for example, the stitching operation. Stitching speed can reach 10 cm/s for a proficient operator, when the stitching machine sounds like "TokTokTokTokTokTokTok…" continuously, as heard besides the workstation. When one operator begins to study stitching a new style of vamp, he should weigh over the thickness and hand feel of the material, stare at the changing spatial curve of vamp leather margin, hold the vamp, and carefully feed it into the presser of the sewing machine. An unskilled operator has to feed little by little, when the on and off "TokTok, TokTokTok, TokTok…" can be heard, and the stitching speed is only 2–3 cm/s. Stitching can speed up with the gradual adaptation to vamp characteristics and gradual familiarity with actions.

We followed and analyzed the operating time of most types of operations in shoemaking, identifying the significant learning curve of most works, which are in the range of 80–90%. In other words, if the accumulated throughput of new model doubles, the operating time of a single piece will be reduced to about 80–90% of the previous work time. The operating time of one operation gradually decreases with the increase of throughput accumulated, as shown in Figure 2.8. However, it is evident that operating time may fluctuate within a short period, which can lead to the unstable running of the production line.

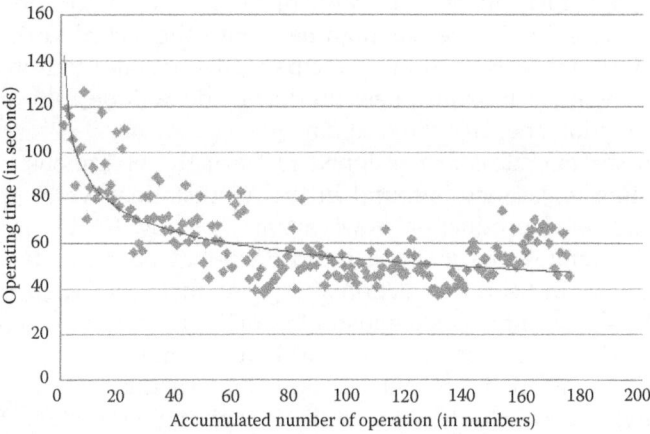

Figure 2.8 Time evolution of shoemaking operation.

Based on this figure, we can understand why the changeover time is long and post-changeover efficiency is very low. The gradually increasing operating skills and production efficiency are the key issue of difficulty in changeover of shoemaking.

The first action we took is "trial production of pilot shoes," which is inspired by the automaker's "pilot car." In the early days of the established new car-making production line, several cars are tested on a daily basis from the first workstation to the last one, in an attempt to test the manufacturing process of each workstation; evaluate the process, quality, and efficiency; identify problems; and make corrections on time. By doing so, each detail of the production line is subjected to a full test, which may establish the subsequent mass production.

In a shoemaking line, people—or operators, to be more specific—are the key to production activities. "Trial production of pilot shoes" means that every operator should know the characteristics of new products, and understand the operating skills and quality requirements, the day before the go-live of a new model. It is usually done by setting a general workstation at one end of the production line, where is laid with machines, tools, and all auxiliary materials or articles. An operator who is making the old model is told to stop and come to the preset general workstation, where the technician will discuss the requirements and skills required for producing new products following the sequence of operations. In general, three pairs of shoes should be finished in the trial production of pilot shoes: (1) For the purpose of "instruction," the technician demonstrates the process of making the first pair, explaining the requirements and basics or key skills. (2) For the purpose of "learning," an operator tries a pair by his/her own hands after being instructed with the basics; then the technician comments on the finished pairs and explains the difference or any problems. (3) For the purpose of "confirmation," an operator confirms the processes and skills in precise alignment with the technician's requirements and guidance. By making three pairs, the operator will master the basics and notes of finishing new products, and will be fully prepared when production goes live the next day.

When the operators are assigned to learn the operations of a new model offline, a vacancy is found in this workstation, which may lead to a discontinued production flow as time goes by. In this regard, an additional operator—basically, a multifunctional operator (termed the "Water Spider" in Lean Production and pronounced as "Mizumushi" in Japanese)—should fill in as a substitute during the vacancy and ensure that production continues to run smoothly in normal takt.

The second action is "relay-race parallel instruction." During the previous line changeover, only one technician is giving instructions from the first to the last operation, but the instruction is slower than the normal production, which would lead to "waiting in downtime" at the downstream of production line. To end the downtime of downstream workstations as soon as possible, we concentrated several technicians and multifunctional operators in one workshop to aid the production line to "run through" the line's changeover, in order to stabilize its running. The first technician begins giving his instructions in Workstation 1; when he goes to Workstation 4, the last pair of an old model is flowing through Workstation 14. At this moment, the second technician is giving his instruction in Workstation 14, then the downstream ones. When the second technician goes to Workstation 18, the last pair of an old model is flowing through Workstation 23. At this point, the third technician begins to give his own instruction in Workstation 23, as shown in Figure 2.9. Thus, three technicians worked together to give work instructions. At the moment the first technician has finished instructing at Workstation 13, the second technician is also finished instructing at Workstation 22, just as the third technician is winding down with his instruction in Workstation 30. In this way, the three technicians shortened the "run through" time to about 2 h, compared to 4–5 h previously. The effective utilization of work hour in the whole production line has been significantly enhanced.

As a matter of fact, there is a lot that needs to be improved during production changeover. Model difference, before and after the changeover, should be small as much as possible, which is considered the better status, as it may quicken the changeover and throughput ramping up. To this end, products of the same category should be assigned to the relatively fixed production line when scheduling the production. Operators will eventually know the processes well and improve the efficiency as well. For example, in one vamp plant, there are 20 vamp lines in total, of which 10 lines are reserved for producing long or short boots, eight lines are for closed shoes, and two lines are for sandals for orders from domestic and

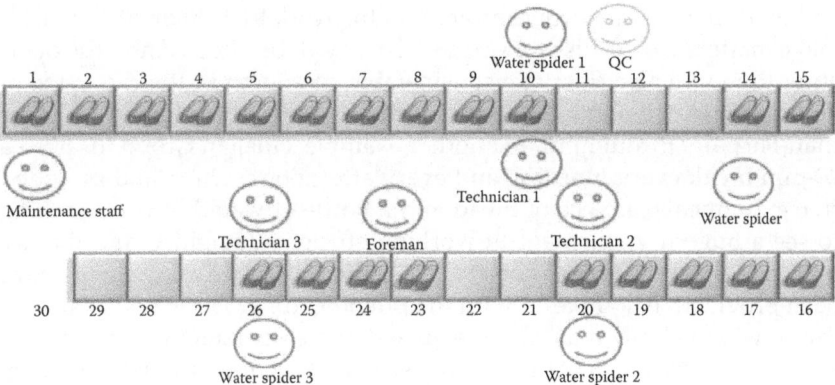

Figure 2.9 Several technicians instruct operators in a parallel way as in a "relay race" during the changeover.

Southern Hemisphere customers, during the busy summer and autumn seasons. Through appropriate and specialized assignments, each production line can perform according to its own strength and ensure the overall productivity.

Production scheduling should be carefully arranged. For example, production orders of the same structure, and different colors and materials should be scheduled to one production line and operated in sequence as much as possible. In this way, process content, sequence, and operation actions can be maintained, where only notes to color and material characteristics slightly change. The production changeover like this means no change, where all orders can smoothly transit to the next in a "seamless" manner, causing no loss of efficiency and quality. Occasionally, several small orders are combined to form a big one, which is steadily processed in batches and for a long time.

2.4   Dynamic balancing of production lines

The operating time of each operation indicates the significant learning-curve effect as described in the preceding section. We also realize the drastic variation of operating time, as exhibited in Figure 2.8. Even if 80–100 pairs of this model of shoes have been made, the operating time fluctuates drastically within a short period (e.g., 10 or 20 min), which may roughly translate to a change of about 10% or 20%, although it seems that the averaged operating time tends to be stable. In other words, this fluctuating process is treated as stationary, or there is no significant tendency of change.

The facts in a shop floor cannot be purely described as "learning curve + random variation." For example, a new model is issued to production. After more than 10 h of hard work, the operator is exhausted, and is unlikely to stay concentrated on his work by 9:00 or 10:00 P.M. To make matters worse, his working efficiency will be affected and the operating time will also fluctuate considerably, as shown in the right side of Figure 2.8, which exhibits the distribution of operating time with more than 140 pairs throughput. In another example, Huajian Group inserted a 20-min break every morning and every afternoon in 2009, and provided free extra meals (a piece of bread and a bottle of water). It is heartening to see a huge improvement in working efficiency, stability, and quality within the same course after the rest period and free extra meals have been provided, if compared with the previous time. These two examples reveal what is known as "fatigue curve," which is intertwined with the learning curve. This means that physical and mental fatigue may affect the employees' working efficiency and lead to an aggravated variation of operating time with the increase in continued working time, followed by an unstable and declining production quality. In the third example, the production scheduler may assign some types of production orders to a certain line as much as possible, when he receives a new order of a product resembling a former product model. This decision is prompted by the fact that this particular line has accumulated the experience of producing a similar category or type of product. Although half a month or 2 months may have already passed, or even if one-fourth or one-third of the senior workers may have already left, the majority of operators in this line, especially the foremen and technical experts, can still count on their memory and past experience to help them handle the operations and quickly ensure a stable production. This is also related to the interval between producing two similar products. If only a few days or 1 week has passed, the workers can easily resume their past production capability; however, if an interval of 2 or 3 months has already passed, it will take them a much longer time to refresh their memory—it is almost like they are trying to handle a new model again. This phenomenon can be interpreted by mankind's memory characteristic of learning knowledge, which is termed the "forgetting curve." Simply put, the acquired knowledge and skills can be quickly recalled and past capabilities can be recovered; however, knowledge and memory of skills are likely to fade off after a long time has passed.

To sum up, in the shoemaking production process, the operating time of each operation is a complex concept—"learning curve + fatigue curve + forgetting curve + random variation"—wherein "random variation" is not presented with a stable variation in the characteristic, and its amplitude will always evolve with the "learning effect," "fatigue effect," and "forgetting effect."

When it comes to production reality, how do we respond to the unidentified variation of shoemaking operations?

One action we can take is to allow WIP buffer stock in some quantity between two workstations (termed "on-hand WIP quantity" in Lean Production) in the production flow of unfixed takt. These buffers avoid any downtime and ensure the continued operation of a production line, even when there is a large difference of operating time between upstream and downstream workstations.

Based on instinctive experience, foremen understand this concept. After we investigated an excellent production line, which was awarded for good production efficiency by HuaJian Group, we identified a common characteristic: WIP stocks in large quantity have been input, consciously or unconsciously, where WIP shoes at each workstation may be 3–5 or 7–8 pairs. By doing so, the production line can continue operating nonstop, and keep the utilization rate of working time at almost 100%, thereby ensuring high throughput per hour. "High throughput cannot be possible without the substantial WIP stocks," added the foreman, when he realized it. Unfortunately, underperforming production managers may not be able to ensure a high throughput, even if a large amount of WIP stocks are accumulated in the production line.

Through analytic calculation and simulation, the effectiveness of WIP quantity (on-hand WIP quantity) at each workstation was examined. And it can be concluded that operators at each workstation can deal with the fluctuating operating time between upstream and downstream workstations, provided there is an additional pair of WIP buffer besides the pair being processed at hand, in normal cases. Even in some special cases, if there is no significant systematic difference in operating time between upstream and downstream workstations although the variation amplitude of operating time may reach 30–50% of the average rate, 2–3 pairs of WIP buffer should be sufficient.

In this regard, why does line break frequently happen even though there is an average of 2–3 pairs of WIP buffer at each workstation in the production line?

As shown in Figure 2.10, if the operating time of Workstation 3 is always longer than that of any other workstation in the same line within a specific period. After a long time, own product being tended to by the Workstation 3 operator will eventually accumulate. Although Workstation 4 had 1–2 WIP stock earlier, that task is soon finished because the WIP that transferred to Workstation 4 is slower to arrive than the processing at Workstation 4, which will result in downtime at Workstation 4. If this problem cannot be resolved, downtime will gradually spill over to Workstation 5 or even to the other downstream operations.

From the viewpoint of production system analysis, Workstation 3 is the bottleneck of this line or is the workstation with the least production capacity. If we organize the production along with the bottleneck operation, according to Goldratt's Optimized Production Technology (OPT) theory, where we adopt push and pull flow, and apply the longest cycle time of a single workstation to the takt cycle of the whole line, then a huge portion of the operating time will be wasted. In contrast, according to Goldratt's Theory of Constraints (TOC), we should let go of the bottleneck workstation, thereby increasing the throughput and releasing the total throughput in the whole line.

Certainly, a multifunctional operator should walk along the production line from time to time and "aid" the workstation that accumulates WIP stocks to some extent (e.g., more than five pairs) until the WIP

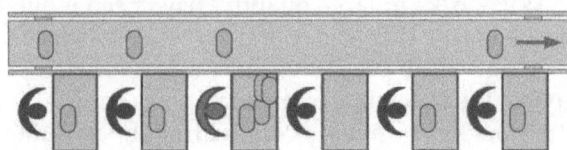

Figure 2.10 "Line break" caused by bottleneck workstation.

quantity is reduced to a more manageable level (e.g., two pairs). After this, he will again search for another bottleneck (workstations with more WIP stocks) and repeat the procedure, as shown in Figure 2.11.

However, bottlenecks do not always necessarily happen in a specific workstation; instead, it will "move around" along the production line with the go-live of new model. For example, in the early phase of production, it may "go through" the line after the changeover, such that if Workstation A is currently the bottleneck, Workstation B may become the bottleneck afterward. If a large order is ongoing for several days, which tends to be stable 2 days after the go-live, Workstation C may turn out to be the new bottleneck.

This is primarily attributed to the degrees of difference in the learning effect among various operations. For some complex operations, the average working time can be greatly reduced with the increase in throughput accumulated, as shown in Figure 2.8, which can be referred to as Type A learning curve. As shown in Figure 2.12, the bottom curve indicates an 80–85% learning rate. It is also possible that an operator may master the basics of some simple operations after processing three or five pairs of shoes; moreover, the average operating time may stabilize after 10 pairs have been processed. In Figure 2.12, the upper curve can almost be fitted

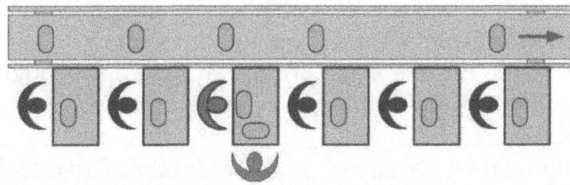

Figure 2.11 Multifunctional operator extends "aid" to handle WIP accumulation at bottleneck workstation.

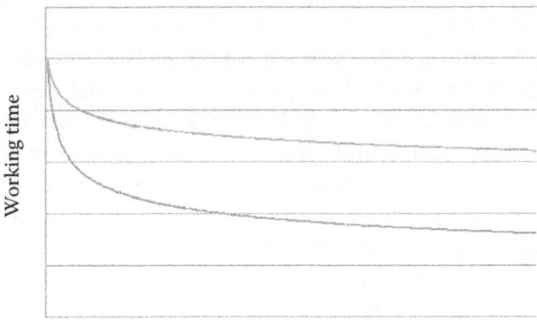

Repeated number of the operation

Figure 2.12 Different learning curve types.

to a straight line, indicating a learning rate of more than 95%. The learning effects of other operations fall between Types A and C, indicating that the learning rate is about 90%.

When assigning the go-live operations of a new product, the foreman estimates the static value of operating time until the production line runs steadily, which can be used for "line balance." In contrast, if the actual operating times of a production line at different phases are averaged (e.g., operating time average of 1 h), it also dynamically varies along the learning curve.

Generally, in the early phase, Type A operations are susceptible to bottlenecks, because the actual operating times in this type of operation at this phase is far from stable, and may even be 2- or 3-fold longer. The production line remains in the phase of "ramping up" throughput, although production takt is below the target capacity, when the relative speed of Type A operation is slow, as indicated by the bottleneck workstation.

A Type A operation usually has to process more than 100 pairs of footwear before the production stabilizes, when its actual operating time decreases to the value at or over the ideal capacity, along the steep learning curve. By contrast, the actual average operating time of a Type C operation within a certain course decreases a little, which does not show a large difference compared with the early phase. Under such circumstances, the Type C operation will become the bottleneck of the production system.

For a similar reason, a Type B operation in medium learning effect usually stays in the middle stage of the ramping-up throughput. When processing the 50–80th pair of the new model, this kind of operation may result in a bottleneck workstation.

"Changing against variation" is the only alternative to counter the "moving around" bottleneck. Within a short time (e.g., 0.5–1 h), WIP stocks accumulate in the bottleneck workstation, which can be addressed by the multifunctional operator's "aid" as shown in Figure 2.11. If one workstation is found to drag the whole line within the long course, the multifunctional operator is bound to come to its aid frequently, which indicates that the total workload assigned to this workstation within this period has surpassed the prevailing takt cycle. In this case, the "aid" cannot address the problem, so the overloaded assignment at the bottleneck workstation should be reallocated to other workstations, either upstream or downstream. Some independent small operations at the bottleneck workstation may be transferred to the previous or the next workstation. If a task at the bottleneck operation takes a long time to finish, then it can be divided into smaller work steps. Dividable, easy, and short preparatory steps (e.g., "marking" or "gluing") or ending work steps (e.g., "cutting threads") can be reallocated to some upstream or downstream workstations. By doing so, the workload of the bottleneck workstation is reduced, allowing the team to align with the total takt—where the "bottleneck" is broken and "dynamic balancing" along the whole line is ensured.

Multifunctional operators can untangle bottlenecks and handle accumulated WIP stocks, not only along one line, but also within several production lines of a workshop or a plant. They may be reallocated according to the actual needs of production processes. For example, when a line is inclined to be stable, the operating time of each workstation also strikes a balance by adjusting the work content, thereby minimizing the variation of "on-hand" WIP in one or two pairs in balanced production. In this situation, a multifunctional operator can spare the time to aid the changeover of new product in another production line, or hand over the early ramping-up throughput immediately after the line changeover. Such appropriate use of multifunctional operators is grounded on exercising the "talent," which ensures the dynamic balance among several lines, thus improving the overall performance of one production unit.

2.5 Production scheduling based on actual capability of lines

Besides the capability difference among operators, preferred product category, production efficiency, and quality level also differ widely among tens of production lines in a large-scale shoemaking plant. This also applies to people in all walks of life. With a view to military forces, for instance, national armed forces can be classified into central major forces, local armed forces, and miscellaneous armed forces; in central troops, some are good at assaults, some are good at guerrilla warfare in mountains, whereas others are good at amphibious operations, etc. Capability difference can be revealed in the past production experience and performance. The historical data of product category, production efficiency, production quality, or delivery lead time, of one production line designated by customers to make high-end goods are provided to us as evidence of the differences in production lines. However, these differences do not rank tens of lines, but expose the various strengths and weaknesses of these workers.

If we analyze the inherent mechanism of line capability differences, it can be treated as individual difference and common difference, which are related to products. Individual difference with products means that different lines are tailored for various product categories. A production line of long riding boots is usually loaded with orders of riding boots throughout the years, where the operators can expertly process all kinds of materials and various styles. Nevertheless, in early spring, there is almost no order of riding boots, so this line can also be loaded with orders of sandals. These employees may feel as if their expertise (processing riding boots) is being disregarded or inappropriately used for making sandals with very simple vamps. This explains why they are inactive and inattentive, which in turn affects their efficiency and quality.

Meanwhile, common difference means the differential efficiency, quality, or other key performance indicators in several production lines for the same product category. It can be attributed to the nature of the labor-intensive industry: personal difference, including individual difference and group difference. As this has already been discussed in the preceding paragraph, individual difference will not be discussed further. The performance of the production line greatly depends on the on-site production foreman (line leader and assistant). As a typical example, if two multifunctional operators are assigned to a regular line and then promoted as the new line leader and his assistant, after the resignation of the former line leader and assistant, this line will become stagnant, and vulnerable to quality-related problems or late delivery. As a result, the operator's income falls, and the other senior guys have no alternative but to resign, thus worsening the situation within more than 1 month. The managerial capability of on-site foremen is a catalyst to the production performance of the entire line. A good production foreman can understand the strengths and weaknesses of the line operators. In particular, the foreman knows how to utilize the strength of each operator, make real-time and dynamic adjustments as needed, and ensure that the whole line is running in a balanced state. Otherwise, the working atmosphere and the team morale will deteriorate. Another problem is the cooperation between employees and foremen. Each operator will actively cooperate if he trusts and understands the action taken by the line leader, thereby smoothing the production flow, whereas if there is any doubt about the action, they will be less productive, create trouble, and be generally difficult, which may lead to overall underperformance. Briefly, grooming a team of competent and dependable production foremen and building a good relationship between leaders and employees are keys to enhancing the company's overall competitiveness in a labor-intensive industry.

The differential actual capacity of production lines cannot be neglected in terms of production planning and scheduling, which should be highly noted in assigning production. The following factors should be taken into account, as they may influence the rules of production scheduling:

1. Which line is well tailored for this model?
2. Which line is processing or will process the product of the same category? If combined, can it reduce the loss arising from line changeover?
3. If this order is loaded on one line, will the delivery be guaranteed according to the actual production capability?
4. Does the quality level of the production line align with the required quality as provided in the order?
5. Load this order to the production line that the customer relies on, in order to enhance customer loyalty.

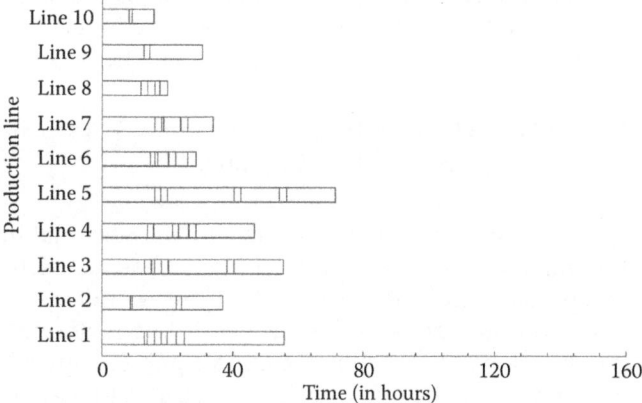

Figure 2.13 Production scheduling of several lines.

Figure 2.13 shows the outcome of production scheduling. In this figure, the gray part, the white part, and the shadowed part represent the order production time, changeover time, and deferred time, respectively. It is evident that Lines 6 and 7 have been loaded with a large order, and that they can maintain a high level of average production efficiency and stable quality, if the delivery is not urgent. Meanwhile, other orders with small quantities and tight delivery schedules will be loaded to other lines. The scheduler will assign operational tasks to each line in this figure, according to the predefined rules, following the receipt of a new order. The objective is to finish these tasks while keeping the total cost as low as possible, provided that the delivery time can be guaranteed.

2.6 Economic production way and scale

The traditional design method of a production line usually presents a production capacity (e.g., annual production throughput) for each line, or infers the throughput takt of production line in accordance with order quantity and the remainder of delivery time. For automakers, for example, an auto assembly line has an annual throughput of 200,000 cars, as provided in the defined production capacity. If calculated by 250 working days, double 8-h shifts per working day, 7.5-h effective time per shift, then the takt time of this assembly line should be presented as

$$(250 \text{ days} \times 2 \text{ shifts} \times 7.5 \text{ h} \times 3600 \text{ s/h}) \div 200{,}000 \text{ cars} = 67.5 \text{ s/car}$$

For another example, a shoemaker receives an order of 3000 pairs of shoes, which is scheduled to make a delivery within half a month. Apart from preprocessing development, material procurement, and mass production preparation, only 4 days are available for production, of which only 3 days are available for stitching vamps. If a line adopts 8 h of regular work and 3 h of extra work, then the takt time should be presented as

$$[3 \text{ days} \times (8 \text{ h} + 3 \text{ h}) \times 3600 \text{ s/h}] \div 3000 \text{ pairs} = 39.6 \text{ s/pairs}$$
$$(\text{rounded as } 40 \text{ s/pairs})$$

If this production instruction is issued to the shoemaking workshop, the shop manager is likely to regard it as infeasible. However, the workshop will probably load this order to five lines simultaneously in an attempt to make on-time delivery, because each line is normally preset with 25–30 workstations, as shown in the middle part of Figure 2.5 (vamp stitching line). In general, the takt time is 150–250 s, but it may vary with the complexity of models. These five lines can stitch 3000 pairs within about 3 days.

In the Donggaun and Ganzhou Plant of the HuaJian Group, a vamp stitching production line is set with 25–30 workstations, which are arranged in "U" shape. In contrast, the Ganzhou and Ji-an Vamp Plant adopted the production layout as parallel workstations arranged at the two sides of a long conveyor belt, where the number of workstations at each line is set at 40–50, as shown in Figure 2.14.

For shoes of the same model, the total working time of all operations accumulated is a fixed value, so the takt time of throughput is inversely related to the number of workstations along the production line in theory. In general, the takt time of the long flow line of the Vamp Plant is approximately 100–160 s.

Figure 2.14 Production method of long production line in vamp plant.

Concerning assignments against order, material transportation and information exchange are inconvenient as the Vamp Plant is remotely located from the main plant, so the orders assigned are always in large quantities (e.g., 3000–5000 pairs) with long time of delivery. As such, employees are more likely to have accumulated production experience; proceed on a later period of learning curve; ensure higher quality and efficiency; and reduce the total production cost—provided that the Vamp Plant can assign works in more detailed job divisions, enable continued production time, and accumulate higher throughput.

The vamp stitching line of the main plant should make a quick response to the urgent order in small quantity (e.g., several hundreds of pairs), as in most cases. On one hand, "line changeover" will take a shorter time during the product change because of fewer workstations. On the other hand, the work assigned to each operator may be diversified; when one can manage two or three operations, the variation in operation time will be offset to some degree, and the effective utilization rate of every takt time will be a little higher. It is considered as the strength of a short line.

Historically, the long line of the Vamp Plant can be traced back to the golden age of shoemaking, which is supposed to be well tailored for orders of tens of thousands of pairs with long delivery time. In the main plant, however, a short line is more appropriate to accommodate an order of thousands of pairs of shoes. Should the vamp stitching line be downsized to 12–15 operators, when orders for "hundreds of pairs" prevail (at the moment)?

The small range and short-term test question its feasibility. Two factors can demonstrate this: one is personnel—that is, when an operator handles multiple operations (e.g., 5–6 operations), his proficiency will be significantly affected. The frequent changeover between different operations or actions slows down the flow, which is supposed to be the primary reason. The other factor is facilities. Many specific operations should be finished by using a special machine or tool, and the equipment or facility required for one workstation may occupy a large area, which was used up by 2–3 workstations before. The operator has to gradually walk to these seats in each takt. Obviously, standing up, turning back, walking, and sitting down result in additional time, thereby affecting his working efficiency. A clever guy wracks his brains to enhance his working efficiency by making 4–5 pairs in each seat, then move to the next seat, and so on, until he returns to the first seat; then he repeats this procedure. However, this procedure may increase the batch size to be transferred, increase WIP quantity, and cause downstream "downtime."

To sum up, product complexity, order size, delivery time, and technological or spatial limitations of the production facility should be considered for production flow, in an attempt to define the most appropriate length of production line.

For orders of 100–200 pairs or even 10–100 pairs, the flow production will "underperform" woefully, because the preparation and warmup costs are too high. In this regard, we reallocated several multifunctional operators to the tailored small-order workroom. The production cell approach, the production method used in a small batch trial that was usually applied in product development, is used instead of the traditional flow line production, as shown in Figure 2.15.

Each cell consists of a production team, where assignments and coordination are available as led by each team leader. The quantity and operation status are handled by the team leader, who is responsible for mapping production through good coordination, although there is no definite production takt. Within the production cell, every operator works from time to time, and the downtime that is frequently seen in flow production scarcely happens. Moreover, equipment and facilities including sewing machines, gluing workbenches, and hot presses are shared by multiple operators, by time segment and in operation sequence. In this way, working sequences and stands are more flexible, thereby saving working space.

The scale of production cell is likely to change with the complexity and process characteristics of a product category. With repeated trials, it can be concluded that the production cell assigned to riding boots should be appropriately allocated with seven to eight operators, because it has many parts and complex operations. The production cell of close shoes should be appropriately allocated with five to six operators, whereas the production cell of sandals and slippers should be appropriately allocated with three to four operators, because it only has a few parts and simple procedures.

In the cell production approach, the total production cycle time of an order lasts for 1–2 or 5–6 days depending on the order quantity. It is a rational time course, which enables operators to accumulate experience and expertly perform their assigned tasks.

Figure 2.15 Production cells of shoemaking in small batches.

Reality can resolve the controversy of production methods. We prefer applying the new production approach and the flexible scale of the production line as a response to market volatility, product, as well as staff or process technology and equipment change.

2.7 Summary

In a response to the list of problems raised in the beginning of this chapter, we may draw the following conclusions:

1. Newcomers should be trained appropriately so as to screen for the most appropriate candidates for each operation, regardless of labor shortage and recruitment difficulties.
2. The adaptive line balancing method can form an effective production force, called the "Dream Team," by grouping those operators whose skills are below the required standard.
3. In a response to highly volatile market demand, "quickness against change" production changeover should be performed to ensure that high efficiency and quality are maintained, regardless of market volatility.
4. In a response to volatile process change, "change against volatility" should be performed to strike a good balance, regardless of market volatility.
5. To reduce the total cost and improve production efficiency, it is necessary to harness the strength and weakness of each production line.
6. In our humble opinion, there is no optimal level but the most appropriate production way and scale.

chapter three

Culture

Management style comparison

Qin Gao

Contents

Visual management (VM) is a management technique that uses visually stimulating signals, symbols, and objects to convey important information in the working environment. It has many forms, such as figures, tables, and signs, and comes in different colors and sizes. Typical VM tools include production kanban, safety signs, and standard operating procedures (SOPs). VM can be implemented in a top–down or a bottom–up manner. However, the systemic approach from top to down is more effective to improve the organization's performance. Most companies have implemented VM within their organization, but not all of these VM systems work effectively in these companies. The reasons why VM has not proved successful in these companies will be discussed in this chapter. In addition, as the development of global economics rolls on, more and more companies develop their businesses at abroad. One of the common problems these companies encounter is that their successful management experience does not work well in foreign countries. The reason for this is that cultural differences are ignored. Although the level of industrialization and economic development plays an important role in management, the influence of culture on management is profound. In this chapter, we will first introduce VM and the problems with implementing VM in practice. Then, the influence of national culture on management is discussed. Third, based on the results of our field study, we will compare VM across six factories from four different countries.

3.1 VM

VM is an important tool for supporting effective workshop management. The objective of VM is to improve productivity, safety, quality, on-time delivery, profits, and employee morale (Liff and Posey, 2004). The elements used in VM include figures, tables, signs, and other means that can be perceived by workers easily and directly. VM uses these visual stimuli to highlight, report, clarify, and integrate mission, vision, values, and culture into an organization's operating systems and performance requirements. Since the 1990s, VM and 5S (Sort, Set in Order, Shine, Standardize, and Sustain) have been widely accepted as cornerstones of lean thinking. VM is often implemented as the first step in a lean transformation, and some companies adopt it as the central theme in their "lean" production system.

Why are VM and 5S so readily accepted? It happens that good house-keeping and workplace organization is perfectly compatible with the mass production economies-of-scale paradigm. This implies that a company can fully endorse VM and 5S without embracing any other aspect of lean thinking. For example, if tools and materials are conveniently located in uncluttered work areas, operators spend less time looking for items or stumbling around unneeded materials and equipment. This leads immediately to higher workstation efficiency, a fundamental goal in mass production. Additionally, VM provides a clear and common understanding of goals and measures. It allows people to align their actions and decisions with the overall strategic direction of the company. It is also an open window to factory performance, and it provides the same unbiased information to everyone, whether owner, manager, operator, or visitor.

The VM system has the following functions:

- It translates critical organizational requirements into visual stimuli that cannot be ignored.
- It uses these visual stimuli to highlight, report, clarify, and integrate mission, vision, values, and culture into an organization's operating systems and performance requirements.
- It creates an environment that enhances employee commitment to the success of the organization by ensuring that the work environment and culture directly support the mission and values of that organization.
- It presents key data and information through the use of compelling sensory messages that reinforce what is important to the organization.
- It addresses performance issues and keeps people focused on the real mission and goals of the organization.
- It provides a mechanism for continuous improvement through system alignment, goal clarity, and engagement of people in the process, and improved communication and information sharing throughout the organization.

3.1.1   Classification of visualization tools

The production activity contains five elements: man, machine, material, method, and environment. Based on these five elements, we classify VM tools into eight categories:

Man—Performance management
Machine—Equipment management
Material—Material management, Quality management
Method—Operation management, Continuous improvement, Production control
Environment—6S (Sort, Set in Order, Shine, Standardize, Sustain, and Safety)

- Visualization of performance management
  The visualization of performance management refers to visualizing the assessment result of personal performance, employee's skill level, and so on. Typical visual tools include table of skill level, honor board, employee's behavior norms, and table for performance assessment.
- Visualization of equipment management
  Visual tools for equipment management are used to represent the working status of equipment, instructions of operating equipment,

and performance indicators related to equipment. Popular visual tools include operating instructions, instructions for emergency, monitoring chart of equipment utilization ratio, and monitoring chart of abnormal equipment stopping.

- Visualization of material management

    Visualization tools for material management refer to signboards of raw materials, work in process (WIP), and products, operation board in the inventory, and statistics table of material level. Visualizing material management can keep the material in good order, which can help workers search for material and under the volume of material easily. Some instances for visualizing material management are signboards for visualizing the name, number, location, and appearance of material and product, and inventory status kanban.

- Visualization of quality management

    Visualization of quality management is used to visualize standards for inspecting, categorization of deficient products, quality status of production, quality performance indicators, and so on. The purpose of visualizing quality management is to demonstrate quality requirements and quality status, which can assist managers to find the quality problems quickly and take measures to solve problems.

- Visualization of operation management

    Visualization of operation management refers to visualizing the operation procedure, methods, and related considerations of operation. Tables, figures, and text are used to visualize the detailed operation procedure to operators and assist them to accomplish the job. SOPs are a widely used visual tool in factories for visualizing detailed operation procedure, operation time, and related considerations.

- Visualization of continuous improvement

    Visualization of continuous improvement is used to exhibit the results of continuous improvement and related performance indicators of workers—for example, the number of continuous improvement proposals and profits of continuous improvement. The most common visual tool is kanban for continuous improvement. The precondition for visualizing continuous improvement is that the factory conducts continuous improvement activities in the factory.

- Visualization of production control

    Visualizing tools for production control can transmit information about production planning and production control across managers and workers. Usually, the workers who engage in production activities are the source of basic production information. However, managers have to grasp the production information to make decisions in daily work. Then the visual tools for production control can visualize the production information and performance indicators to transmit the information between managers and workers. The visual tools include the electronic production kanban, production planning kanban, task assignment sheet, and so on.

* Visualization of 6S

    Area identification includes area classification for materials, tools, and equipment, which is an important method for field management and also the requirement of 5S. Visualization of safety and 5S includes safety warnings, safety signs, and safety bars. Some examples include the track for tools position, area division in the workshop, safety checklists, and safety records.

3.1.2   Problems with VM

Although the VM system is widely used in factories, there are still numerous problems associated with it. These problems are the reasons for the failure of VM in practice. The problems of VM can be classified into three large categories:

* Problem with information

    Some failures of VM are caused by improper information being visualized. Usually, the visualized information is not critical to the production management or not useful for an individual's work. Another problem is that the presented information is incorrect or out of date owing to lack of maintenance.

* Problem with representation

    This type of problem is concerned with the improper visualization style of information. The representation style is not tailored to users' cultural background, and the cultural difference is ignored. The ergonomic and psychological aspects are not taken into consideration. As a result, the representation is not easy to understand.

* Problem with implementation

    One of the most common problems with VM implementation is that the executive committee does not give inadequate introduction and training to users. The worst case is that the target users even do not know that the company has established visual tools. Another common problem is that the executive committee does not conduct continuous improvement activities for VM.

3.2   Cross-cultural management

Culture plays an important role in management style, although management style also depends on education, organization culture, economics, and technology. With the globalization of the economy, more and more companies are expanding their businesses in foreign countries. The communication style, management style, and problem-solving techniques are varied in different cultures. Consequently, the management and leadership style that prevails in the home country are very likely to fail in another country. In-depth understanding of the cultural differences and taking adaptation measures are imperative for cross-cultural management.

3.2.1   Hofstede's cultural dimension

According to Hofstede (1980), culture is the collective mental programming of the human mind that distinguishes one group of people from another. He conducted a comprehensive study of how values in the workplace are influenced by culture. The data were collected from more than 116,000 IBM workers, who hailed from more than 70 countries. A four-dimension model was proposed based on the data, which differentiated different national cultures and different values of people. The four dimensions are *Individualism versus Collectivism, Power Distance, Uncertainty Avoidance,* and *Masculinity versus Femininity.* In 1991, a fifth dimension was added, which is called *Long-term Orientation.* In 2010, using World Values Survey data, Minkov generated another two dimensions. One was a new dimension, and the other was similar to the fifth dimension. Then, the fifth dimension was adapted and called *Pragmatic versus Normative* (PRA). The sixth dimension is *Indulgence versus Restraint.*

- Individualism versus Collectivism

    Individualism (IDV) refers to the relationship between the individual and the collectivity that prevails in a given culture. Along the continuum of individualism and collectivism, an individualistic country places primary importance on the needs and goals of the individual, whereas a collectivistic country places emphasis on the needs and goals of the collective. People from individualistic cultures assume that individuals look after themselves. Independence, self-reliance, autonomy, and individual achievement are valued in an individualistic culture. In contrast, collectivist cultures are mostly concerned with in-groups. Group harmony and welfare, interdependence, and relationships are valued in a collectivist culture. East Asian cultures are more collectivistic than Western cultures, which tend to value individualism. Some studies suggest that both approaches may foster organizational cooperation, but in a different way (Chen, Chen, and Meindl, 1998; Wagner, 1995).
- Power Distance

    Power distance refers to the degree of inequality that the population of a country considers normal. Hofstede (2001) defined power distance as the extent to which the less powerful members of institutions and organizations within a country expect and accept that power is distributed unequally. High Power Distance Index (PDI) indicates that unequal power and financial conditions are approved by a society. By contrast, low PDI emphasizes on minimizing the differences on power and wealth between individuals. A society with high PDI tends to focus on high autocratic leadership and centralization of authority.

- Uncertainty Avoidance

    Uncertainty Avoidance Index (UAI) refers to the extent of uncertainty and ambiguity that a society can tolerate (Hofstede, 1980). Cultures with high uncertainty avoidance tend to minimize unstructured situations. These societies are rule-oriented and institute laws and regulations in order to reduce the extent of ambiguity. By contrast, countries with low UAI have more tolerance and are more willing to accept changes.

- Masculinity versus Femininity

    This dimension refers to the extent of role divisions between genders. Hofstede (1980) asserts that masculine cultures are those that insist on maximum distinction between the roles of men and women in societies. High masculinity countries place emphasis on culture with a high domination of males in the society with competitive, assertive, and ambitious trains. By contrast, feminine cultures care more about quality of interpersonal relations and quality of working life.

- Pragmatic versus Normative

    This dimension describes how people in the past or today, relate to events that are happening around us and accept the fact that they cannot be explained. In societies with a normative orientation, most people have a strong desire to explain as much as possible. People in such societies have a strong concern with establishing the absolute truth; they are normative in their thinking. They exhibit great respect for traditions, a relatively small propensity to save for the future, and a focus on achieving quick results. In societies with a pragmatic orientation, most people do not have a need to explain everything, as they believe that it is impossible to understand fully the complexity of life. The challenge is not to know the truth but to live a virtuous life. In societies with a pragmatic orientation, people believe that truth depends very much on situation, context, and time. They show an ability to adapt traditions easily to changing conditions, a strong propensity to save and invest, thriftiness, and perseverance in achieving results.

- Indulgence versus Restraint

    Indulgence stands for a society that allows relatively free gratification of basic and natural human drives related to enjoying life and having fun. *Restraint* stands for a society that suppresses gratification of needs and regulates it by means of strict social norms.

Hofstede (1980, 1991) suggested that managerial style is influenced primarily by the cultural context in which an organization exists. The six cultural dimensions all have influence on the management style. Key's (2000) study has found that individualism was positively related with participative management style. In participative management style, all levels of employees are encouraged to participate in the decision-making process. Power distance is an important determinant of relations between managers and employees in a company. In high power distance cultures, the most widely used management style is the autocratic style, in which subordinates are under closely supervised control and decisions are made by managers. Managers in the masculine work environment are more decisive and assertive. In feminine cultures, managers are intuitive and insist on general agreement (Jandt, 2006). In cultures with high uncertainty avoidance, subordinates prefer that goals, assignment, policies, and procedures are carefully detailed and pronounced. By contrast, in cultures with low uncertainty avoidance, subordinates can tolerate unclear descriptions of the goals and processes. Table 3.1 gives each country's six dimension indexes.

China ranks high for both power distance (PDI) and Pragmatic/ Normative (PRA). Unequal relationships are accepted in China. Elders naturally receive respect from young people, and the same goes for the relationship between seniors and subordinates. In terms of IDV, China ranks lowest among the four countries. China's strong collectivist orientation leads to the belief that everyone takes responsibility for the members of their group. The Chinese management style is authoritative and directive. The managers are expected to make decisions on behalf of the group. Managers' orders are usually given to the immediate subordinates and then systematically passed down the hierarchy.

The United States ranks highest for IDV among these countries. Its power distance (PDI) and uncertainty avoidance (UAI) ratings are also lower than those of other countries in the table. The relationships between different management levels are generally cooperative, so as to achieve a common goal. Similar to China and Germany, its Masculinity index is moderate. This indicates that the country supports a higher degree of gender differentiation.

Table 3.1 Hofstede's cultural dimensions

Country	PDI	IDV	MAS	UAI	PRA	IND
China	80	20	66	30	87	24
USA	31	100	63	37	21	68
Germany	35	67	66	65	83	40
Japan	46	47	100	81	82	42

Note: IDV, individualism; IND, indulgence versus restraint; MAS, masculinity index; PDI, power distance index; PRA, pragmatic versus normative; UAI, uncertainty avoidance index.

Germany ranks lowest in the dimension of power distance among the countries listed in Table 3.1. The relationships between managers and employees are often close. German companies concentrate more fully on product quality and product service. Both managers and employees believe that working together can create a good product. Germany's individualism index is also high. Germans value people's time and freedom. Competitiveness, assertiveness, and ambition are also valued by Germans.

Japanese culture places high emphasis on the male's role (it scored highest on Masculinity index). Compared with China, its PDI is lower and its IDV is higher. Japanese workers deem equality as a way of maximizing cohesion, which impacts productivity. Both supervisors and employees are required for making decisions. Rather than being a source of authority, the top management is seen as the facilitator and responsible for maintaining harmony between employees. Japanese companies invest substantial resources in their employees' training and career development. The employees are expected to stay in one company for their entire working career. In contrast, Americans usually jump from one company to another for higher salary and to gain more experience.

3.2.2 Cultural influence on communication style

Communication style also varies widely between cultures. Through communication, managers' plans and decisions are conveyed to subordinates, and in turn, managers learn about their subordinates' ideas and requirements. An appropriate communication style determines the effectiveness of the management measures. The communication style is especially important for VM. One of the implementation steps for VM is to promote and execute the VM tools among managers and subordinates. Usually, VM tools are developed by one group of people and used by another group of people. How to effectively communicate the purpose and use of VM tools to users is important.

Hall and Hall (1995) presented a clear distinction between *high-context culture* and *low-context culture*. A high-context culture is one that is highly dependent on the context—that is, many aspects of the culture are only understood by those living within that culture—the "in-group" so to speak. In a high-context culture, people have had similar experiences and so many things are left unsaid. High-context cultures are more common in Eastern and Middle Eastern countries, such as Japan, China, Egypt, and Saudi Arabia. In contrast, in a low-context culture, many more things are "explicit" in the environment because members of the culture come from a wide variety of backgrounds and traditions. In low-context cultures, people tend to have many loose connections of shorter durations. Because of their heterogeneity, such cultures can change significantly from one generation to the next. Some examples of low-context cultures are the United States, Germany, UK, Canada, Denmark, and Norway.

3.3   Comparison of VM across cultures

We conducted a field study to investigate VM implementation in practice. Six factories were investigated. These factories represent different countries and different cultures. The investigation time for each factory lasted 2 or 3 days. The investigation procedure for each factory was as follows: background information collection, field observation, manager interview, and worker interview. Background information about the investigated factory includes history, products type, scale of the factory, manufacturing mode, organization hierarchy, and information system. During field observation, we observed what visual tools were used and how these tools are used in daily operations. The VM tools used in the factory were recorded photographically for further analysis. During the observation, the supervisors in the workshop were asked to interpret the functions and usage of visual tools in the field. Next, managers and workers were interviewed about VM in practice. We attempted to interview staff members from the same department and also covered different levels of the organization (e.g., department manager, section manager, monitor, and workers). Through this process, we can identify how VM tools were established, how these tools were implemented in the workshop, and the attitudes of the managers and workers toward these tools. The number of interviewees and their education level are shown in Tables 3.2 and 3.3, respectively.

After the field study and the data collection phase were completed, we conducted an overall assessment across the investigated factories. We attempted to discover the good cases of VM in China and dig out the reasons why they were successful. VM used in these factories is compared

Table 3.2 Number of interviewees across investigated factories

Factory	A	B	C	D	E	F	Total
Manager	4	5	7	6	7	6	35
Worker	3	4	5	2	3	2	19

Table 3.3 Education level of interviewees

Factory	A	B	C	D	E	F
Above bachelor's degree	3	4	4	2	4	6
Junior bachelor's degree	1	1	3	0	3	0
Middle school	2	2	4	6	2	2
Polytechnic school	1	2	1	2	1	0

horizontally and vertically. From the horizontal perspective, the type of VM tools, the expected users of these tools, and the effectiveness of these tools were compared. From the vertical perspective, the planning, design, implementation, and control process of VM in each factory were compared. Additionally, a best working map for execution of VM is summarized from these cases.

3.3.1 Overview of six investigated factories

The investigated factories cover the electric industry, transportation industry, and shoe industry. Different industries have different types of products and production modes. For example, automobile plants usually adopt flow line production. The following background information will be explained and compared among the six factories: industry of the factory (the industry domain to which the factory belongs), history (development history and the state owner), scale (production capacity and economic scale), and production management and products (including production organization, planning, scheduling, and control type of products). The overview of the factories is depicted in Table 3.4.

- Industry

    Factory A (Germany) designs, manufactures, and sells signaling products and systems for main line, mass transit, special railways, other rail bound transportation systems, and related engineering and services. Other business activities cover system installation, execution, and subsystem development. Its component products include electric point machines, light-emitting diode (LED) signal light units, axle counters track circuits, and point controllers. System products include interlocking system, train control system, and other systems.

    Factory B (USA) designs, produces, and sells four-wheel and all-wheel drive torque management systems and provides auto companies with the optimum four-wheel and all-wheel drive system

Table 3.4 Overview of investigated factories

Factory	Industry	Country
Factory A	Electric	Germany
Factory B	Automobile	USA
Factory C	Electric	Germany
Factory D	Clothing	China
Factory E	Automobile	Japan
Factory F	Machinery	Japan

solution. The factory not only develops strategies and products to manage engines for fuel efficiency, reduced emissions, and enhanced performance, but also develops interactive control systems and strategies for traditional mechanical products.

Factory C (Germany 2) produces and sells medium-voltage vacuum interrupters, vacuum circuit breakers, and gas insulated switchgears, as well as after-sales service. These products are used in the electric industry.

Factory D (China) produces shoes for several famous international brands and some local brands, although the factory has its own brand. In addition to shoemaking, the factory also manufactures tanning and dies.

Factory E (Japanese) is an automobile plant. It is a joint venture of Japanese and Chinese automobile manufacturers. It mainly produces passenger cars including compacts, medium size cars, full size cars, multipurpose vehicles, and sports utility vehicles.

Factory F (Japanese) produces pneumatic components, covering high-pressure products, high frequency, high speed response, long service life products, fluororesin products, air preparation, and airline equipment, and stainless products. To meet the specific requirements of some customers, the plant set up a design and production department for nonstandard products. Besides standard products, the factory also produces nonstandard products and integrated systems such as pneumatic control cabinet based on specific requirements from customers.

- History

    The history of the factory has an impact on the adoption of VM, especially regarding its acceptance. The factory's history influences the attitude of both its managers and workers. One prevalent phenomenon we found during investigation is that when worker interviewees were asked about the establishing process of the visual tool, usually their answer was that the tool was already there before they entered the company and it continues to exist even until now. Several factories just imported the visual tools from parent companies, without localizing them in the Chinese factory. Once the employees become familiar with tools, it is difficult for them to abandon the old tools and accept new ones. Japanese factories dedicate a large portion of their resources to field control and VM, which are affected by the style of Toyota Production System (TPS), whereas European and American factories take field management less seriously. When we conducted our field investigation, we found that Japanese factories have had several years of experience of implementing VM, whereas the European factories have not yet conducted systematic VM in the entire factory. As a result, the attitude toward the adoption of VM

varies considerably between Japanese and European and U.S. factories. Japanese organizations have a long tradition of making a neat
and clear production field, whereas European organizations may
not even intend to do so.

Factory A (Germany) is a Sino–German joint venture that was
established in December 1995. The Germans hold a 70% share of the
factory, whereas Chinese stakeholders own 30%. The parent company, which is located in Germany, brings advanced technology
to the factory. Factory B (USA), a joint venture between Chinese
and U.S. companies, was established in early 2001. Factory C
(Germany 2), a joint venture of German and Chinese companies,
was established in November 1997. Factory D (China) is a local
plant that was established in 1996. Factory E (Japanese), a Japanese-
invested automation plant, was formally established in 2003.
In 2004, the factory started to manufacture its products. In 2005,
the factory brought in the Japanese production mode and began
to promote Total Productive Maintenance. At present, the factory
can produce about seven types of cars simultaneously, which is a
highly mixed production mode. Factory F (Japanese) is Japanese
solely invested corporation that was established in 1994. In 1996, the
factory started to manufacture its products. In 1997, a second plant
was established. The factory then obtained the ISO9002 (Quality
System Authentication) certification. In 2004, the research center of
the corporation was set up.

* Scale

The scale of the factory is measured by the volume of its sales
each year, the size of its workforce, and the number of equipment. The scale is related to the characteristic of the products.
Generally, the larger the scale of the factory, the more systematic VM
becomes. The large-scale factory always conducts VM across different departments of the factory. An execution committee is set up and
takes charge of the promotion of VM across the factory. The basic
templates for VM tools are also created, and then each department
can modify the templates to set up their own tools. In a small-scale
factory, each department may set up their own visual tools without
referring to other departments. Different departments do not learn
from each other. In terms of scale, Factory D, Factory E, and Factory F
are larger than Factory A, Factory B, and Factory C. In addition,
large-scale factories have the means to procure some special tools to
monitor the status of production, for example, setting many monitors for the unmanned production field.

Factory A (Germany), which has a registered capital of 41.4 million RMB, occupies a total land area of 34,266 m^2. The area of the
workshop is 2165 m^2, and the office area is 1192 m^2. Factory B (USA)

has a registered capital of $3.76 million. American stakeholders own 80% of the company, whereas Chinese shareholders own the rest (20%). Factory C (Germany 2), which has a registered capital of $26 million, is mostly owned by German shareholders. Factory D (Chinese) has a workforce of about 20,000, of whom 80% are field workers. At present, the factory has 47 modern production lines. The factory has set up several subsidiary factories. Each year its sales volume is about 20 million pairs of shoes. Factory E (Japan 1) occupies an area of about 1,075,000 m², of which the green area accounts for 301,000 m². The factory has a staff of more than 3000 employees. Its annual production capacity is about 360,000 cars. Factory F (Japan 2), which has about 4200 employees, has a registered capital of 41 billion yen. The factory has four subsidiary plants in Beijing. In total, the plants occupy about 450,000 m². Its products are mainly sold to foreign countries (e.g., Europe, America, and Japan).

- Production management and products

VM is also related with the production model and types of products. For example, the electronic kanban is not well tailored for small batch and multitype production, because the daily output is very easy to count. However, the electronic kanban is very effective for large batch and flow line production mode, which can monitor the real-time production output. For small batch and multitype production mode, VM should focus on visualizing production information (e.g., production procedures, setup time of equipment) of each processing unit. In addition, usually quality inspection in small batch and multitype production takes full product inspection. VM tools for recording the number of deficient products are also not very appropriate. With respect to flow line production, it is necessary for the factories to establish visual tools for recording the number of output products and deficient products.

Factory A (Germany) produces signaling products and systems for main line, mass transit, special railways, other rail bound transportation systems, and related engineering and services. Factory B (USA) has several production lines for different types of products. It is not a typical flow line production. Factory C (Germany 2) operates on typical small batch and multitype production mode. The extent of customization is relatively higher compared with other investigated factories. Daily production plan is achieved only about 50% of the time, because there are many urgent orders every day, which have not been originally planned for. Factory D (China) is a labor-intensive factory, and the level of automation is relatively low compared with that of a carmaker or an equipment maker. Most processes are done by hands with the assistance of very simple tools or equipment. In 2008, the factory began to learn from Lean Manufacturing and this brought substantial changes to its production lines and manufacturing philosophy. Factory E (Japanese) uses a typical flow line production system. The extent of customization of products is low. The general manufacturing process includes forging and pressing, welding, resin and painting, and assembling. Factory F (Japanese) is a typical make-to-stock type. Its production is inventory oriented. Daily production is used to supplement inventory and not for market sales. The production process mainly involves three steps: parts and component production by machines, washing and cleaning, and assembling.

3.3.2 Comparison of VM across culture—A horizontal view

3.3.2.1 Overall assessment of VM across factories

Based on the classification discussed earlier, an overall assessment of VM is conducted across six factories. In order to assess how well VM is implemented in each factory, we developed a four-level evaluation criterion. The first level, also the best level, is that the factory has set up a complete set of tools and the tools are functioning very well in the factory. The second level is that this type of tools exists and functions, but is not complete. The third level is that this type of tools is not complete and also does not function. The fourth—and worst—level is that the factory has not yet set up this type of tools. Table 3.5 gives the assessment results for the investigated factories.

In the following discussion, we will provide more details on the implementation of VM in each factory. According to the above classification, VM is compared across the factories for each category. The assessment covers the following aspects:

- Types of visual tools used in each factory
- Effectiveness of visual tools, which is assessed on the basis of the employee's perception toward the tool, motivation for using the tool, and behaviors shaped by the visual tools

Finally, the best practice for this category of visual tools will be shown.

3.3.2.2 Visualization of 6S

Under 6S, the intention is to integrate safety with the traditional 5S. The purpose of VM of the workplace and 5S is to attract the attention of workers and regularize their behaviors. The ultimate goal is to decrease the probability of making mistakes and increase profits. From the interview

Table 3.5 Overall assessment of visual management
across investigated factories

Factory	Germany 1	USA	Germany 2	China	Japan 1	Japan 2
Performance management	◌	○	○	○	◎	◌
Equipment management	◌	◎	◎	○	◎	◎
Material management	○	◎	◎	◎	◎	◎
Quality management	○	○	○	○	●	◎
Operation management	◎	◎	○	◎	●	◎
Continuous management	◌	○	◌	◌	◎	◎
Production control	◌	◎	○	◎	●	●
5S, safety, area identification	◎	◎	○	○	●	◎

● Complete and function very well

◎ Incomplete but existing tools function

○ Incomplete and not functional/do not function very well

◌ Do not exist

results, we find that the visualization tool of workplace management and 6S are universally established in the investigated factories. However, the extensity and intensity levels during the process of conducting VM vary across factories. We summarize the visual tools used by the investigated factories in Table 3.6.

Not all factories use the visual tools listed in Table 3.6. Most of the factories only set up about four kinds of tools. The visual tools used by each investigated factory are listed in Table 3.7. Generally, more than half of the established tools are useful in investigated factories. Some tools should be collaborated with supervision measures to take effect. Tools of S01 (Area division) and S02 (Area identification board) are usually effective to regularize the behavior of employees after they are set up. The problems for implementing S01 and S02 are that the factory usually does not set up enough and appropriate tools for workplace management and 5S activity. When establishing S04 (Safety warning board) and S05 (Safety signboard), the manager should monitor the behaviors of workers, and may set up key performance indicator (KPI) to regularize their behaviors. Checklists (S08) are a good and useful tool for shaping the behaviors of workers.

Table 3.6 Summarization of visual tools for 6S in investigated factories

Category	Code	Visual tool	Function
1. Area identification	S01	Area division	Distinguishing different areas for material, product, equipment, transportation vehicle, or other facilities
	S02	Area identification board	Identifying different processing units
	S03	Tool position	Setting track to place tools in fixed position
2. Safety signs	S04	Safety warning board	Warning staffs to be careful
	S05	Safety signboard	Demonstrating safety requirements
3. Monitoring execution of 6S	S06	5S standard and knowledge	Demonstrating the standards for conducting 5S activity
	S07	Safety accidents statistics	Monitoring and controlling safety accidents
	S08	5S and safety checklist	Checking the safety-related items according to checklist
	S09	5S assessment	Marking across different departments on the 5S activity

- Best practice for visualizing of workplace management and 6S

  Among the six factories we have investigated, Japan 1 does best in the visualization of workplace management and 6S. The factory has established the most complete and systematic VM system. All types of tools are covered by the factory. When designing the styles of visual tools, the factory also adopted several novel elements to attract the attention of workers, for example, a cute foot symbol. Moreover, the factory has set up several rules to assure the effectiveness of the tools. For example, managers supervise the workers to encourage them to walk in blue-coded areas. Once the workers violate the rules, they will be issued warnings, and may even have their salaries deducted if they commit serious mistakes.

  Good examples of this concept include (Figure 3.1):

1. Area identification for tools. This is used to identify the positions of different tools. With this measure, workers can quickly find the tools they need. Also, the tool that is missing can be quickly found.

Table 3.7 Evaluation of used 6S tools

Factory	Evaluation
Germany 1	• Established tools: S01, S02, S04, and S06. • General evaluation: The established tools are effective, but lack other advanced visual tools, such as S03 and S07.
USA	• Established tools: S01, S02, S04, S05, S07, and S08. • General evaluation: Visual tools S01 and S04 are effective and regularize the workers' behavior; tools S06, S03, and S07 can motivate the workers to some extent. The visual tools are also incomplete, but better than those of Germany 1.
Germany 2	• Established tools: S01, S02, S04, and S06. • General evaluation: The factory lacks many useful tools and the established tools do not function very well.
China	• Established tools: S01, S02, and S04. • General evaluation: Tools S01 and S02 are widely used and effective; tool S04 is too simple and does not function very well; the factory also lacks many other useful tools.
Japan 1	• Established tools: S01, S02, S03, S04, S05, S07, and S08. • General evaluation: The array of visual tools in the factory is almost complete, and supervision measures are also adopted to assure their effectiveness.
Japan 2	• Established tools: S01, S02, S03, S04, and S07. • General evaluation: The extent of visual tools is effective; other tools should be set up in the future.

(1) Area identification for tools   (2) Safety signboard   (3) Safety checklist

Figure 3.1 Good examples of 6S visualization.

2. Safety signboard. This is used to demonstrate the safety requirements for operations. It clearly shows which protection tools are required for a specific operation and which type of behavior is prohibited in the workshop.
3. Safety checklist. This lists the items that should be checked for safety purposes.

3.3.2.3 Visualization of operation management

The most widely used visual tool for operation management is SOP. SOP is especially important to novice workers, who are not familiar with the job. One popular phenomenon we found during the investigation is that the senior workers do not always refer to SOPs after grasping the operation skills. This is not good for the continuous improvement of operations. SOPs also need regular updates and maintenance. There are two levels of SOPs. One is the decomposed version, on which very specific operation procedures are listed. The other is the compiled version, in which only the key procedures are shown. The decomposed version is used to train novice workers, whereas the compiled SOP is referred to by senior workers. Each worker owns a copy of the compiled SOP, and the production line owns one or several copies of the decomposed SOP. For a well-designed SOP, figures are usually used to illustrate the procedures. In addition to SOPs for repeated operation jobs, there are also nonrepeated SOPs, which are mainly used for emergency or irregular daily jobs (e.g., equipment breakdown). The nonrepeated SOPs instruct users how to handle irregular problems correctly and efficiently. Different factories achieve different levels of SOP execution. The monitoring chart of SOP fulfilling rate is another tool for operation management. The SOP fulfilling rate is a kind of KPI for the individual worker; it is used to measure to what extent the worker operates according to SOPs. The visual tools for operation management across the investigated factories are summarized in Table 3.8.

All investigated factories have created their own SOPs. However, some factories have set up a very complete and effective SOP system, whereas

Table 3.8 Summarization of visual tools for operation management in investigated factories

Category	Code	Visual tool	Function
1. Standard operation procedure (SOP)	O01	SOP	Demonstrating operation procedure with figures illustration for repeated jobs
	O03	Nonrepeated SOP	Demonstrating the handling procedure for nonrepeated jobs
2. Standard operation time (SOT)	O02	Standard operation time	Demonstrating the standard operation time
3. Monitoring fulfilling rate of SOP	O04	Monitoring chart for fulfilling rate of SOP	Showing the personal KPI of fulfilling rate of SOP

others have come up with very simple and ineffective ones. The factory that has set up a monitoring chart for the fulfilling rate of SOP finds it effective for motivating workers, whereas some factories believe that it is not feasible to visualize personnel performance. The visual tools used by each investigated factory are listed in Table 3.9.

- Best practice for visualizing operation management

    The best practice for visualizing operation management is the one established by Japan 1. The factory has set up a highly complete SOP system. To increase productivity and decrease production costs, the operation procedures are continuously improved. The factory also takes measures to make workers operate according to their SOP. This can induce the established SOP to take effect in practice.

Table 3.9 Evaluation of used visual tools for operation management

Factory	Evaluation
Germany 1	• Established tools: O01 and O02. • General evaluation: The established tools are effective, but they are not regularly updated. The tools are not enough compared with excellent example.
USA	• Established tools: O01, O02, and O03. • General evaluation: The established tools are effective, but the tools are also not regularly updated. Supervision measures can be adopted to assure their effectiveness.
Germany 2	• Established tools: O01. • General evaluation: The number of tools is far from enough. The existing tools are also not effective. No corresponding supervision measures are adopted.
China	• Established tools: O01 and O03. • General evaluation: The visual tools are far from enough. Across the whole factory, only several work stations have set up SOP. The established SOP has no clear operation requirements.
Japan 1	• Established tools: O01, O03, and O04. • General evaluation: The factory does better than its peers. A system of SOPs has been set up, including repeated and nonrepeated, decomposed and compiled SOPs. Furthermore, the SOP system is regularly updated. The managers take many measures to shape the behavior of workers. The visual tools are effective.
Japan 2	• Established tools: O01 and O03. • General evaluation: The established tools are effective. Managers inspect the operation of workers every day. Color figures are also used for designing tools.

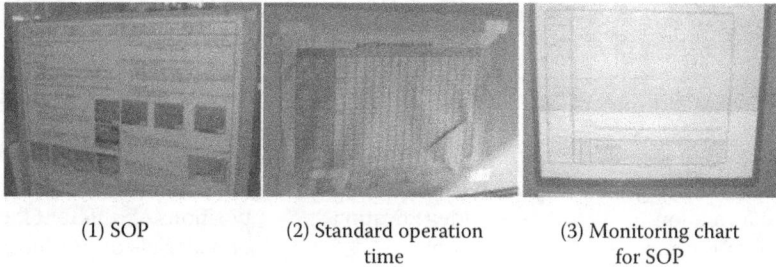

| (1) SOP | (2) Standard operation time | (3) Monitoring chart for SOP |

Figure 3.2 Good examples of operation management visualization.

Good examples (Figure 3.2):

1. SOP. This describes the operation procedures for repeated jobs. The key steps are illustrated with pictures. This is especially effective for novice workers.
2. Table of standard operation time. This table gives the standard operation time for each step. It can be used as the benchmark for workers' operation time.
3. Monitoring chart for fulfilling rate of SOP. This chart shows the fulfilling rate of SOP for workers. How well the workers operate according to SOP can be perceived easily and directly from the chart. Workers are encouraged to operate according to SOP.

3.3.2.4   Visualization of product/material management

Visual tools for material management include different kinds of labels, which indicate the name, ID, amount, appearance of materials, and inventory status. The status kanban can easily show the current volume of materials and products in the inventory, from which the worker can easily find out whether he should generate orders to suppliers. The material identification signs are similar across investigated factories. The better ones are combined with figures and different colors. Table 3.10 summarizes the visual tools for material and product management.

The visual tools used in each factory are shown in Table 3.11. Among the investigated factories, Factory A (Germany 1) performed worst, and others had their own features. Germany 1 only used some very simple material labels, whereas other types of tools were not established. The American factory has set up a relatively complete set of visual tools for material management. Different colors are used by the American factory to mark the different entry time of materials or products. The visual tools used by Germany 2 functioned very well. Germany 2 not only used material labels, but also set up boards to visualize the position of materials. Germany 2 used the SAP system (ERP system), and the labels had barcodes

Table 3.10 Summarization of visual tools for material management

Category	Code	Visual tool	Function
1. Material label	M01	Material label	Marking different types of material, WIP, and products with figure illustration
2. Indicating logistic information (location and volume)	M02	Location identification kanban	Demonstrating the positions of different kinds of materials or products in the inventory
	M03	Volume kanban	Showing WIP or product's volume in a workstation
	M04	Kanban table	Transmitting information on dwindling supplies of materials and setting order on the kanban table to suppliers
3. Differentiating different parts	M05	Parts differentiate card	Demonstrating the slight differences between similar parts

Table 3.11 Evaluation of visual tools for material management

Factory	Evaluation
Germany 1	• Established tools: M01. • General evaluation: The established tools are too simple. Some important visual tools are lacking.
USA	• Established tools: M01, M02, and M03. • General evaluation: The tools are relatively complete. Colors are effectively used to mark different entry time of materials. Visual management for material is complete and effective.
Germany 2	• Established tools: M01, M02, and M04. • General evaluation: Overall, the visual tools are complete and effective in the factory.
China	• Established tools: M01, M02, and M03. • General evaluation: Visual management for material management is relatively complete and effective.
Japan 1	• Established tools: M01 and M05. • General evaluation: Visual management for material management in the factory is effective and relatively complete. Pictures are widely used in marking labels.
Japan 2	• Established tools: M01 and M03. • General evaluation: The established tools are effective. Different colors are also used to differentiate distinct types. M02 can be added for further improvement.

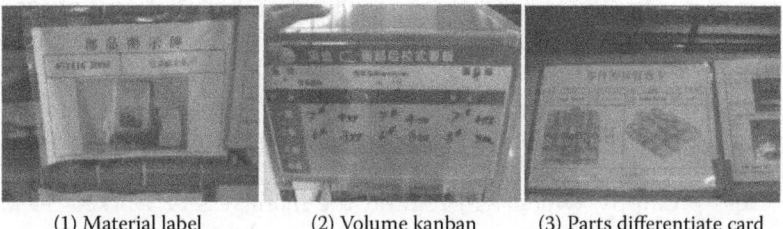

(1) Material label   (2) Volume kanban   (3) Parts differentiate card

Figure 3.3 Good examples of material management visualization.

that can be scanned easily. Moreover, Germany 2 also established a kanban table, which was used to visualize the materials in short supply. After the worker finds out that the material or certain parts were going to be used up, he would put the corresponding boards on the kanban table. Then the material controller orders the lacking materials from suppliers. Japan 1 used figures to illustrate the materials that facilitate the searching tasks of workers. The visual tools in Japan 2 were effective, although it did not have a complete set of visual tools.

- Best practice for visualizing of material management
  Except Germany 1, each investigated factory has its own unique features. Consequently, the best practice is to combine the tools described earlier. Considering the special situations, some tools can be omitted. Pictures and colors should be appropriately used to assist the daily tasks of workers.
  Good examples (Figure 3.3):

  1. Material label. This uses pictures to demonstrate the shape, name, and code of parts. Different types of material can be easily identified with this label.
  2. Volume kanban. This shows the volume of WIP or products in a workstation. This kanban can be used repeatedly. The old information can be easily erased, and then new information can be recorded.
  3. Parts differentiate card. This clearly shows the differences between similar parts. This card is very useful for complex and similar parts or materials.

3.3.2.5 Visualization of production control

Visual tools for production control are used to transmit control information between workers and managers. Managers allocate tasks to workers by demonstrating production planning information. Workers present the production situations to managers by updating the production output kanban. There are real-time updated kanbans, hourly updated kanbans, daily updated kanbans, and monthly updated kanbans. Different types of production kanban serve different functions. Daily and monthly updated kanbans are usually used to compute weekly or monthly production KPIs (e.g., production output and labor efficiency), whereas real-time updated kanban and hourly updated kanban are used to monitor the production. These two types of kanbans can highlight the problems of the production field promptly, which will remind the managers or workers to take the appropriate measures (e.g., equipment breakdown or quality problems). The summarization of visual tools for production control is shown Table 3.12.

The production control tools used in each factory are shown in Table 3.13. Overall, Japan 1 did best in VM for production control. The factory has set up a very complete VM system and has also taken measures to ensure that the visual tools can take effect in practice. Managers in Japan 1 factory can easily acquire the information they need about the production field from the visual tools. Workers are likewise motivated to work hard to accomplish their tasks by visualizing the production output. The visual

Table 3.12 Summarization of visual tools for production control

Category	Code	Visual tool	Function
1. Indicating production output	P01	Real-time updated production output kanban	Demonstrating current production output and goal
	P02	Hourly updated production output kanban	Visualizing hourly output and causes of unaccomplished tasks
2. Allocating daily tasks to workers	P03	Production planning sheet	Demonstrating weekly or monthly production plans
	P04	Task allocation kanban	Showing daily task assignments among workers
	P05	Production information sheet	Demonstrating parts requirement for each product
	P06	Production indicating light	Lighting lamp demonstrating current processing product type
3. Indicating production KPI (e.g., labor efficiency, hourly output)	P07	Daily updated production kanban	Demonstrating production daily KPI, including daily output goal, daily labor efficiency
	P08	Monthly updated production KPI kanban	Demonstrating monthly statistics of production output

Table 3.13 Evaluation of visual tools for production control

Factory	Evaluation
Germany 1	Established tools: P03 and P04.
	General evaluation: Types of tools used by the factory are far from enough. Production output information needs to be visualized to managers, which can also motivate workers to perform better.
USA	Established tools: P01, P02, P07, and P08.
	General evaluation: The factory has set up four types of visual tools. Colors are also used to attract the attention of users. However, workers do not care much about the general information kanban. The kanban just stays there for visitors.
Germany 2	Established tools: P03 and P08.
	General evaluation: The tools used by the factory are far from enough. Much production output information could be visualized. The production plan kanban also does not take into effect. Daily plan is fulfilled only about 50% of the time.
China	Established tools: P01, P02, P03, P04, and P07.
	General evaluation: The factory has set up five types of visual tools. Most of the tools are effective.
Japan 1	Established tools: P01, P02, P03, P05, P06, and P08.
	General evaluation: The factory has set up a very complete visual system for production system. Most of the visual tools are effective. The factory also takes measures to assure that the tools take effect in practice.
Japan 2	Established tools: P01, P04, and P07.
	General evaluation: The tools for production control in the factory are almost complete. The visual tools are also effective.

tools used by Japan 2 factory are almost complete and are effective. The visual tools in Germany 1 and Germany 2 are far from adequate to serve their purpose. Although Category 1 tools are not well tailored to the two factories, Categories 2 and 3 are also lacking. The incomplete visual tools for production control in Germany 1 and 2 are related with the production mode of these two factories. Different from the flow production line of Japan 1 and 2, the production mode used by the two German factories is small batch and multitype production mode. The daily output of each factory is less than about 50. Each product has to be processed by many procedures. There is no need to set up many visual tools.

- Best practice for implementing VM of production control
  The two Japanese factories can serve as good examples for best practice. The two factories have covered the three categories of tools. The production output kanban demonstrates the following information: goal output, current output, and the gap between goal and

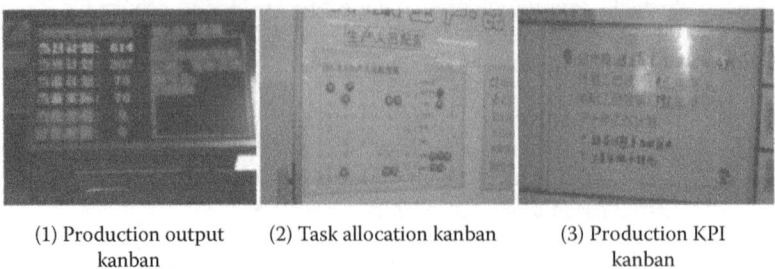

| (1) Production output | (2) Task allocation kanban | (3) Production KPI |
| kanban | | kanban |

Figure 3.4 Good examples of production control visualization.

actual outputs. The hourly updated kanban presents the information about the causes of an incomplete goal. Japan 2 sets up a feasible task allocation tool (P04) for assigning tasks. Every day, the managers just need to move the chore magnet, which is pasted with workers' names, to distribute tasks on the tool. Japan 2 updates and maintains the KPI tool on time and also discusses with the workers about the situation of production in the meeting. These measures induce the tools to take effect in practice.

The following are good examples (Figure 3.4):

1. Real-time updated production output kanban. This shows the production output and goal in real time. Both workers and managers can easily evaluate the current production status from this kanban.
2. Task allocation kanban. This is used to allocate daily production tasks for workers. Each worker's name is pasted on a magnet. In this manner, the allocation sheet can be used repeatedly. This is an effective method.
3. Daily updated production KPI kanban. This kanban demonstrates the production KPI for each day, including output goal, labor efficiency, and plan.

3.3.2.6   Visualization of quality management

Quality visualization is an important aspect of VM. The visual tools for quality management are summarized in Table 3.14. The hierarchy of visual tools is shown in Figure 3.5. There are two types of visual tools: visualization of quality KPI and visualization of quality assurance measures. The tools for visualizing quality KPI are used to monitor the quality status of the factory and support the decision making of managers. The quality KPIs includes the number and percentage of deficient products and the number of customer complaint events. Different factories have their own particular indicators. The tools of visualizing quality assurance measures are established for training workers or motivating workers to improve product quality.

Table 3.14 Summary of visual tools for quality management in the factories

Category	Code	Visual tool	Function
1. Indicating quality KPI (e.g., number of deficient products, percentage of deficient products)	Q01	Quality target and target decomposition kanban	Visualizing the quality target of workshop and the decomposition to each working unit
	Q02	Quality KPI kanban of organization	Recording and demonstrating the current quality status of working unit
	Q03	Individual quality records	Recording the number of deficient products for individual
2. Controlling manufacturing quality	Q04	Visual tool of quality assurance guaranty	Showing the items of assurance and making workers do as it allows
	Q05	Quality standard and policy kanban	Visualizing the quality standards and policies of the workshop or factory
	Q06	Resolution for quality deficiency	Visualizing the method to resolve the deficiency, also the importance of the deficiency

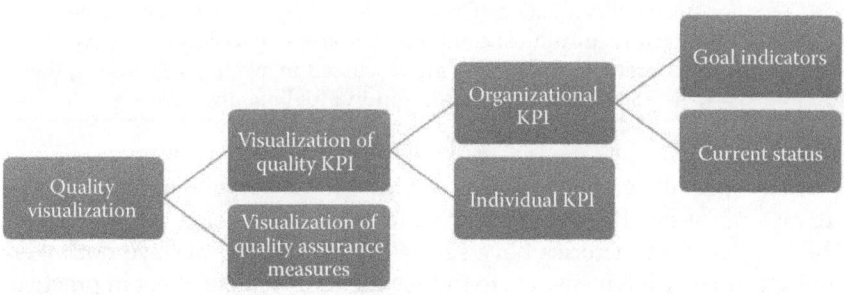

Figure 3.5 Hierarchy of visual tools for quality management.

The visual tools used in each factory for quality management are shown in Table 3.15. Japan 1 has established a very complete and useful VM system for quality control. The factory also adopts many quality control measures, and the workers take the issue of quality seriously. Japan 2 also established many tools for controlling quality, and the tools are useful. However, the factory lacks individual tools for quality management, which can motivate workers to pay more attention to quality control. The Chinese factory did not take adequate measures to control

Table 3.15 Evaluation of visual tools for quality management

Factory	Evaluation
Germany 1	Established tools: Q03. General evaluation: The tools used by the factory are far from enough. More tools are needed to improve the quality.
USA	Established tools: Q01, Q02, and Q05. General evaluation: The factory has set up several visual tools for quality management. However, the workers do not care much about their content.
Germany 2	Established tools: Q02. General evaluation: Very few tools are established for quality management. Furthermore, only a few interviewees are aware of the existence of these tools.
China	Established tools: Q02 and Q05. General evaluation: The factory did not set up many tools to monitor the quality. The assurance of quality is mostly through inspections. Workers did not care much about the existing tools.
Japan 1	Established tools: Q01, Q02, Q03, Q04, and Q05. General evaluation: The factory set up a very complete visual system for quality management. Most of the visual tools are effective. The factory also took measures to assure the effectiveness of the tools in practice (e.g., awarding, punishing, reminding).
Japan 2	Established tools: Q01, Q02, Q05, and Q06. General evaluation: The tools for quality management in the factory are almost complete. These visual tools are effective. The content on the kanban are discussed in meetings. However, there is no visual tool to monitor individual quality.

quality, especially visualization measures. Moreover, the established tools are also not very effective. The information on the tool is not very clear. The two German factories have set up several tools, but have not taken enough supervision measures to induce the tools to take effect in practice. Some workers even did not know the existence of these tools and did not care much about quality.

- Best practice for visualizing quality control
  The two Japanese factories performed best among the investigated factories. The two factories have set up the most complete set of tools, which cover almost every type listed in Table 3.14. In addition to the organizational and individual quality tools, Japan 1 also uses trend charts to clearly represent the trend of quality. The managers can easily find out from the charts whether or not the targets have been achieved. The tools are updated and maintained almost in real time.

(1) Quality target and decomposition kanban

(2) Quality assurance guarantee

Figure 3.6 Good examples of quality management visualization.

The factory also used supervision measures (e.g., awarding, KPI) to make the tools take effect in practice. Japan 2 visualized many quality assurance measures (e.g., resolution for quality deficiency).
Good examples include (Figure 3.6):

1. Quality target and decomposition kanban. This presents the quality target of the workshop and the decomposition of target to each processing unit. It clearly demonstrates the quality target and hierarchy of quality index.
2. Quality assurance guarantee. This demonstrates the key points for quality. Workers need to understand these key points, and sign a guarantee to assure the quality. This is effective in encouraging workers to pay attention to product quality.

3.3.2.7 Visualization of equipment management

Visual tools for equipment management include three types: equipment indicators, equipment operation instructions, and equipment maintenance method. The tools are summarized in Table 3.16. Equipment indicators are used to indicate the working status of equipment (normal or abnormal). The performance indicators (KPI) of equipment can also be visualized (e.g., equipment cost, stopping time). The visualization of performance indicators are used to monitor the equipment performance and motivate the managers and workers to improve the equipment performance. Equipment operation instruction is similar to SOP, and it is used to instruct workers how to operate equipment. Equipment maintenance is an important part of equipment management. Equipment checklists are widely used in factories. They allow workers to maintain the equipment by themselves without relying on engineers. Equipment maintenance records are used to record the maintenance details, including time, reason, and engineer. Equipment breakdown handling method is another equipment operation instruction; it tells the workers how to handle equipment breakdowns. The visualization of equipment breakdown analysis reports is used to train workers to pay attention to the same mistakes.

Table 3.16 Summary of visual tools for equipment management in the factories

Category	Code	Visual tool	Function
1. Indicating equipment status and KPI	E01	Equipment indicating lights	Indicating current status of equipment: normal or abnormal
	E02	Equipment KPI	Demonstrating equipment KPI, for example, stopping time
2. Equipment operation instructions	E03	Equipment operation instructions	Demonstrating how to operate equipment
3. Maintaining equipment	E04	Equipment checklist	Listing inspecting items for equipment
	E05	Equipment maintenance records	Recording maintenance details
	E06	Equipment breakdown handling methods	Demonstrating how to handle with equipment breakdown
	E07	Equipment breakdown analysis reports	Demonstrating the analysis results for equipment breakdown

The visual tools used in each factory for equipment management are shown in Table 3.17. Japan 1 did best among these factories. The factory did not only set up a very complete VM system for equipment, but also took measures to assure the effectiveness of these tools. Japan 2 also set up a relatively complete set of visual tools for equipment management, and most of the tools are effective. Because of its labor-intensive manufacturing style, the Chinese factory did not set up many tools for equipment management. The tools for visualizing the equipment performance indicators KPI can be established to increase the utilization ratio of equipment. Germany 1 used several types of tools, such as operation instruction, checklist, and maintenance records, and most of these tools have been effective. However, the factory also could have set up some tools to visualize performance indicators to motivate its workers. Germany 2 used very few visual tools, and most of the tools were ineffective. The workers rarely used them, and some were not even aware of the existence of these tools.

- Best practice for visualization of equipment management

    The best example for visualization of equipment management is Japan 1. The factory has established a very complete VM

Table 3.17 Evaluation of visual tools for equipment management

Factory	Evaluation
Germany 1	Established tools: E03, E04, and E05.
	General evaluation: The factory has set up several types of tools, The established tools are effective. The tools are used extensively by workers in daily operation.
USA	Established tools: E02, E05, and E06.
	General evaluation: Most of the tools are useful. The factory demonstrates substantial information about the equipment KPI, including stopping time and stopping number. However, the workers do not care much about the KPI information. They are not motivated by the content.
Germany 2	Established tools: E02 and E03.
	General evaluation: The visual tools for equipment management are very few. Also, some of the tools were not very well designed. The interviewed workers did not even know about the existence of these tools.
China	Established tools: E03 and E04.
	General evaluation: The factory did not create many tools for equipment management. This is related to the fact that most jobs in the factory are accomplished by hands instead of equipment. The workers did not care much about the equipment.
Japan 1	Established tools: E01, E02, E03, E04, E06, and E07.
	General evaluation: The factory has set up a very complete visual system for equipment management. Most of the visual tools are effective. The factory also took measures to assure that the tools are effective in practice (e.g., awarding, punishing, and reminding). The workers conducted equipment inspection and maintenance by themselves.
Japan 2	Established tools: E01, E04, and E07.
	General evaluation: The tools for equipment management in the factory are almost complete. These visual tools are effective.

system for equipment, including equipment indicating lights, equipment KPI, equipment operation instructions, equipment checklist, equipment maintenance records, equipment breakdown handling methods, and equipment breakdown analysis reports. It also used trend charts and pictures to illustrate the content, which facilitated the identification of problems. The workers performed inspection and checked the equipment every day. In addition, the monitors explained the content of equipment management kanban to workers.

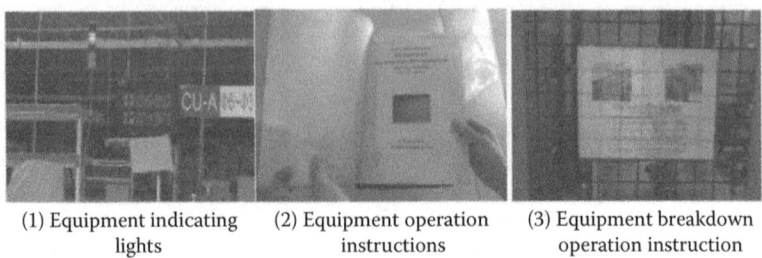

| (1) Equipment indicating lights | (2) Equipment operation instructions | (3) Equipment breakdown operation instruction |

Figure 3.7 Good examples of equipment management visualization.

Good examples include (Figure 3.7):

1. Equipment indicating lights. This clearly shows the current status of equipment. If it flashes yellow or red, the operator should stop using the equipment right away. If it is green, this means that the equipment is in normal status.
2. Equipment operation instructions. They demonstrate how to operate the equipment. They are particularly useful to novice users and are especially useful for equipment that are not often used.
3. Equipment breakdown operation instruction. This demonstrates how to operate the equipment in case of an equipment breakdown. With this instruction, the operator does not need to wait for the technician to solve the problem when the equipment fails to work.

3.3.2.8   Visualization of performance management

Visual tools for performance management are not widely used by factories. The dispute on visualizing the workers' performance continues. Some people believe that visualizing personnel performance may have a negative impact on some workers, who prefer not to show their performance to the public. Some believe that it can actually motivate workers to work harder, and make the managers' assessment process open to workers. Some of the investigated factories have set up several tools for performance management, which are shown in Table 3.18.

Among the investigated factories, Japan 1 showed the most concern about employee morale in the factory and has also established several visual tools to facilitate performance management. Most of the established tools are effective. Managers and workers care about the content a lot. The American factory has also set up many tools for performance management. However, these tools are not very effective. The factory made efforts to visualize performance assessment standards, but in reality the managers assessed the workers' performance subjectively. Japan 2 also set up some tools to motivate workers, for example, honor board and behavioral norms. The managers in the factory strongly disagreed with

Table 3.18 Summary of visual tools for performance management
in the factories

Visual tool	Code	Function
Behavioral norms	A01	Demonstrating the requirements for workers' daily behaviors
Performance assessment results	A02	Demonstrating the performance assessment standards and results
Honor board	A03	Rewarding staffs who do good jobs on the honor board
Punishment measures for violating disciplines	A04	Demonstrating the punishment measures to disciplines violating workers
Multiskill development	A05	Showing the development plan of multiskill workers
Current status of staff distribution in the workshop	A06	Demonstrating the status of staff distribution

the visualizing of personal performance assessment results. In their view, this public display of results may disappoint or discourage some workers. The Chinese factory has set up several tools for performance management. The honor board motivated some workers to work harder. The two German factories did not create many tools for performance management. No tools were found in Germany 1. Germany 2 set up one tool to visualize the requirements for multiskill workers. However, the workers did not care much about it (Table 3.19).

- Best practice for visualization of performance management

  More measurable indicators can be established for performance management in order to increase the morale of workers. For example, Japan 1 set up an indicator for managers regarding the number of people being trained on specific operation skills in a season. The factory also used trend charts to demonstrate the results. As for visualizing the performance assessment results, although Japan 2 disagreed with this, these tools were effective in Japan 1. Setting an honor board or hosting a competition is another method to increase workers' morale.

  Good examples include (Figure 3.8):

  1. Honor board. This is used to reward those workers who performed well in their jobs. This is especially effective for Chinese workers and can motivate them to work harder.
  2. Current status of staff distribution in the workshop. This demonstrates the status of staff distribution, including skill level and performance.

Table 3.19 Evaluation of visual tools for performance management

Factory	Evaluation
Germany 1	Established tools: Null. General evaluation: None of the tools are established in the factory.
USA	Established tools: A01, A02, A03, and A04. General evaluation: The factory has created many tools for performance management, but the tools are not effective. Workers do not care much about the content on the tools.
Germany 2	Established tools: A05. General evaluation: The factory only visualizes the requirements for multiskill workers. No other tools are used.
China	Established tools: A03 and A05. General evaluation: The factory uses two types of tools. Some workers are motivated by the honor board.
Japan 1	Established tools: A02, A03, A05, and A06. General evaluation: The factory has set up a relatively complete visual system for performance management. Most of the visual tools are effective, which can improve morale in the factory.
Japan 2	Established tools: A01, A03, and A04. General evaluation: The factory has set up several tools for performance management. Generally, the tools are effective to workers.

 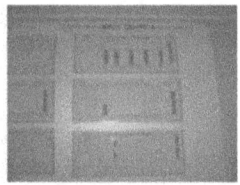

(1) Honor board

(2) Current status of staff distribution in the workshop

Figure 3.8 Good examples of performance management visualization.

3.3.2.9   *Visualization of continuous improvement*

Continuous improvement activities are prevalent in Japanese-style factories. Continuous improvement originated from the philosophy of *Lean Production*. The improvement covers all production activities, including operation, equipment, quality, material, and 5S. Everybody is required to participate in the continuous improvement activity. The workers can propose any ideas about improving the production efficiency and reducing production costs. The two Japanese factories and the American factory have conducted continuous improvement activities. The workers are required to come up with one or two proposals every 2 months. The factories also hold competitions about continuous improvement among workers. The visualization of continuous improvement mainly demonstrates the good cases of improvement instigated by workers, which can serve as a good example to other workers. Table 3.20 makes a comparison of the visual tools used by the investigated factories.

- Best practice for visualization of continuous improvement

    The best practice for visualization of continuous improvement is to set up good cases kanban and performance indicators. When setting up good cases kanban, pictures can illustrate the content and attract users' attention. The managers should also train the workers about the content of the kanban. Performance indicators such as the number of proposals each month can motivate the workers to work hard to come up with ideas.

Table 3.20 Evaluation of visual tools for continuous improvement

Factory	Evaluation
Germany 1	Established tools: Null. General evaluation: None of the tools are established in the factory.
USA	Established tools: Continuous improvement cases kanban. General evaluation: Workers in the factory did not care much about the content on the established kanban.
Germany 2	Established tools: Null. General evaluation: None of the tools are established in the factory.
China	Established tools: Continuous improvement projects and policies kanban, continuous improvement cases kanban. General evaluation: The established tools are not very effective. The workers did not care much about the content. The kanbans were not updated in time, which stayed there for the visitors' benefit.
Japan 1	Established tools: Continuous improvement cases kanban. General evaluation: Every department in the factory has set up continuous improvement cases kanban. Although the workers did not care about the content on the kanban, the managers trained the workers on the content of the kanban during meetings. Generally, the tools are effective in the factory. However, the tools for the factory were slightly redundant, which required a lot of updating work.
Japan 2	Established tools: Continuous improvement cases kanban. General evaluation: The tools were effective in the factory. The managers trained the workers about the content of the kanban.

(1, 2) Continuous improvement cases kanban

Figure 3.9 Good examples of continuous improvement visualization.

One good example is continuous improvement cases kanban (Figure 3.9). This presents the successful cases of continuous improvement. If the managers can lead the workers to learn the content on the kanban, this tool will be more useful to workers. Otherwise, the workers will not care much about its content.

3.3.3   Comparison of VM across cultures—A vertical view

From a vertical perspective, establishing the process of VM involves four steps: planning, design, implementation, and control (Figure 3.10). In the planning phase, the objectives of establishing visual tools, the target users, and the information to be visualized should be determined. In the design phase, the media used to deliver information and the information display styles are selected. The visual tools should be designed to satisfy the users' requirements. The implementation stage is critical to the success of VM. Promotion and training let the users know about the purpose of the tool and how to use it. Updates and regular maintenance enable the tools to become effective in practice. The control stage is also an important step for the effectiveness of VM. Because users are always reluctant to change their habits and use new things, the managers should adopt supervision measures to ensure that the users use the tools properly. In addition, the

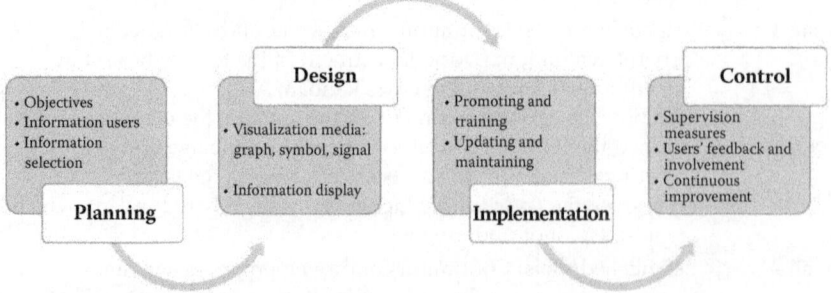

Figure 3.10 Vertical view of visual management: establishment process.

managers should pay attention to users' feedback and involvement to further improve the visual tools. In the following discussion, the establishing process for VM will be compared across factories.

From the view of the whole factory, the best practice for establishment of VM is to conduct VM across departments. It is better to establish an execution committee to conduct VM. The committee investigates the requirements of managers and workers and determines the information that can be visualized in different departments. Although the requirements and information in different departments are not exactly the same, most of them are similar according to our investigation results. Then the committee can create templates for the visual tools, which are used by different departments. Next, each department can adapt the template for its own purpose. The committee also provides overall promotion and training to the rank and file. Later on, the committee will assess the effects of conducting VM and give suggestions regarding the activity. The advantage of setting up an execution committee is that it allows the implementation of VM to become more systematic and enables the company to highlight the effects of VM. Each department can also identify its own requirements and establish visual tools.

3.3.3.1 Plan for VM

Before setting up visual tools, the manager should identify the information to be visualized. VM covers eight categories, and each category covers different types of information. Some information is coming from workers, which should be demonstrated to managers. Some information is transmitted within workers or managers. Others are mainly used by managers to convey orders, tasks, or other types of information to workers. When making plans for VM, the target users should be considered, because the position and education level of users influence the effectiveness of VM. After determining the information and the users, the objectives and expected effects should be pointed out for later evaluation and improvement.

Across the six investigated factories, Factory E (Japan 1), Factory F (Japan 2), Factory B (USA), and Factory D (China) established a committee or team to execute VM. These factories promoted VM in the factory and encouraged department staff members to use VM in their daily tasks. Before implementing VM, the departments first identified the information to be visualized and figured out the purpose of visualization. However, according to the interview results, we found that almost all investigated factories did not figure out the target users very clearly before implementation. The two German factories did not implement VM systematically across departments. Some departments established visual tools for their own use, but managers in other departments did not know the existence of these tools. This situation is not good for the development of the whole factory.

3.3.3.2   Design of visual tools

After the information to be visualized is identified, the executive committee begins to design the visual tools. First, the format should be determined, and options should include label, symbol, signal, line, print, board, etc. Different types of formats are tailored to different types of information and functions. Second, the information representation patterns should be designed. Considerations for the patterns to be used should include text, color, chart, graph, layout, location, etc. The patters of representation have an impact on the visibility, legibility, and readability of visual materials. Text is better for accuracy, whereas graphics is better for speed and intuitive understanding. Usually, it is suggested that the combination of text and graphics be used. A total of five categories of formats were used in the investigated factories.

- Signal lights
    Signal lights are widely used for indicating the status of equipment, which can inform the engineers or managers about the current state of the equipment. Signal lights are also sometimes used to indicate current production information. In the design of signal lights, the color, flash rate, light intensity, and location should be considered.
- Sign
    There are two types of signs for VM: symbols and verbal signs. Symbols are superior to verbal signs in conveying an intended message, because it is easier for users to perceive the meaning of symbols. Symbols are used widely in the factory for safety promotion activities. The three criteria for designing symbols are: *Recognition*, *Matching*, and *Preference*. Recognition means that users can accurately recognize the meanings of symbols. Matching refers to the idea that users can quickly and correctly match the meaning with the corresponding symbol among several alternatives. Preference means that users' preferences and opinions should be considered during design.
- Text
    Hard copy text is often used in VM (e.g., safety signs, SOPs, quality standards, control measures, and analysis reports). Visibility, legibility, and readability should be considered for text design. Visibility refers to the quality of a character that makes it distinctly visible from its background. Legibility refers to the attribute of characters that makes it possible for each one to be identifiable from the rest. Readability refers to the quality that makes it possible to recognize the information content.
- Line
    Line is widely used in the factory to divide distinct functional areas in VM—for example, operation area and logistics area. Different colors have different meanings. Yellow, red, and zebra lines with black

and yellow are used most often. Generally, yellow is used to indicate the passageway and the inventory area. Red is used to indicate waste material, faulty products, or dangerous material(s). Zebra lines with black and yellow interval are used for warning, fire extinguishers, qualified product and waste containers.
- Graphics (chart)

  Graphics is widely used for presenting KPI in VM (e.g., bar charts, pie charts, and line charts). Line chart is good at demonstrating the trend of indicators and the difference between goal and reality. Another popular chart is flow chart, which is used to demonstrate operation procedures, such as SOPs and equipment operation instruction.

Among the investigated factories, Factory E (Japan 1) performed best in designing visual tools. Trend charts, rather than tables, were widely used to demonstrate performance indicators, such as equipment KPI and quality KPI. The goal and the actual value were both presented on the chart. Through this method, the managers can easily find out whether or not the goal has been achieved. A bad example for the design of visual tools is that used by Factory F (Japan 2), which used too many colors to encode different production tasks. This exerted a large memory load for workers.

3.3.3.3   Implementation

According to the results, we found that although many factories have set up many visual tools, VM in the factories is still largely unsuccessful. The key point is that those factories did not take proper and effective measures for their implementation. The implementation process entails introducing visual tools to users, training users on how to use these tools, and regularly updating the information on visual tools. When introducing and promoting VM in the factory, the executives should clearly demonstrate the functions of visual tools with adequate examples. They should provide training to users, or at least to their managers or monitors. Then the managers and monitors, in turn, should provide training to workers on how to use these tools. The tools for visualizing dynamic information should be updated and maintained regularly. Otherwise, the information presented by these tools is invalid, and the problems cannot be discovered and solved in time. Finally, the factories should take measures to supervise the usage and implementation processes. The measures include monitoring, awarding, and punishing. The implementation process for the eight categories of VM is similar. Figure 3.11 presents the implementation process.

Across the investigated factories, Factory F (Japan 2) performed best in terms of implementation. It conducted VM across the whole factory (the

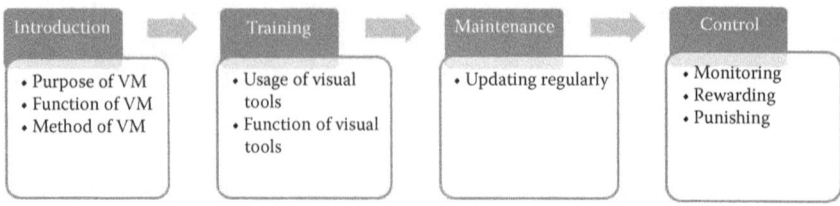

Figure 3.11 Implementation process of visual management.

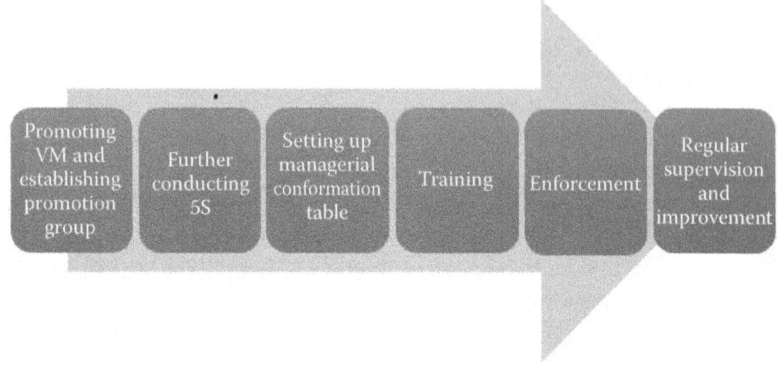

Figure 3.12 Implementation process in Factory F (Japan 2).

implementation process used is shown in Figure 3.12). At the beginning of VM, the factory set up a promotion group to promote VM. Then by conducting 5S activity, the factory identified the information to be visualized. Next, visual tools were designed and created. The managers conducted training during morning meetings on how to recognize and use the tools. The managers also used the information presented on the tool to motivate workers. Finally, the managers regularly supervised the behaviors of the workers and the usage of the tools. In contrast, Factory C (Germany 2) did not handle the implementation stage well. During the investigation, we found that many workers did not even know the existence of the tools. The managers also did not train the workers on how to use the information on the tool. Many tools that should have been updated were not updated for several months.

3.3.3.4 Management with visual tools

After the visual tools are established, feedback from users should be collected for further improvement. There are two methods used for collecting users' feedback in the investigated factories. One is that the users express their opinions or preferences about the visual tools directly to managers. Many factories hold morning meetings every day. During this time, the manager or monitor can encourage the workers to express their opinions. The other method is to provide a suggestion box. The workers can write down their opinions anonymously and place them in the suggestion box.

Although most of the investigated factories have established mailboxes as a feedback mechanism (from their employees), they were not really very effective in practice. The workers are usually reluctant to send in their suggestions, because they do not care much about things that have nothing to do with salary. Factory E (Japan 1) did relatively better than the other factories in this regard. The managers talk regularly with their staff regarding their demands and opinions. Through this method, the workers can provide feedback to the managers.

chapter four

Design of a new production line

Li Zheng and Chenjie Wang

Contents

4.1   Basic information and requirements

4.1.1   Design requirements

The second phase of the construction project of JLRD Company is to design and plan a transmission manufacture assembly line with a production capacity of double shifts of 150,000 sets based on the optimization of the data of technological process and standard hours of the first phase in the condition of original factory building area and technological process. Consider the operation rest interval to be 10%, the Takt time is 87 s. During the planning stage, people need to analyze the standard labor time and design the layout of the workstations and corresponding distribution route.

4.1.2   Basic information of products

Auto continuously variable transmission (CVT) is one of the main parts in automotive powertrain. It is a safe, energy-saving, environmental-protecting high-tech product that integrates machine, hydraulic, and electronic technology in one. CVT technology uses driving belts cooperating with drive and driven wheels with variable working diameters to transfer power. Its electronic control system can obtain speed ratio modification automatically based on the driving cycle and the operation intention of the driver to obtain the continuous transmission ratio to reach the best match of the power trained working condition of the drive engine so that the best economic efficiency and dynamic property can also be achieved. The common CVT has two types: hydro mechanical transmission CVT and Van Doorne's Transmissie (VDT)-CVT.

4.1.3   Product type

Currently, JLRD Company mainly develops and produces X series VDT-CVT, which is the best-fit, cost-effective CVT for the domestic independent brands. The appearance of the product is shown in Figure 4.1. It can be equipped in economy cars with 1.3–1.6 L engines. The fuel economy is the same or 3% lower than the manual transmission of the same type; the dynamic property is the same or even better than the manually operated transmission of professional drivers. It has excellent drivability and seating comfort, low cost, and 50,000 km trouble-free distance of running.

Now, JLRD Company produces mainly two products, X-A and X-B, provided to two car manufacturers A and B, respectively. The existing production line uses batch production mode: produce a batch of a product and then switch to another product according to the market demand. The current switching frequency is once a month and the switching time is half a day.

Although the two products are similar in overall processing and assembly process, one of them is a front wheel drive and the other is a rear wheel drive, which leads to differences in some workstations. Thus, it is planned to respectively produce the two products after the new production lines are designed: the existing product line will be responsible X-A, as its demand is small and requirements can be met by the capacity of the current assembly line, while the new assembly line will be responsible for the production of X-B products and its possible transformation front wheel drive products, as the demand for front wheel drive CVT is larger, which should be produced by the new production line.

The design of a new assembly production line considers B as the main product. Product B (mainly supply B car factory), according to the bill of material list, has a total of 41 standardized parts and 207 components and is divided into 11 subassemblies as follows: input axis assembly, driving axle assembly, driven axle assembly, intermediate axis assembly, differentiating device assembly, front housing assembly, intermediate housing assembly and rear housing assembly, oil sump assembly, hydraulic valve block assembly, and electronic control assembly.

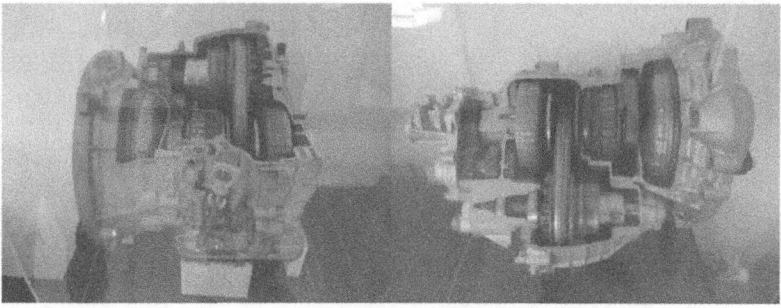

Figure 4.1 Appearance of transmission.

4.2　Time and motion study

The measure and standardization of the technological process and working hours are the basis of the production line design. In this chapter, the existing technological processes of JLRD will be analyzed and the MOD method will be applied to determine standard technology time data.

4.2.1　Time analysis principle

Time analysis can be generally divided into two types: direct observation method and indirect observation method. According to the operation type, nature and analysis purpose of the subjects, the two types can be subdivided into more methods. As shown in Figure 4.2.

The current general assembly production line of JLRD Company is an assembly line designed for an annual output of 50,000 sets, but in actual operation, the assembly capability is about 20,000–30,000 sets only. So the current time cannot accurately reflect the operation time of full

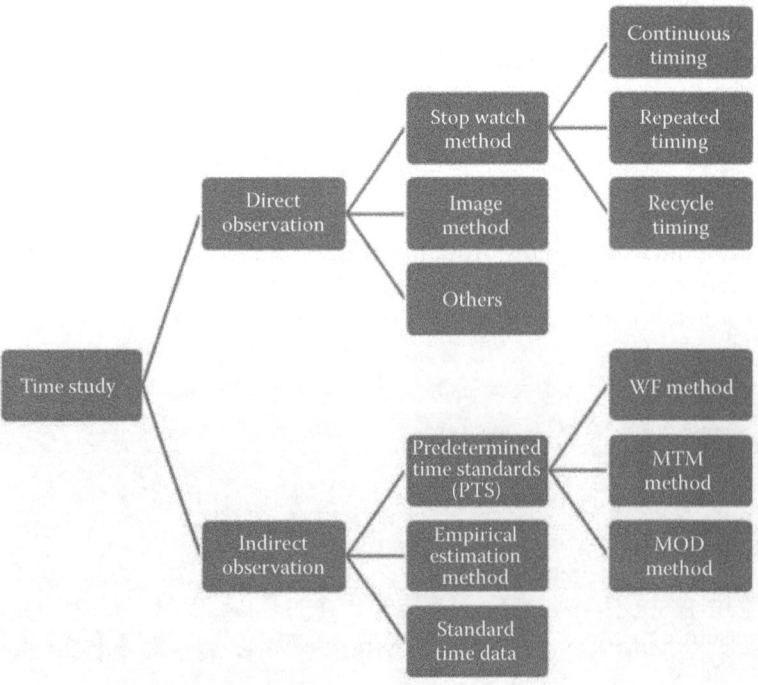

Figure 4.2 Time analysis method.

production. And considering the 150,000 sets of annual production capacity of the new assembly line, its design and layout will be improved and changed compared with the current line. Therefore, we use the predetermined time standards (PTS) method to conduct time analysis for each working procedure and use the stopwatch for assisting correction.

The PTS method is different from direct observation method. It decomposes the motions that form the working unit into several basic actions, and those basic motions are observed carefully, then a standard time form of basic actions is created. When determining actual working time, decompose the work assignment into basic motions, check the standard time of each one in the form of standard time, and then add them up to get the normal working time, which will turn to standard working hours when added with allowance time.

In 1966, Dr G.C. Heyde of Australia created the Modular Arrangement of Predetermined Time Standard, hereinafter referred to as MOD method, based on the long-term research of various predetermined time standard methods and integrated research results in the aspect of human factors engineering. This is a kind of omitted PTS method that is simple and easy to master, combining actions and time while the precision is not lower than in the traditional PTS technology. As a result, we will use the MOD method to analyze the process of assembly line (Figure 4.3).

The MOD method is based mainly on the following assumptions (basic principle):

1. All manual operation actions should include some basic actions. The MOD method summarizes the operations in actual production into 21 basic actions.
2. When people are doing the same basic action (under the same operating condition), the time required is almost equal (about 10% of error).
3. When different body parts do the actions, the ratio of the time needed by the fastest speed and the normal speed is almost the same.
4. When different body parts do the actions, the time needed for the action is proportional to each other.

4.2.2   Analysis of action time of assembly process

Use the MOD method to conduct time analysis on the operational time of each workstation in the existing assembly line; the time analysis takes net assembly time into major account and balances the time needed for fetching material. In practical analysis, record video of the production line at first and analyze DV based on MOD time. Figure 4.4 shows the time analysis of installation of belt wheel and oil cylinder.

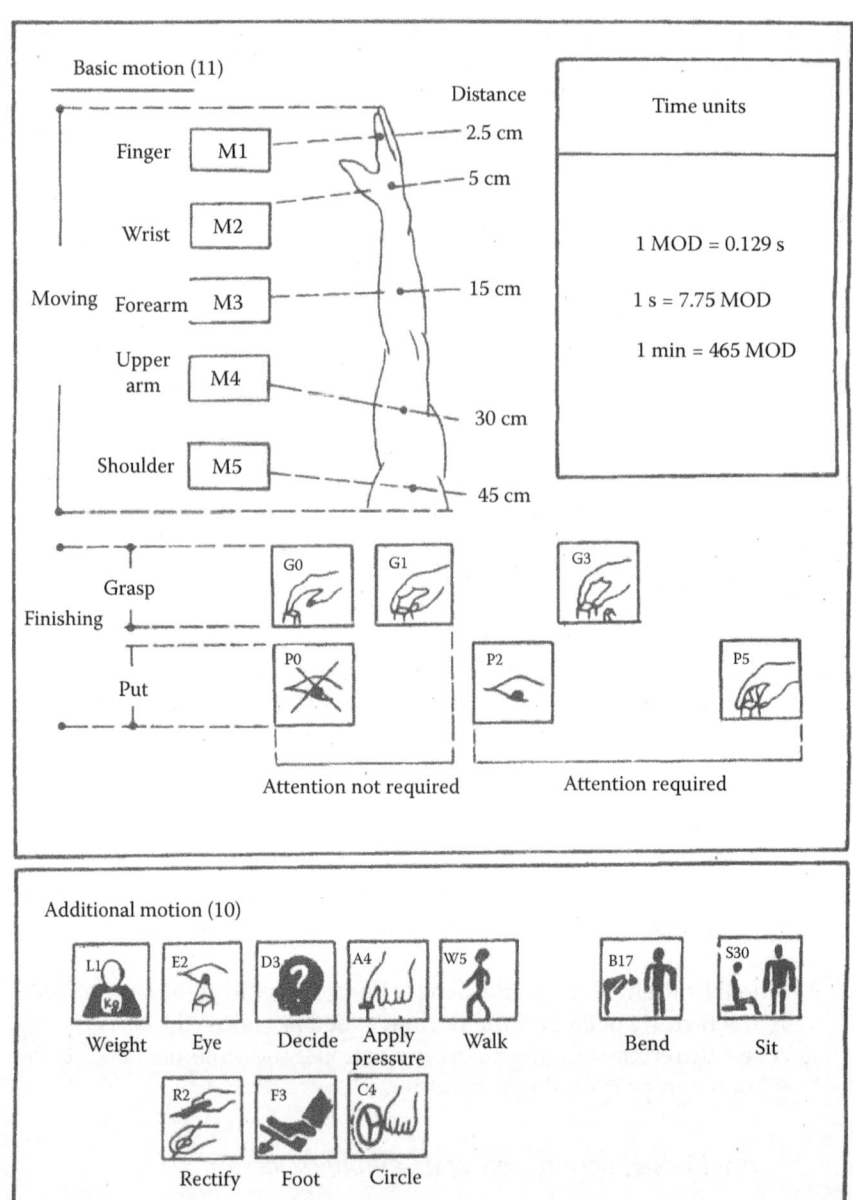

Figure 4.3 Basic principle diagram of the MOD method.

No.	1	Analyzer	Wang Chenjie	Update time	July 9, 2011			
Process no.	F02							
Process name	Install belt wheel and oil cylinder							
Workstation no.	1							
Time benchmark	0.129							

NO	Action of left hand		Action of right hand			MOD value	MOD value conversion (s)
		MOD		MOD	Mark		
	Get drive belt wheel	M4G1	Get drive belt wheel	M4G1	M4G1	5	0.645
	Place drive belt wheel	M4P5	Place drive belt wheel	M4P5	M4P5	9	1.161
	Press-fit drive belt wheel	R2A4	Press-fit drive belt wheel	R2A4	R2A4	6	0.774
	Get drive oil cylinder	M4G1	Get drive oil cylinder	M4G1	M4G1	5	0.645
	Place drive oil cylinder	M4P5A4	Place drive oil cylinder	M4P5A4	M4P5A4	13	1.677
	Place bolts	H	Get 4 M6 bolts	M4+4(G3M2)	M4+4(G3M2)	24	3.096
	Place bolts	H	Get 4 screw adhesive	M4G1	M4G1	5	0.645
	Place bolts	H	Bolt gluing	M4+4(M2P5M2)	M4+4(M2P5M2)	40	5.16
	Place bolts	H	Install 4 M6 bolts	4(M3G3M3P5+5M2)	4(M3G3M3P5+5M2)	96	12.384
	Get pneumatic impact wrench	M4G1M4	Get pneumatic impact wrench	M4G1M4	M4G1M4	9	1.161
	Tighten 4 bolts	4(M2L1P5M1)	Tighten 4 bolts (machine time)	4(M2L1P5M1)	4(M2L1P5M1)	36	4.644
	Tighten 4 bolts (machine time)	4UT	Tighten 4 bolts (machine time)	4UT	4UT	6	6
		BD	Get torque wrench	M4G1	M4G1	5	0.645
	Confirm 4 torque bolts	4(M2P5M2)	Confirm 4 torque bolts	4(M2P5M2)	4(M2P5M2)	36	4.644
		BD	Put back torque wrench	M4P0	M4P0	4	0.516
							43.797

Figure 4.4 Belt wheel and oil cylinder installation: MOD analysis of workstation.

Divide the 24 existing workstations into 98 processes and arrive at the operational time of each process of general assembly line through MOD analysis.

4.2.3   Analyze standard operation based on MOD to reduce unreasonable operation waste

It is found through analysis that the current production line has lots of things to improve, for example:

- Install oil pump workstation; four times of moving back and forth from the workbench and the assembly line, 12W5 = 7.74 s.
- Valve block installation workstation, valve block access tray, move and bend 6W5 + B17 = 6.063 s.
- Invalid moving time for other workstations: install rear end cover, 3.87 s; install differential, 7.74 s; install intermediate housing, 3.87 s; hydraulic torque converter oil filters, 9.933 s. Frequent movements and bending over increase muscle load and fatigue degree, which also require more broad time indirectly.

Actions can be optimized to improve production in the following four ways:

Improvement method 1: Add online work bins, as shown in Figure 4.5, to reduce invalid actions.

Improvement method 2: Some product packagings are removed in the dispatching process to reduce invalid action in the bottleneck process (Figure 4.6).

Improvement method 3: Adjust the operating mode; make full use of both the left and right hands.

Improvement method 4: Standardize the action of turning screws.

Figure 4.5 Add online work bins.

Figure 4.6 Invalid action: unpacking.

In many bottleneck processes, for example, in oil pump assembly, a lot of actions of screw turning are repeated. The path of screw turning should be standardized to improve the efficiency of screw turning.

4.2.4 Allowance time and standard time

After analyzing the normal time, considering reduction in working speed caused by an interruption, delay, or fatigue in actual work, a broad time should be added before obtaining a operation standard time (Figure 4.7).

The types of time allowance include mainly fatigue allowance and special allowance. Fatigue allowance refers to the time that workers need to recover from the fatigue caused by working or by the working

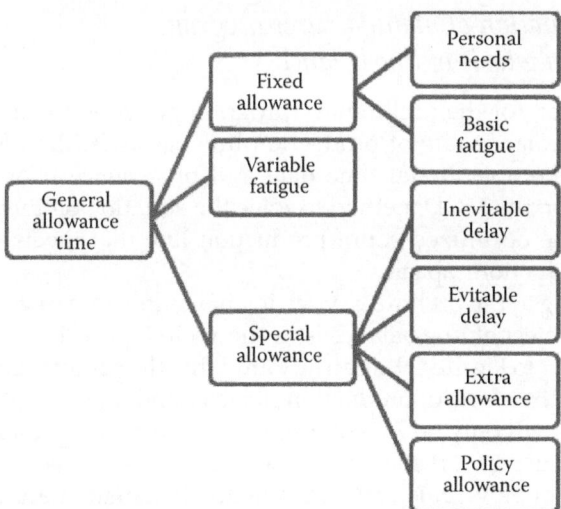

Figure 4.7 Time allowance.

environment, while special allowance includes many factors such as inevitable delay related to process procedure, equipment, and raw materials.

In the general assembly line of CVT, the main time allowance considers mainly personal needs, such as drinking water or going to washroom; basic fatigue, which considers mainly the time loss caused by the monotonicity of work for the workers in the assembly; and inevitable delay, such as interruption of work caused by the shortage of raw materials, shift supervisors, or others.

According to the allowance standard recommended by the International Labor Organization, the allowance value of individual demand is 5%, basic fatigue is 4%, and inevitable delay is 1%. Therefore, the total allowance is 10%.

Standard time is converted based on the normal time: standard time = normal time × (1 + allowance ratio).

Based on the previous time analysis, further divide the working procedure of those workstations into indivisible procedure to get the operation time of each procedure and then multiply by an allowance to get the standard allowance time of operation. Please refer to the appendix for the specific information.

4.2.5   Analysis of precedence relationship

Precedence relationship provides key foundation for balance and layout of production line. The precedence relationship diagram of transmission assembly is shown in Figure 4.8.

4.2.6   Evaluation of manufacturer program and analysis of line balance

In this chapter, for the preliminary program given by the manufacturer, analyze the balance rate of production line based on the MOD method before; use the operational time data and precedence relationship diagram based on the MOD method to solve the equation of line equilibrium and to get the optimized set of production line that meets the requirements of production capacity.

According to the preliminary setting of workstation of general assembly line, the effect of line balance is as shown in Figure 4.9.

According to Figure 4.9, it can be found that the production line cannot meet the requirements of production capacity and takt so that the production capacity will only be 73% of prediction. Effective operation time and complexity among workstations are seriously imbalanced: The complexity and time of valve block and gear switch installation are greater than those in workstations installing hydraulic torque converter and fastening

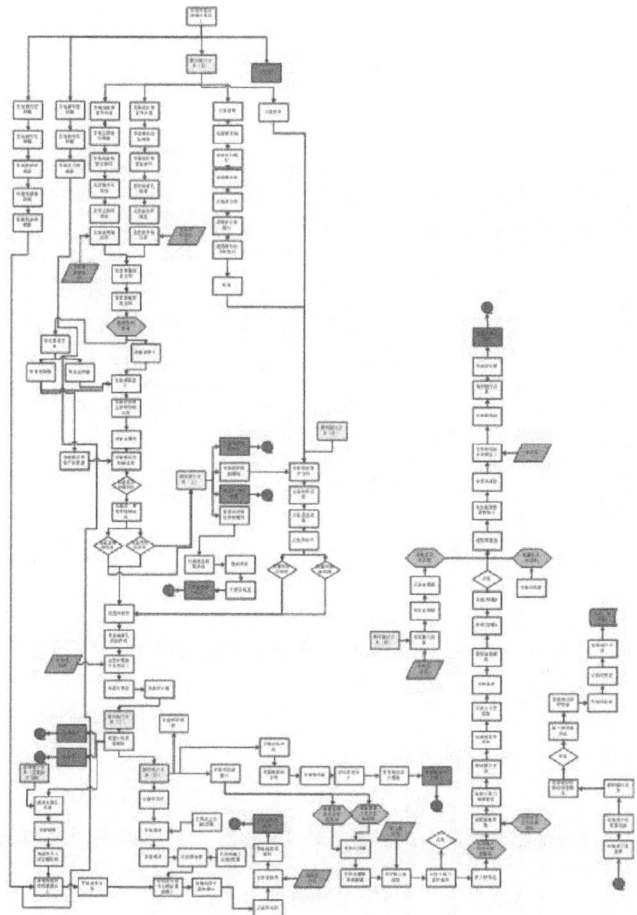

Figure 4.8 Precedence relationship of transmission assembly.

driven axle nuts. The balance rate of production line of transmission is only 64.5%.

Based on the precedence relationship diagram line, calculate line balance equation and readjust stations and production line balance; the balance effect of production line of general assembly line is shown in Figure 4.10.

The balance rate of production line of transmission assembly after readjustment of workstation increases from 64.5% to 84.9% and can reach the takt of 87 s, thus achieving the predicted production capacity of two shifts and 150,000 sets.

In addition to the rearrangement of line balance equation, some of the workstations are also adjusted technically, including the following:

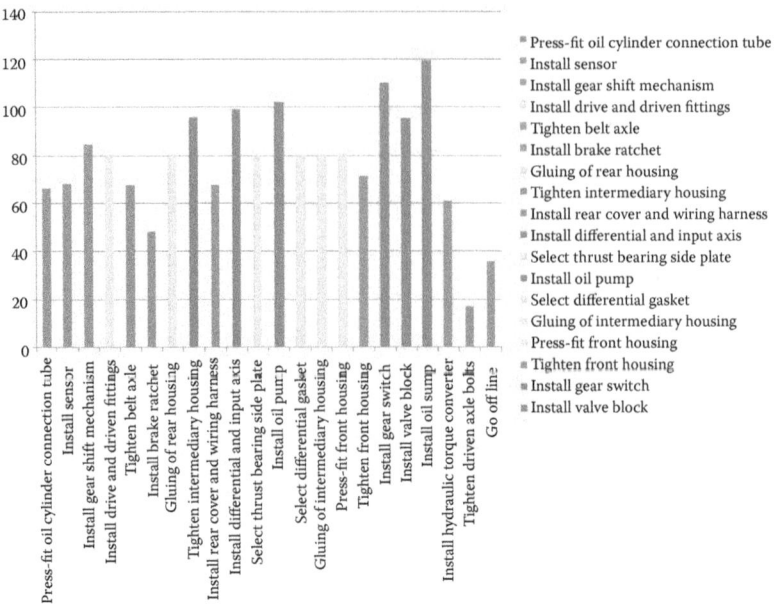

Figure 4.9 Line balance analysis: manufacturer program.

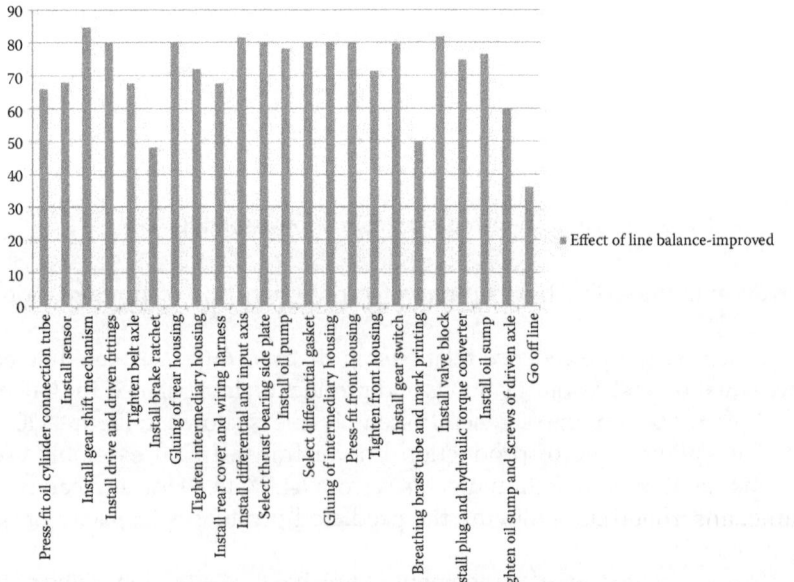

Figure 4.10 Effect of line balance after adjustment.

Adjustment 1: The installation of intermediate housing is a bottleneck as the time spent on turning screws takes a large proportion. The 12 screws in the intermediate housing need to go through three steps, including being tightened manually in advance, by electric tightening machine and torque wrench, which costs 81.786 s in total, with 6.8155 s needed for each screw turning, on average. For the tightening of front housing, only 71.466 s is spent on 16 screws, because in the process of tightening, torque function of electric tightening machine rather than torque wrench is used for tightening directly. At the same time, the order of preinstallation and tightening of screw influences the path of actions, which has a big effect on time consumption. By changing the tightening equipment, improving the efficiency of tightening, and planning the action path of tightening screws by operators in a better way, the time spent for tightening each screw can thus be reduced from 6.8155 to 4.8155 s, and the time for tightening screws is reduced at the same time.

Adjustment 2: Considering that the artificial measurement may be affected by the psychological condition of people, which may lead to errors or even mistakes, using machines for measurement is suggested. The manufacturer also can omit the measurement step by constraining the dimension precision of suppliers.

Adjustment 3: Nine screws need to be tightened in oil pump installation. Increase the efficiency of tightening by improving the tightening devices to save a lot of time. In addition, remove the packing of oil pump in the process of handling, set online work bins, and place seal rings or spacer to reduce turning action.

4.3   Analysis and design of workstation layout

In this section, the balance of production line is adjusted and the setting of workstation is determined according to the results of the analysis of the action time and the requirements of takt. In this chapter, the layout and corresponding logistics equipment of manufacturers will be analyzed, and the final corresponding layout scheme will be given.

4.3.1   Existing logistics equipment

The current logistics storage equipment used by the manufacturer are as follows: medium turnover box, auxiliary work bin, small and big cleaning frame, and roller rack. The roller rack has three to four layers. Logistics transportation equipment is a platform cart. Carts include valve block cart, oil pump cart, housing cart, oil sump, etc. These carts are similar in structure and size but different in the method of place of the parts. The cart can be two or three layers.

Table 4.1 Size of basic logistics equipment

	Benchmark metric	L	W	H
Medium turnover box	cm	55	40	30
Auxiliary working bins	cm	20	12	15
Small cleaning frame	cm	50	25	25
Big cleaning frame	cm	55	40	30
Roller rack (interlayer)	cm	117	90	45
Platform cart	cm	130	80	90

Table 4.2 Accommodation relationship of logistics equipment

	Platform cart level	Rack level	Medium turnover box
Medium turnover box	4	4	
Auxiliary work bin	24	24	6
Small cleaning frame	6	6	
Big cleaning frame	4	4	

The size of the logistics equipment is shown in Table 4.1.

Logistics equipment have a certain nested accommodate relationship, as shown in Table 4.2.

4.3.2   Analysis and setting of logistics channels

The logistics channels of production line include mainly cart channel and walking channel. Cart channel is a two-way channel, which is the main channel of distribution. The formula of the width of the two-way channel is as follows: $W = 2W_P + 2C_0 + C_m$. W stands for the width of two-way channel, W_P stands for cart width, C_0 is the size margin cart side, and C_m is the minimum distance when passing. In this factory, W_P is 80 cm, C_0 is 10 cm, and C_m is 20 cm. Thus, the minimum width of a two-way channel is 2 m. There is no large turnover in the manufacturer. There is mainly short-distance reclaiming and walking by operators. According to the standard, 0.75 to 1 m is advisable. Here, 1 m is set.

4.3.3   Analysis of storage and material flow quantity

Before setting the workstation layout, analyze the storage capacity and material flow quantity of each workstation. Draw the forms of distribution batch, storage quantity, and material flow quantity; the format is as shown in Table 4.3 (only a part is selected for sample, and the complete distribution information table is provided in the appendix).

Among them, the data in bold in the table stand for large items that are delivered directly by cart (housing, valve block) or that use a large cleaning frame (driven axle, driving axle); the distribution period is less than 1 h. The data in italics stand for the auxiliary parts, such as screw gasket, which are placed in the work bin, whose distribution period is 2 or 4 h. For the stations with larger consumption (1 takt needs 14), the distribution period is 1 h. The rest of the data in the table stand for common parts; the general distribution period is 1 to 2 h according to the different sizes of the parts.

According to the distribution batch and frequency table, the material flow quantity and storage of each current workstation can be drawn as Table 4.4.

The basic unit of the material flow quantity and storage content is small cleaning frame. The material flow quantity of rear housing installation is 9; namely, there are nine small cleaning framing flows into the workstation per hour. The storage quantity is 13; that is, according to the current distribution frequency, the storage of pipeline inventories needs to occupy the volume of 13 small cleaning frames.

To analyze the material flow and storage content more directly, draw Figures 4.11 and 4.12 as follows.

By analyzing the figures, it can be seen that material flow and storage concentrate mainly on the workstations of large parts, such as housing assembly and driving and driven wheel axle assembly; on the other hand, a large amount of material flow concentrated at the intersection of a subassembly line and the general assembly line, for example, the intersection of drive assembly and driven assembly or intersection of intermediate housing and general assembly line. If these material flows are totally delivered manually, it will cost a lot of time. Therefore, when considering the layout of the workstations, try to use subassembly lines, especially for subassembly line with large material flow, which could be directly connected by assembly line, thus avoiding manual distribution.

4.3.4   Evaluation of existing layout of second phase workstation

The manufacturers provide the preliminary program of the second phase of planning, as shown in Figure 4.13.

From the aspect of material flow, in the program, the drive and driven belt axle assembly, intermediate housing assembly, differential assembly, input axle assembly, front housing assembly, and so on, are directly connected to the general assembly line, which decreases distribution largely and is consistent with the idea of minimizing material flow. From the point of view of material flow channel and distribution, the main channel with a length of 2.3 m can accommodate turns of cart, which is a reasonable double-channel.

Table 4.3 Distribution batch and frequency

Workstation storage	Workstation material flow quantity	Process no.	Process name	Part	Consumption/takt consumption/takt	Consumption/h	Container	Container volume container accommodation	Turnover/h	Distribution batch	Distribution cycle (h)	Number of rack/cart level accommodation container
13	9	ZZ01	Install rear housing	Rear housing	1	40	Cart	60	2/3	1	11/2	1/2
			Install rear housing	Joint tube (including four seal rings)	2	80	Small cleaning frame	80	1	1	1	6
		ZZ02	Preinstall sensor	Lip rocker arm radial sealing ring	1	40	Auxiliary working accessories	80	1/2	2	4	24
			Preinstall sensor	Revolving speed sensor	2	80	Small cleaning frame	80	1	1	1	6
			Preinstall sensor	M6*16 Bolt M6*16	2	80	Auxiliary parts cases	80	1	2	2	24
	1.875		Preinstall sensor	Pressure sensor	1	40	Small cleaning frame	80	1/2	1	2	6
3		ZZ03	Install shifting mechanism	Control lever	1	40	Small cleaning frame	80	1/2	1	2	6

(Continued)

Table 4.3 (Continued) Distribution batch and frequency

Workstation storage	Workstation material flow quantity	Process no.	Process name	Part	Consumption/ takt consumption/ takt	Consumption/h	Container	Container volume container accommodation	Turnover/h	Distribution batch	Distribution cycle (h)	Number of rack/cart level accommodation container
			Install shifting mechanism	Control stick	1	40	Small cleaning frame	80	1/2	1	2	6
			Install shifting mechanism	Fan-shaped plate fitting	1	40	Small cleaning frame	80	1/2	1	2	6
			Install shifting mechanism	Rocker arm	1	40	Small cleaning frame	80	1/2	1	2	6
			Install shifting mechanism	Elastic pin	2	80	Auxiliary parts cases	80	1	2	2	24
			Install shifting mechanism	Locating plate fitting	1	40	Small cleaning frame	80	1/2	1	2	6
6	3		Install shifting mechanism	M6*12 Bolt M6*12	2	80	Auxiliary parts cases	80	1	2	2	24

Table 4.4 Analysis of material flow and storage

Process no.	Section	Process name	Storage	Flow (small cleaning frame/h)
ZZ01	General assembly line	Install rear housing	13	9
ZZ02	General assembly line	Preinstall sensor	3	1.875
ZZ03	General assembly line	Install shifting mechanism	6	3
ZZ04	General assembly line	Axle fitting of rear housing and driving and driven	27	52.25
ZZ05	General assembly line	Fix driving and driven axle	1.25	0.625
ZZ07	General assembly line	Install brake ratchet	4.5	2
ZZ08	General assembly line	Install intermediate housing	13.5	25.5
ZZ09	General assembly line	Install rear end cover	4.5	2
ZZ10	General assembly line	Install differential, input axle	28	25.25
ZZ11	General assembly line	Measure and select side plate	0.5	0.25
ZZ12	General assembly line	Install oil pump	21.5	21.875
ZZ13	General assembly line	Measure and select differential pad	0	0
ZZ15	General assembly line	Press-fit prehousing	17	26.25
ZZ16	General assembly line	Mark permanent identification	0	0
ZZ17	General assembly line	Install gear switch and breathing brass tube	11	6
ZZ18	General assembly line	Install valve block	15.25	13.75
ZZ19	General assembly line	Install sump	13.75	13.75
ZZ20	General assembly line	Install hydraulic torque converter, go offline	12	24

(Continued)

Table 4.4 (Continued) Analysis of material flow and storage

Process no.	Section	Process name	Storage	Flow (small cleaning frame/h)
FZ01-01	Drive axle subassembly line	Drive axle assembly	7	13.41
FZ01-02	Drive axle subassembly line	Install oil cylinder	3.75	4.75
FZ01-03	Drive axle subassembly line	Assemble and test steel ball	3	1.5
FZ01-04	Drive axle subassembly line	Install bearings and locking nut	1	0.5
FZ02-01	Driven axle subassembly line	Install driven axle	5	17
FZ02-02	Driven axle subassembly line	Install oil cylinder	3.5	5.29
FZ02-03	Driven axle subassembly line	Install and test steel ball	3	1.75
FZ02-04	Driven axle subassembly line	Install piston	6.5	10.63
FZ02-05	Driven axle subassembly line	Tighten the nut	1	0.5
FZ03-01	Input axle subassembly line	Weld part of input axle	5.25	5.88
FZ03-02	Input axle subassembly line	Weld part of sun	15	13
FZ04-01	Intermediate housing subassembly line	Press fitting oil seal of intermediate housing	13	12.5
FZ04-02	Intermediate housing subassembly line	Press-fitting reverse gear piston	4	2
FZ04-03	Intermediate housing subassembly line	Select and match pressing plate of friction plate	5	3.5
FZ04-04	Intermediate housing subassembly line	Press-fit roller bearings	3.25	1.625
FZ04-05	Intermediate housing subassembly line	Install peripheral parts	4.75	2.625

(Continued)

Table 4.4 (Continued) Analysis of material flow and storage

Process no.	Section	Process name	Storage	Flow (small cleaning frame/h)
FZ05-01	Differential subassembly line	Pretighten differential bolts	25.25	25.25
FZ05-02	Differential subassembly line	Press bearing	2	1
FZ05-03	Differential subassembly line	Install planetary mechanism	3.75	3.75
FZ06	Preassembled of front housing	Front housing assembly	21.5	23.5

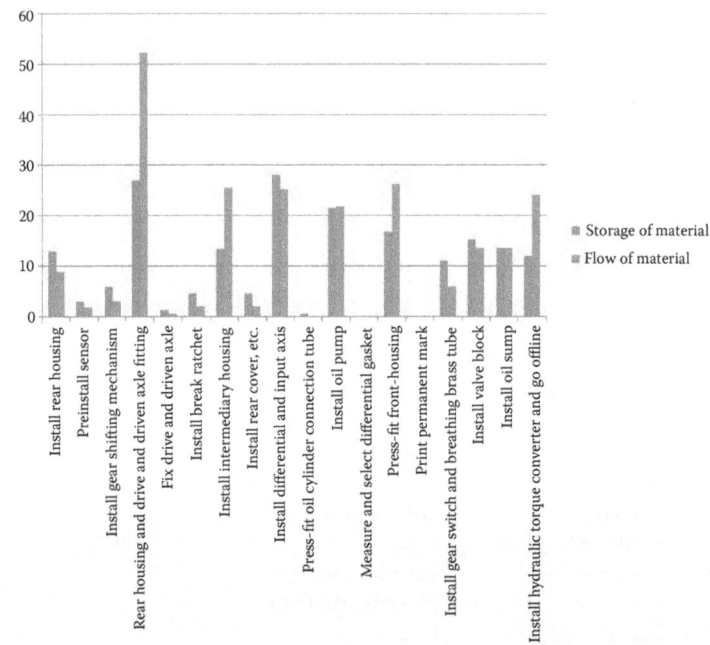

Figure 4.11 Material flow and storage distribution of general assembly line.

But for the line side material flow, the program still needs to be improved (see Figure 4.14).

Problem 1: Does oil cylinder connect to the channel between the servo press and driven axle assembly line? The horizontal distance is only 1.4 m, and the vertical distance is only 0.75 m, while the length of the small carts is 1.3 m and the width is 0.8 m; it is difficult to go through the channel. In

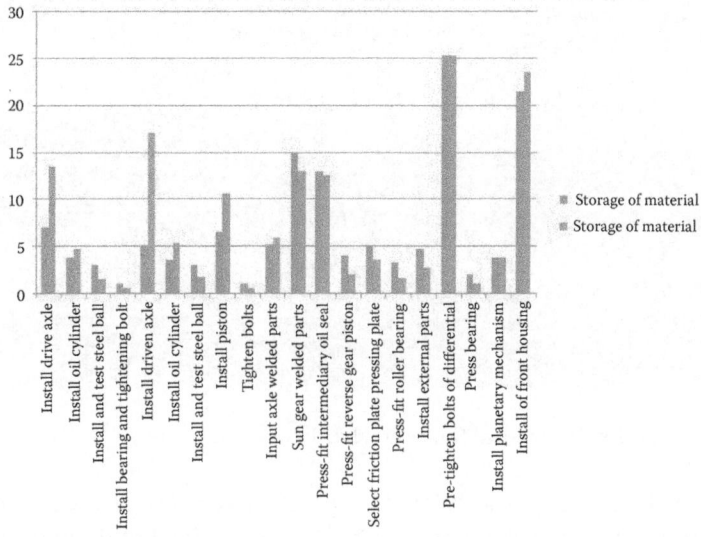

Figure 4.12 Material flow and storage distribution of subassembly line.

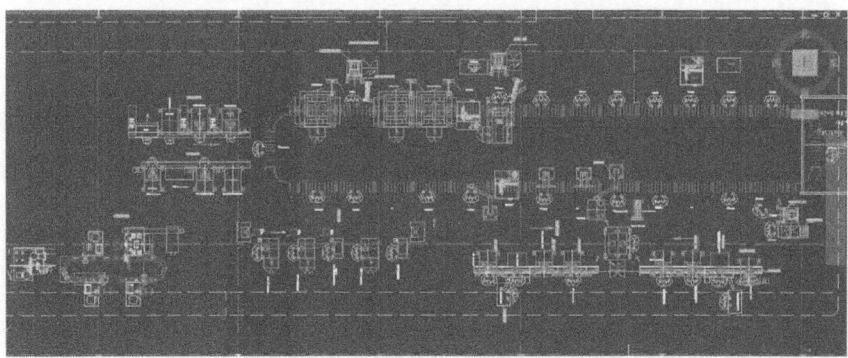

Figure 4.13 Preliminary planning of the second phase, provided by the manufacturer.

addition, the setting of roller rack will make cart transferring even more difficult. On the other hand, the sensor of distribution process 2, install sensors; gear shift mechanism of process 3, gear shift mechanism; and metal belt of processes 4, installation of drive and driven wheel axle need to go through the channel. Thus, the setting of the material flow channel

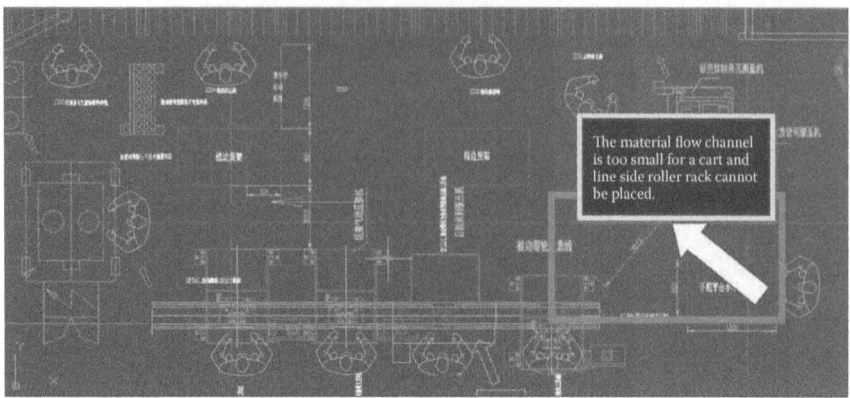

Figure 4.14 Line side material flow and channel status.

needs to be improved. In addition, the space between the processes of sensor assembly and drive and driven axle is so small that the standard roller rack cannot be placed.

Problem 2: The distance between the rear cover installation workstation and the subassembly line of intermediary housing is 1.3 m. Considering the possibilities of placing racks and the reasonable moving range of people, the distribution by carts has the same problem.

Problem 3: As shown in Figure 4.15, the space between the installation of rear cover process and the subassembly line is so small that standard roller rack cannot be placed.

Problem 4: There is no space between the differential and input axis subassembly lines for placing the roller rack.

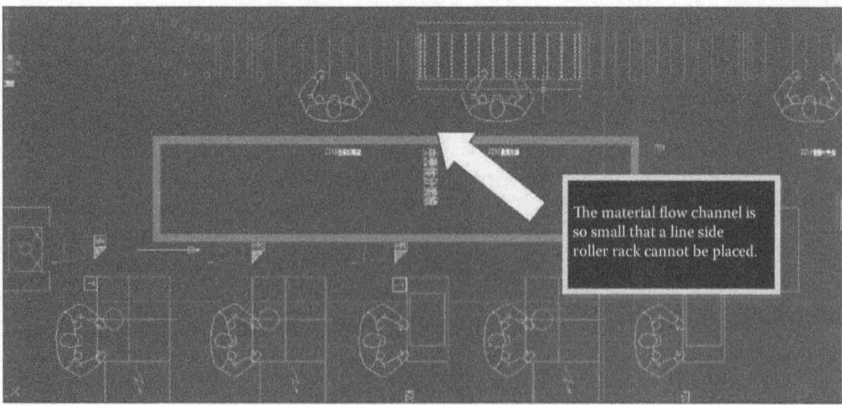

Figure 4.15 Line side material flow channel status 2.

Problem 5: As shown in Figure 4.16, the front housing subassembly line is very close to the main material flow channel, which leads to the lack of space for placing the roller rack at the side of the front of the housing subassembly line to store the respective 11 kinds of assembly parts (prehousing, intermediate axle, middle wheel, framework oil seal, etc.).

Problem 6: Turnover cases of line side cannot be placed at the side of the subassembly line 1, drive belt axle assembly, and subassembly line 2, driven belt axle assembly, process 2, workstation of oil cylinder installation.

Based on the previous analysis, the readjusted layout of workstation is as in Figure 4.17. The detailed proposal for revision is referred to in the comments of the figure. See the detailed AutoCAD file in "Assembly Line Program-Revised.dwg."

Instruction 1: See detailed CAD file in "Assembly Line Program-Modified version.dwg." The specific material flow equipment will be analyzed in Chapter six.

Instruction 2: Adjust the position of the oil cylinder connecting the tube installation machine to enlarge the channel between the position and the right side of the subassembly line of the driven axle assembly.

Instruction 3: Move down the subassembly line of intermediary housing for 20 cm.

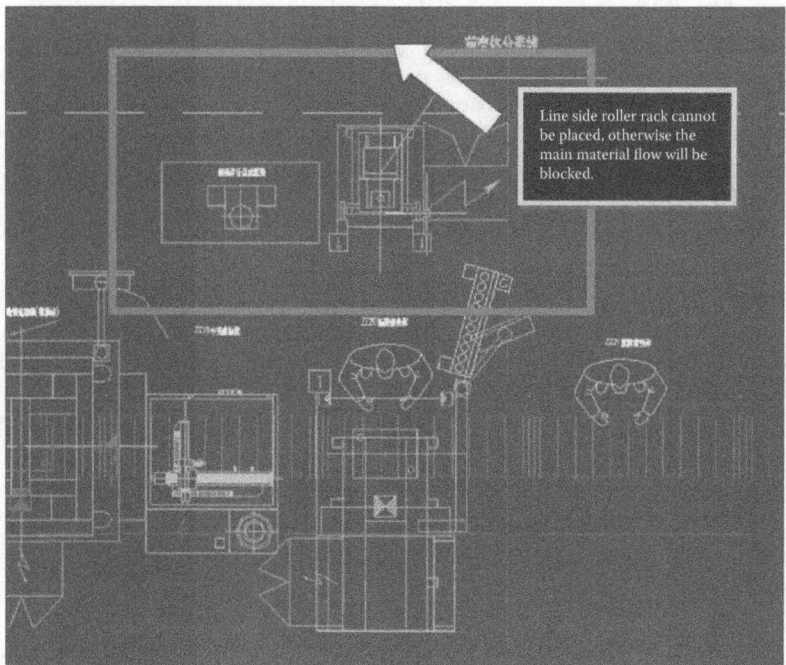

Figure 4.16 Problem of front housing subassembly material flow channel.

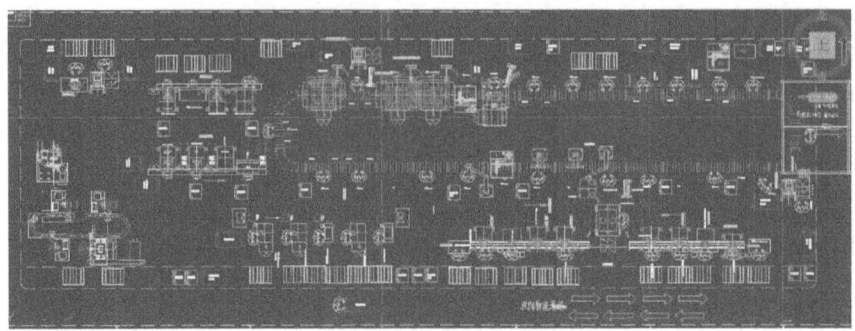

Figure 4.17 Layout of the second stage—adjusted.

Instruction 4: Change the direction of the operators of the differential subassembly line and input axis subassembly line to provide enough space behind the operator for storage rack.

Instruction 5: Move the differential subassembly line to the bottom of the graph, and the move the input axis subassembly line to the top of the graph. As the material flow and inventory of input axis are relatively larger, if it is placed below, there will not be enough roller racks because of planet gear assembly line; thus, the line side stock cannot be guaranteed. The material flow and stock of differential are relatively smaller, and the effective time of some workstations is short and can assist material acquisition. As a result, the operators assemble differential first, then install input axis, which is the reverse of the sequence of the original plan.

Instruction 6: Enlarge the distance between the differential and workstation of input axis subassembly line to 1 m, which ensures the passage of withdrawable cart and operators.

Instruction 7: The direction of the differential subassembly line is adjusted as from left to right in the layout figure and joins into the general assembly line.

Instruction 8: Workstation—install sensor, gear shift mechanism, and brake ratchet storage and use withdrawable cart as container, to meet the limitation of channel and storage space.

Instruction 9: For the first process of differential corresponding to differential housing and differential gear ring, with these two parts with large material flow and storage, a double box cart is used; for the flowing working procedure, relative to the small material flow and storage, a withdrawable cart is used to meet the limitation of space and material flow channel.

Instruction 9: The space of oil pump installation is not enough, so that the oil pump cart is placed far, which leads to invalid movement waste. It is suggested that the matching thrust bearing side plate instrument and measurement differential pat instruction be placed a bit away from the workstation of oil pump.

Instruction 10: The front housing subassembly line is moved from the side of the transmission assembly of prehousing installation to the space on the left side of the production workshop (vertical view). Considering that the subassembly line of the front housing needs to install 12 parts, among which front housing, intermediary axle, and intermediary well take large spaces for storage, it will cost a lot of line side stock, while placing them at the original location will influence the material flow channel. So the front housing subassembly line is moved to the left side of production workshop (vertical view).

Instruction 11: Adjust the assembly tables of assembly line 1, drive belt axle assembly, and process 2 of assembly line 2, driven belt axle assembly–oil cylinder installation, to the position that is parallel to the subassembly line and place them at the first workstation of the subassembly line rather than the vertical position. Therefore, the subassembly is a straight line that facilitates distribution. At the same time, there will be enough space near the workstation of oil cylinder for storing corresponding parts.

4.4   Design of material flow distribution

4.4.1   Basic rules of distribution

1. In principle, the distribution should be conducted according to distribution cycle; the different cycles are 0.25, 0.5, 1, 2, and 4 h.
2. Divide the distribution route according to regional division distribution within the distribution batch divided by cycles.
3. Some special distribution principles are as follows:
   a. Deliver by multiples of 15 min to reduce the memorial burden of distribution staff and reduce distribution error. Some of the parts are delivered not according to the bearing material flow installation capacity, including the following:
      i.   One big cleaning frame can contain at most 28 drive belt wheels and driven belt wheels. Here, each cleaning frame contains 20 sets and the distribution cycle is 1/2 h.
      ii. Rear housing cart: A cart can contain 60 rear housings at most; distribution cycle is 1.5 h. To save space and to adapt to the width of channels between process 2 and 4, a smaller cart can be used that can load 40 rear housings each. Distribution cycle is 1 h.
      iii. Drive axle: A big cleaning farm contains eight drive axles. To facilitate distribution, each big cleaning frame contains five drive axles. The distribution cycle is 1.5 h.

iv. Drive and driven cylinders: A big cleaning farm contains 14 drive or driven cylinders. To facilitate distribution, each big cleaning frame contains 10 sets and the distribution cycle is 15 min.

v. Oil pump: In the original cart design by the manufacturer, a cart can contain 36 oil pumps. To match the consumption of 40 overall per hour, properly increase the loading area of an oil pump cart, making one cart contain 40 oil pumps.

b. The batch of distribution of a few parts is different:

i. Driven axle: A big cleaning frame contains four drive axles. To facilitate distribution, three sets are distributed for each batch, and the distribution cycle is 18 min. To reduce the complexity of memory of the distribution of the distribution staff, the first 15 min per hour is for three cleaning frames and the second 15 minutes is for two cleaning frames, which takes turns.

c. Consumable materials with low value, such as bolts and steel balls, are difficult to count, which should be delivered in their original package.

4.4.2 Parts distribution batch and cycle analysis

In accordance with the previous discussion, analyze all the parts need distribution and sort out as in Table 4.5.

In Table 4.5, not all parts in the material flow and storage are included. Parts such as drive and driven belt wheel assembly, intermediary assembly, and differential assembly do not need to be delivered as the subassembly line is connected to the general assembly line in design. In the table, the data in bold stand for large parts, which are usually stored by cart, large cleaning frame, and medium turnover box; the data in italics stand for the auxiliary parts, usually placed in a medium part case with auxiliary parts case or put into a medium parts case for distribution; and the rest of the data use a small cleaning frame for distribution (a small cleaning frame can deliver parts that do not need to be cleaned, such as sensors).

4.4.3 Rule of cycle of delivered parts and analysis of parts type

There are five distribution cycle time: 0.25 hour, 0.5 hour, 1 hour, 2 hours, and 4 hours in total. For the distribution cycle of 0.25 hour, four types of part are distributed to the assembly line. They are driving axle, driving cylinder, driven axle, and driven cylinder, distribution containers are large cleaning frames, which are transported by carts. The start of the distribution is the exit of cleaner and the end of distribution is the roller rack behind the subassembly line of driving and driven belt axle assembly.

Table 4.5 Example of distribution batch and cycle

Process name	Part	Takt consumption quantity	Consumption/h	Container	Quantity	Turnover container/h	Distribution batch	Distribution cycle (h)
Gear switch	Control stick weld	1	40	Small cleaning frame (not clean)	80	1/2	1	2
Install gear switch	Line card	1	40	Auxiliary parts case	80	1/2	2	4
Install gear switch	Control stick wire drawing support	1	40	Auxiliary parts case	80	1/2	2	4
Install gear switch	Wiring harness bracket	1	40	Auxiliary parts case	80	1/2	2	4
Install breathing brass tube	Pipe clamp bracket	1	40	Auxiliary parts case	80	1/2	2	4
Install gear switch	Pipe clamp	1	40	Auxiliary parts case	80	1/2	2	4
Install breathing brass tube	Breathing brass tube	1	40	Auxiliary parts case	80	1/2	2	4
Install breathing brass tube	Breathing tube	1	40	Auxiliary parts case	80	1/2	2	4

(Continued)

Table 4.5 (Continued) Example of distribution batch and cycle

Process name	Part	Takt consumption quantity	Consumption/h	Container	Quantity	Turnover container/h	Distribution batch	Distribution cycle (h)
Install breathing brass tube	M8X12 hexagon flange bolts	1	40	Bolt packaging	80	1/2	2	4
Install valve block	M6X50-10.9 G hexagon flange bolts	3	120	Bolt packaging	80	1 1/2	6	4
Install valve block	M6X80-10.9 G hexagon flange bolts	1	40	Bolt packaging	80	1/2	2	4
Install valve block	Valve block assembly	1	40	Cart of valve block	40	1	1	1
Install valve block	Oil filter	1	40	Medium turnover case	40	1	2	2
Install oil sump	Oil sump	1	40	Cart of valve block	40	1	1	1
Install hydraulic torque converter, go off-line	Hydraulic torque converter	1	40	Cart for hydraulic torque converter	20	2	1	1/2

For the distribution cycle of 0.5 hour, 13 types of part are distributed to the assembly line. Carts are used only for hydraulic torque converter and differential gear ring directly; the others use large cleaning frames as distribution container. The start of distribution is the exit of cleaning matching and the end of distribution is the roller racks behind the workstation of drive and driven belt axle assembly.

For the distribution cycle is 1 hour, 55 types of part are distributed to the assembly line. The rear housing, oil pump, valve block and oil sump, etc., are placed and delivered by cart directly. Bolt is delivered with original packaging; most of the others, such as control rods, use small cleaning frame as distribution container.

The distribution cycle is 2 hours, 73 types of part are distributed to the assembly line, half of which use a small cleaning frame (parts that do not need cleaning, such as gear switch, can use metal wire frame or medium turnover case, which occupy the same volume). The other half use original packaging, such as bolts.

The distribution cycle is 4 hours, 36 types of part are distributed to the assembly line. All of them are auxiliary parts, usually placed into medium turnover case with original packaging or placed in medium turnover case with auxiliary parts case without packaging.

4.4.4   Design of the place of line side storage rack

Determine the name and quantity of the corresponding parts of turnover case used for storing parts of roller rack, drawer table cart, and platform cart of line side.

A combined roller rack has four layers, in which the first to third layers are the main insulation layers for the material acquisition for operators and the fourth layer is used for placing empty cleaning frames or turnover boxes for the moment. According to the information on parts placing in the following combined roller rack, usually, the second and third layers are used to facilitate the material acquisition of operators; some use the first layer, but the fourth layer will not involve information of parts placing. Four medium turnover cases or large cleaning frames or six small cleaning frames can be placed on each layer of the combined roller rack. The storage of turnover cases in the combined roller rack adopts the double cases rule, namely, case 1 is used as the turnover case in use, while case 2 is used as safety stock. Once the parts of case 1 are used up, the parts of case 2 will be used at once. The distribution staff provides cases with distributed parts to replace the empty case 1 when the operators are using the parts in case 2.

A drawer-type cart, as transport equipment of material flow, can also be used as storage equipment of line side stock. The drawer-type cart is divided into three layers; each layer can contain two medium turnover

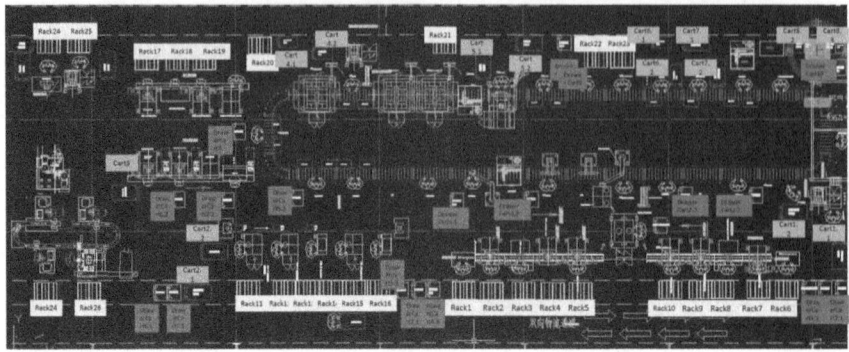

Figure 4.18 Layout of line side rack.

cases or three small cleaning frames. In the aspect of stock and distribution strategies, the drawer-type cart should use the double-boxes rule, namely, case 1 is used as the turnover case in use, and after the parts in the first drawer-type cart are used up, switch to the second drawer-type cart. When the operator is using the second drawer-type cart, distribution staff provides the drawer-type cart with distributed parts to replace the first drawer-type cart. For a few workstations, the drawer-type cart is used only for transferring and is used together with online container to play the role of double boxes.

A platform cart is used as transport equipment of material flow as well as stock equipment of line side stock, which is used to transfer and store large parts of housings and oil sump. In the aspect of stock and distribution strategies, the double-boxes principle should be applied to platform cart.

The layout of a line side rack is shown in Figure 4.18.

4.4.4.1 Combined roller racking
As shown in Figure 4.18, there are a total of 29 combined roller racks, which are concentrated in the lower part of the layout (vertical view). Most combined roller racks are placed mainly beside the subassembly lines, including the drive belt axle subassembly line, driven belt axle subassembly line, intermediate housing subassembly line, and input axis subassembly line. There are some combined roller racks in workstation 17, install gear switch, and workstation 18, install breathing brass tube, of the general assembly line (Figure 4.19).

4.4.4.2 Drawer-type cart
There are 10 spots needing the use of a drawer-type cart. There are seven double-box drawer-type carts used for transferring and online stock;

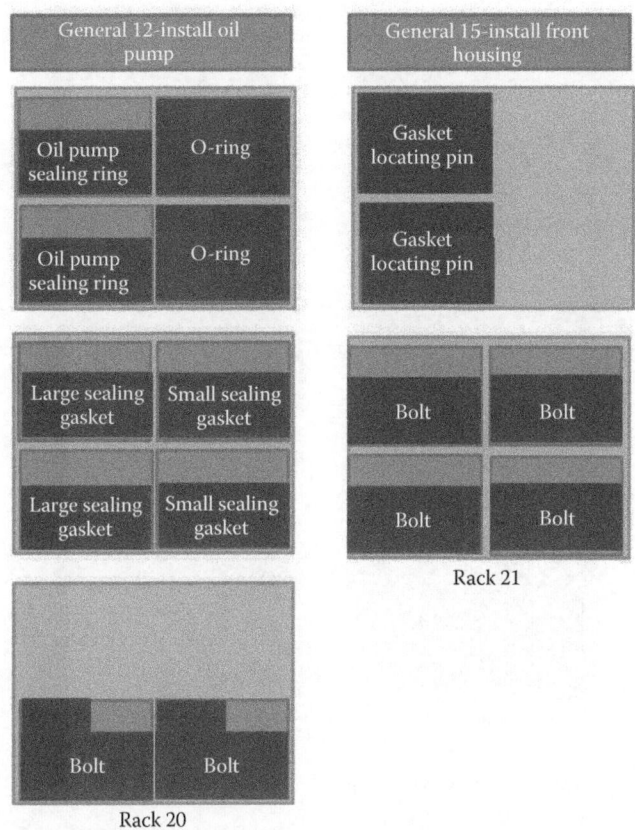

Figure 4.19 Inner layout of roller rack of transmission assembly.

three of them are used only for transferring and for the online container platform. A drawer-type cart can also play the role of double boxes.

Figure 4.20 shows the inner layout of a drawer-type cart.

4.4.4.3 Platform cart
Some platforms need assembly delivery of some small supporting parts of the corresponding spots.

Hydraulic torque converter cart: one distribution cycle, namely, 0.5 h for bolt, 40 bolts.

Differential gear ring: one distribution cycle, namely, 0.5 h for bolt, 200 bolts.

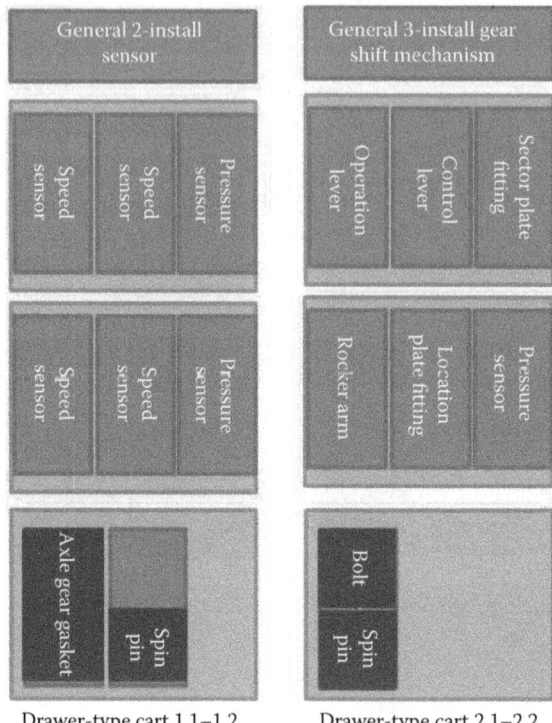

Drawer-type cart 1.1–1.2   Drawer-type cart 2.1–2.2

Figure 4.20 Sample of drawer-type cart.

Rear housing cart: one distribution cycle, namely, 1 h for oil tubing, including four gaskets.

Oil sump cart: one distribution cycle, namely, 1 h for oil return pipe clamp plate, oil return pipe, and magnet.

4.4.5 Set of entrance

There are three entrances of distribution, as shown in Figure 4.21:

- Entrance for cleaning parts (subdivided into entrance of ultrasonic cleaning machine and entrance of passing spray, which are named collectively as entrance of cleaning parts)
- Entrance for channel
- Entrance for housing (set a channel in the partition plate of working storage area and production workshop to facilitate the distribution of housing parts and crankcase oil)

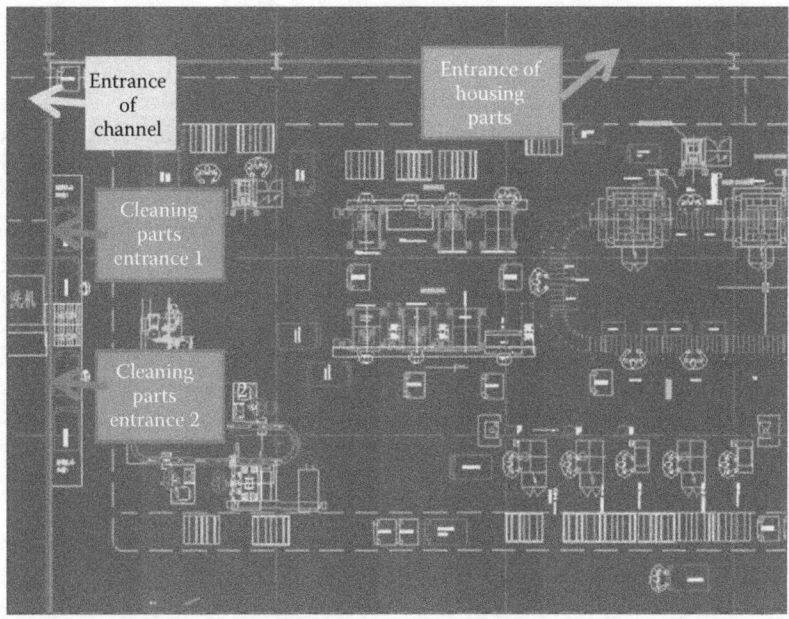

Figure 4.21 Setting of production workshop entrance.

4.4.6   *Analysis of auxiliary operation time of distribution*

Regardless of the preparation process, the preparation staff is responsible for stocking, while the distribution staff is in charge of distribution. The auxiliary actions of the distribution staff can be summarized as follows: Fetch the cart, move the cart, stop and locate, feed material, empty the cleaning frame, move the cart (return).

According to the time analysis of the MOD method, for the parts placed in the cleaning frame, the average time of manual feeding is about 6 s owing to the different quantities of materials; the time for taking off empty cleaning frames is about 4 s. The load and unload time of one cleaning frame is added to 10 s. The time of stopping and positioning is about 2 s.

The speed of carts pushed manually is 1 m/s. From the cleaning frame inlet to the driven belt axle subassembly line and then returning to the backflow of cleaning frame, the moving distance is 90 m, costing 90 s.

For the parts simply placed directly on a special cart, such as oil sump, valve block, housing, etc., they are not controlled by material acquisition and uninstalling of cleaning frame, and the time equals to the time sum total of moving, stopping, and locating time.

4.4.7 Cycle distribution design

In the distribution design, the distribution is generally divided by distribution cycle. Under the same distribution cycle, the division is conducted according to the constraints and cart capacity as well as shortest-route principle. A cart usually has three layers, of which two layers are used to place containers for loading material and the last one layer is used for recycling cleaning frame. Therefore, the ceiling of calculation of distribution is two layers.

A special drawer-type cart is used in the workstation of sensor installation, gear shift mechanism, brake ratchet and rear cover, input axis differential, etc.; thus, the same logistics equipment is used in both distribution and storage. Distribution is not divided by cycle of division, but according to the workstation. In the first part, the distribution of these workstations will be analyzed individually.

4.4.7.1 Distribution of special workstation—Drawer-type cart

Some workstations can neither use roller racks for storage nor use carts for replenish parts because of the limitation of channel and space. Therefore, material flow equipment needs to be replaced. Different from other parts delivered by cycle, its distribution and storage use the same material flow equipment: a drawer-type cart. To facilitate distribution, the distribution is based on the principle of one-time distribution of the same work position, making the distribution cycles of all the parts of the same workstation the same with each other by adjusting distribution batch, namely, complete in a single distribution.

The drawer-type cart of special distribution workstation can load four cleaning frames or four intermediary turnover cases. There are hooks in the rear of a drawer-type cart so these carts can be interconnected, facilitating distribution staff to deliver several drawer-type carts at one time. Each distribution staff is able to move at most four connected drawer-type carts at one time.

Generally, there are three distribution routes.

Route 1: Install rear end cover and wiring harness, etc. Differential subassembly—install gears, install brake ratchet, and install gear shift mechanism; the distribution cycle is 2 h. Four drawer-type carts connected by hooks can be delivered at one time.

Route 2: Differential subassembly 2—install bearing, install bolts of intermediary housing, and preinstall sensor; the distribution cycle is 4 h. Three drawer-type carts connected by hooks can be delivered at one time.

Route 3: Install the bolts of the front housing and oil sump. The distribution cycle is 2 h. Two drawer-type carts connected by hooks can be delivered at one time.

For drawer-type cart, as the workstations of differential and input axis installation are close to the working storage area, the drawer-type carts are fetched by the worker of the workstation.

The distribution analysis of these workstations is shown in Table 4.6; the three distribution routes are shown in Figure 4.22. In the figure of the distribution route, the lines stand for routes, while diamonds stand for spots of allocation and material loading and unloading. The analysis of distribution time is as follows.

Route 1: The distribution cycle is 2 h. The traversing distance is 100 m, and the travel time is 100 s. There is no loading and unloading time for drawer-type carts; stop for allocation six times and for 12 seconds. The general distribution time of route 1 is 112 s, namely, 1.86 min.

Route 2: The distribution cycle is 4 h. The traversing distance is 100 m, and the travel time is 100 s. There is no loading and unloading time for drawer-type carts; stop for allocation five times and for 10 s. The general distribution time of route 2 is 110 s, namely, 1.83 min.

Route 3: The distribution cycle is 2 h. The traversing distance is 80 m, and the travel time is 80 s. There is no loading and unloading time for drawer-type carts; stop for allocation four times and for 8 s. The general distribution time of Route 1 is 88 s, namely, 1.46 min.

Routes 1, 2, and 3 do not move the drawer-type carts to the line side but to the working storage area of drawer-type carts, which is closest to the special workstation. Operators need to move the empty drawer-type carts to the working storage area of drawer-type carts after the carts are used for exchange of the carts full of materials.

4.4.7.2   Example of distribution cycle: 0.5 h

The distribution cycle is 0.5 h, and there are 11 types in total. The differential gear ring and torque converter use the platform carts. The differential gear ring cart needs 0.5 h, namely, 200 bolts, while a torque converter bolts cart needs 40 bolts; the others use a platform cart together with a large cleaning frame as distribution container. The start of distribution is the exit of cleaning machine and the end of distribution is roller rack.

Table 4.6 Distribution of special workstation: drawer-type cart

Process no.	Process name	Part name	Takt consumption quantity	Consumption/h	Container	Accommodate quantity/container	Hour turnover container	Distribution batch	Distribution cycle (h)	Start of distribution	End of distribution
ZZ02	Preinstall sensor	Lip rocker arm radial	1	40	Auxiliary parts case	80	1/2	2	4	Working storage area	ZZ02
	Preinstall sensor	Revolving speed sensor	2	80	Small cleaning frame	80	1	4	4	Working storage area	ZZ02
	Preinstall sensor	M6*16 Bolt M6*16	2	80	Bolt packaging	80	1	4	4	Working storage area	ZZ02
	Preinstall sensor	Pressure sensor	1	40	Small cleaning frame	80	1/2	2	4	Working storage area	ZZ02

Note: Drawer-type cart 1.

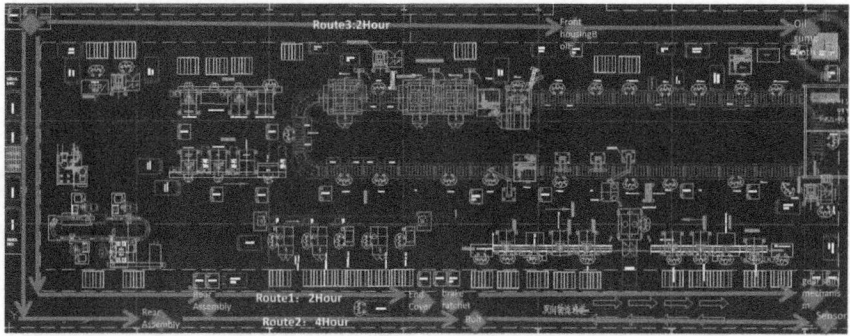

Figure 4.22 Distribution route of drawer-type cart.

Part name	Process name	Container	Distribution batch	Distribution occupation plane	Start of distribution	End of distribution
Differential gear ring	Subassembly of differential	Platform carts	1	2	Cleaning parts entrance	FZ04-01

Cart 1: occupies two layers of platform carts.

Part name	Process name	Container	Distribution batch	Distribution occupation plane	Start of distribution	End of distribution
Drive belt wheel	Drive axle subassembly line	Large cleaning frame	1	1/4	Cleaning parts entrance	FZ01-01
Drive pistons	Drive axle subassembly line	Large cleaning frame	2	1/2	Cleaning parts entrance	FZ01-04
Driven belt wheel	Driven axle subassembly line	Large cleaning frame	1	1/4	Cleaning parts entrance	FZ02-01
Driven pistons	Driven axle subassembly line	Large cleaning frame	2	1/2	Cleaning parts entrance	FZ02-05

(Continued)

Pistons shield	Driven axle subassembly line	Large cleaning frame	2	1/2	Cleaning parts entrance	FZ02-05

Cart 2: occupies two layers of cart platform.

Part Name	Process name	Container	Distribution batch	Distribution occupation plane	Start of distribution	End of distribution
Weld part of input axle	Input axle subassembly line	Large cleaning frame	2	1/2	Cleaning parts entrance	FZ03-01
Sun Gear Weld Part	Input axle subassembly line	Large cleaning frame	2	1/2	Cleaning parts entrance	FZ03-03
Intermediary axle	Preassembled of front housing	Large cleaning frame	2	1/2	Cleaning parts entrance	FZ05
Intermediary wheel	Preassembled of front housing	Large cleaning frame	2	1/2	Cleaning parts entrance	FZ05

Three carts occupy two layers of cart platform.

Part name	Process name	Container	Distribution batch	Distribution occupation plane	Start of distribution	End of distribution
Hydraulic torque converter, go off-line	Install hydraulic torque converter	Platform carts	1	2	Housing parts entrance	ZZ20

Cart 1: occupies two layers of cart platform.

Among the parts with a distribution cycle of 0.5 h, all the parts that pass routes 1, 2, and 3 need to be cleaned, and the start point is the joint queue of cleaning parts entrance. The torque converter entrance of route 4 is the entrance of housing entrance. The detailed route is as follows

Route 1: The distribution cycle is 0.5 h, traverse distance is 30 m, and the travel time is 30 s. Stop for allocation twice; stop for allocation for 4 s. The general distribution time of route 1 is 34 s, namely, 0.57 min.

Route 2: The distribution cycle is 0.5 h, traverse distance is 120 m, and the travel time is 120 s. Load material eight times; the processing time of the material is 80 s. Stop for allocation five times, and for 10 s. The general distribution time for route 2 is 210 seconds, namely, 3.5 min.

Route 3: The distribution cycle is 0.5 h, traverse distance is 60 m, and the travel time is 60 s. Load and unload material eight times; the processing time of the material is 80 s. Stop for allocation four times, and for 8 s. The general distribution time of route 3 is 148 s, namely, 2.46 min.

Route 4: The distribution cycle is 0.5 h, traverse distance is 50 m, and the travel time is 50 s. Stop for allocation twice; stop allocation for 4 s. The general distribution time of route 4 is 54 s, namely, 0.9 min.

4.4.7.3   Distribution route summary

Summarize the cycle distribution route and named as shown in Table 4.7, in which the unit of distribution cycle is in hours, the unit of length of route is in meters, and the unit of distribution tie is in minutes.

4.4.8   Distribution task allocation

4.4.8.1   Preparation and distribution staff and responsibilities within the production workshop

The scope of distribution staff is limited within the production workshop, which is aimed to ensure the cleanliness within the workshop and clarify the duties of distribution and preparation.

There are two distribution staff members. Staff 1 is responsible for delivering parts with a cycle of 0.25 and 0.5 h. Except for the hydraulic torque converter, which enters from the housing part entrance, the others all enter from the entrance of ultrasonic cleaning machine as cleaning parts. Staff 2 is responsible for delivering parts with a cycle of 1, 2, and 4 h. Some of the parts enter from the passing spray machine, while some enter directly from the channel entrance.

The preparation staff is divided into two types; one is staff of internal preparation in the workshop and the other is the staff of the warehouse preparation. There are two staff members within the workshop. One is at the entrance of the ultrasonic cleaning machine and is responsible for carrying the parts that have been cleaned by the ultrasonic cleaning machine to the cart and prepare for distribution staff 1. The other one is at the entrance of the passing spray machine, who is responsible for carrying

Table 4.7 Summary of cycle distribution route

Name	Distribution cycle (hour)	Start of distribution	Route length (meter)	Distribution equipment	Time consumption of distribution (minute)	Parts distribution representative
A1	0.25	Cleaning parts entrance	90	Platform carts + cleaning frame	2.86	Driving and driven axle
B1	0.5	Cleaning parts entrance	30	Platform carts	0.57	Differential housing, gear ring
B2	0.5	Cleaning parts entrance	120	Platform carts + cleaning frame	3.5	Drive belt wheel
B3	0.5	Housing parts entrance	50	Torque converter carts	0.9	Hydraulic torque converter
B4	0.5	Cleaning parts entrance	60	Platform carts + cleaning frame	2.46	Weld part of input axle
C1	1	Housing parts entrance	120	Platform carts	2.06	Rear housing
C2	1	Housing parts entrance	20	Platform carts	0.4	Front housing
C3	1	Housing parts entrance	100	Platform carts	1.73	Intermediary housing
C4	1	Housing parts entrance	60	Platform carts	0.23	Oil pump
C5	1	Housing parts entrance	60	Platform carts	1.06	Valve block

(Continued)

Table 4.7 (Continued) Summary of cycle distribution route

Name	Distribution cycle (hour)	Start of distribution	Route length (meter)	Distribution equipment	Time consumption of distribution (minute)	Parts distribution representative
C6	1	Housing parts entrance	60	Platform carts	1.06	Oil sump
C7	1	Channel entrance junction	100	Platform carts + cleaning frame, etc.	2.93	Drive and driven axle steel belt fitting
C8	1	Channel entrance junction	80	Platform carts + cleaning frame, etc.	3.43	Driven axle spring
C9	1	Channel entrance junction	30	Platform carts + cleaning frame, etc.	2.23	Planetary gear fitting
D1	2	Channel entrance junction	100	Drawer-type carts	1.86	Gear switch
D2	2	Channel entrance junction	80	Drawer-type carts	1.46	Oil sump bolt
D3	2	Channel entrance junction	20	Platform carts + intermediary parts case	1.73	Input axis oil drain valve
D4	2	Channel entrance junction	10	Platform carts + intermediary parts case, etc.	1.93	Lubrication cover fitting

(Continued)

Table 4.7 (Continued) Summary of cycle distribution route

Name	Distribution cycle (hour)	Start of distribution	Route length (meter)	Distribution equipment	Time consumption of distribution (minute)	Parts distribution representative
D5	2	Channel entrance junction	60	Platform carts + intermediary parts case, etc.	2.73	Gear switch
D6	2	Channel entrance junction	100	Platform carts + intermediary parts case, etc.	3.8	Reverse gear piston
D7	2	Channel entrance junction	60	Platform carts + intermediary parts boxes, etc.	2.73	Wave spring
E1	4	Channel entrance junction	100	Drawer-type carts	1.83	Sensor
E2	4	Channel entrance junction	60	Platform carts + intermediary parts case	4.6	O-ring
E3	4	Channel entrance junction	100	Platform carts + intermediary parts case	3.23	Steel retaining ring, etc.

the cleaning parts to the carts and collecting the cleaning parts and non-cleaning parts at the entrance of the same distribution route to one cart according to the cart distribution requirements and preparing material for distribution staff 2.

4.4.8.2   Distribution staff 1
Distribution staff 1 is responsible for delivering parts with a cycle of 0.25 and 0.5 h.

Job description:
> From the 0 to 5th minute of each 30 min, deliver hydraulic torque converter with a cycle of 0.5 h from the housing entrance.
> From the 5th to 10th minute of each 30 min, deliver the drive and driven axle with a cycle of 0.25 h.
> From the 10th to 15th minute of each 30 min, deliver the drive and driven belt wheel with a cycle of 0.5 h.
> From the 15th to 20th minute of each 30 min, deliver the differential gear ring with a cycle of 0.5 h.
> From the 20th to 25th minute of each 30 min, deliver the drive and driven axle with a cycle of 0.25 h.

The specific workflow is shown in Table 4.8.
The distribution route of distribution staff 1 is summarized in Figure 4.23.

4.4.8.3   Distribution staff 2
Distribution staff 2 is responsible for delivering parts with a cycle of 1, 2, and 4 h.

Job description:
> From the 0th to 10th minute of each hour, fetch the carts filled with material from the housing entrance by order and use the carts to deliver six kinds of large parts such as housing.

Table 4.8 Task assignment of distribution staff 1

	Distribution staff 1		
Time (min)	0–5	5–10	10–15
0–15	B3: hydraulic torque converter	A1: drive and driven axle	B2: drive and driven belt wheel
15–30	B4: input axis weld	A1: drive and driven axle	B1: differential gear ring

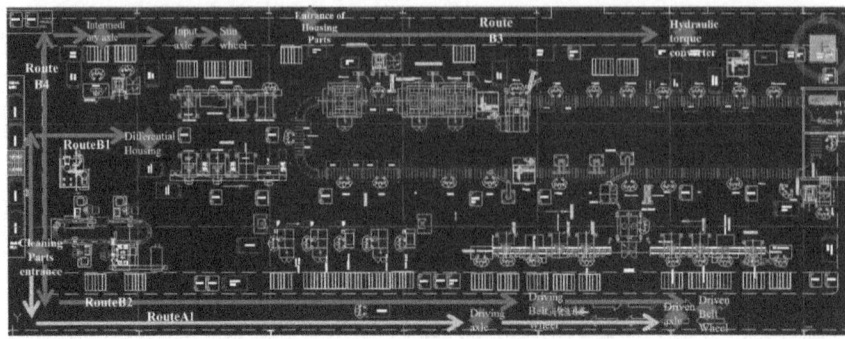

Figure 4.23 Summary of distribution route of distribution staff 1.

Table 4.9 Task assignment of distribution staff 2

Distribution staff 2				
Time (min)	0–10	10–25	25–40	40–50
1	C1–C6: housing, etc.	C7–C9: steel band fitting, etc.	D1–D5: gear switch, etc.	
2	C1–C6: housing, etc.	C7–C9: steel band fitting, etc.	E1–E3: O-rings, etc.	D6–D7: wave spring
3	C1–C6: housing, etc.	C7–C9: steel band fitting, etc.	D1–D5: gear switch, etc.	
4	C1–C6: housing, etc.	C7–C9: steel band fitting, etc.	D6–D7: wave spring	

From the 10th to 25 minute of each hour, distribute the steel band fittings with a cycle of 1 h.

The distribution of parts with a cycle of 2 and 4 h needs to be separated as much possible, making the workload of each hour balanced and facilitating the material preparation by the staff. Specific workflow is shown in Table 4.9.

The distribution route of distribution staff 2 is summarized in Figure 4.24.

4.4.8.4 Efficiency analysis of cleaning machine and cache storage settings

For distribution staff 1, parts of 38 cleaning frames need to be delivered every half hour.

Supposing the parts delivered by staff 1 are all cleaned by ultrasonic cleaning machine, at least 1.3 cleaning frames should be cleaned every minute, namely, the cleaning takt is 47.4 s/frame.

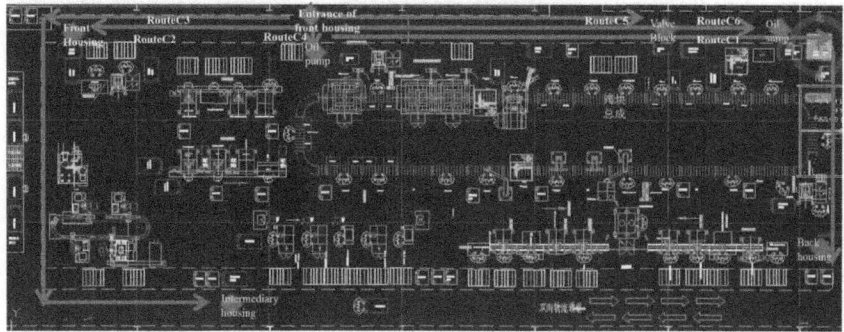

Figure 4.24 Summary of distribution route of distribution staff 2.

Table 4.10 Relationship of takt and cache of ultrasonic cleaning machine

Cleaning takt (seconds)	37.5	40	47.3
Cache settings (cart)	Double-box structure, without caching	1	Unable to meet the distribution requirements

Supposing the cleaning machine cleans 1.5 cleaning frames per minute (takt 40 s), one platform cart loads eight cleaning frames and the time needed for cleaning one platform cart (namely, one distribution route) is 5.3 min. According to the task assignment table of distribution staff 1, the distribution takt is 5 min, which is less than the takt of cleaning a cart platform cleaning frame of 5.3 min. Therefore, the caching of one cart needs to be set. When distribution staff 1 is delivering the hydraulic torque converter, release the caching storage according to the cart's loading cleaning frames needed to be delivered.

Suppose the cleaning machine cleans 1.6 cleaning frames per minute, namely, the cleaning takt is 37.5 s, there is no need to set caching storage, but adopt a double-box mode that can meet the takt requirement of distribution. Ultrasonic cleaning machine and the corresponding caching storage settings are shown in Table 4.10.

4.4.9 Analysis of cart loading and thrust load

It is known that the rolling friction coefficient of asphalt + rolling of wheel is 0.2 cm, and the radius of the wheel is 5 cm. According to Human Factors Ergonomics, the continued thrust of a human being is 200 kg, and the load of cart is calculated to be 1000 kg. According to the current cart that can

be loaded with 12 intermediary turnover cases, the load of each intermediary turnover case, together with parts, would not exceed a maximum
of 480 kg. Reasonably stagger heavy metal parts and light sensor parts to
avoid the maximum continuous thrust of humans and ensure reasonable
load of carts.

chapter five

Optimizing selection of cracking material in petrochemical enterprise

Zuozhuo Zhao and Zhihai Zhang

Contents

This chapter introduces the optimized selection of cracking material in one petrochemical enterprise, which applies the optimization technology. The mathematical model for the entire process of ethylene production is well tailored for the characteristics (e.g., price change, cracking performance, cracking unit) of its particular situation. By optimizing the selection of cracking material and subsequent processes, the optimization management software with user-friendly operation interface was developed. With the implementation of this project, the corporate profit has increased and production costs have been effectively cut down.

5.1   Background presentation

The petrochemical industry is one of the pillars sustaining the stable development of our national economy. The petrochemical industry, which features intensive capital, resource, and technology, high industrial relevance, and large economic aggregate, plays a crucial role in upgrading the relevant industries and spurring the country's economic growth, with petrochemical products widely applied in the national economy, people's living, and national defense technology.

However, in contrast with the healthy growth enjoyed by its foreign counterparts, China's ethylene industry remains underdeveloped in many aspects, including ethylene unit, raw materials, technological level, and product structure. In respect of ethylene units, foreign enterprises keep transforming technologies and increasing their production scale, which demonstrate the advantage of scale economy; in terms of performance, the lifetime of an ethylene unit is 5–6 years in Europe and North America, whereas it is only 2 years in China. Regarding raw materials, the proportion of diesel remains large among the raw materials of China's ethylene units, in contrast with the raw materials composition of ethylene units worldwide, although the past years have witnessed great achievements made by China's ethylene producers, in response to their best efforts to optimize the raw materials in consideration of national conditions and factory realities. This kind of materials will diminish the economic efficiency of ethylene production, which may lead to high consumption of raw materials, additional cost of unit processing, and increased by-products. As for the technological level, foreign ethylene patent companies have made substantial strides in bringing about technological developments in ethylene units in recent years, for example, furnace tube material, cracking furnace type, control technology, separation technology, and environmental protection technology. At the moment, China has no alternative but to buy complete sets of large-, medium-, or small-sized ethylene units from abroad. These hard truths indicate the technological gap between China's ethylene processing industry and its foreign counterparts. In terms of product structure, special materials account for a small proportion. For now, there are more general materials of downstream ethylene products and low-class products, but less special materials and high-end products, as revealed in China's ethylene industry. This restrains the enhancement of petrochemical economic efficiency to a large extent. For example, general plastic products that generate small profits are mass-produced by Chinese petrochemical enterprises, whereas most of the special resins used for producing home appliances, electronics, and other products with high added-value are imported. The low degree of refining–chemical integration and industrial production shaped by conventional thinking are considered as the root causes of irrationalized product structure.

These factors directly impose higher costs on ethylene production for the Chinese ethylene industry, when compared with international levels. To be exact, international advanced and average levels are US$110/ton and US$150/ton, respectively, whereas for a Chinese petrochemical enterprise this carries an average cost of US$182/ton.

We suggest a solution of optimization, which orients at optimizing the selection of cracking materials in one Chinese petrochemical subsidiary, and tailored for the operating status of Chinese petrochemical producers. This solution plays a crucial role in enhancing profitability and cutting operation costs.

This enterprise is a petrochemical giant. It holds an important position in the recorded history of Chinese petrochemical production, witnessing and taking part in several important events in the industry since the founding of the People's Republic of China in 1949. To date, its petrochemical production sector mainly covers two divisions: chemical engineering and oil refining. Some residual historic reasons complicate the product varieties, equipment, and technologies, which pose more requirements for corporate operations and management. When the enterprise adopted Lean Management practices, the operating costs have been decreasing steadily whereas its profitability has been enhanced year after year. Now it ranks among the best of SINOPEC's subsidiaries, a huge leap from its middle or lower level standing several years earlier. However, cost stress is mounting and profitability is drastically diminishing with the rising cost of raw materials and energy sources. Controlling its spiraling production cost is one of the main problems impeding its further development.

5.2   Description of problems

There are two divisions in the enterprise: petroleum refining and chemical engineering. The chemical engineering division tends to be susceptible to the higher stress of profitability and is more sensitive to the volatility of market prices. To cope with these sudden changes, the enterprise can immediately adjust its production processes according to the product or material price; however, this poses corresponding challenges to its operation management. This chapter introduces how its chemical engineering division can effectively optimize the selection of cracking materials, thereby enhancing its profitability and paring down its production costs.

5.2.1   Cracking and main flow of subsequent processes

The main flows of the cracking and subsequent processes in chemical engineering are shown in Figure 5.1. Chemical engineering can be done in two steps: cracking and subsequent processes of cracking products in sequence.

Cracking materials can be sourced either internally (produced by the refinery plant) or externally (purchased from another company). Cracking

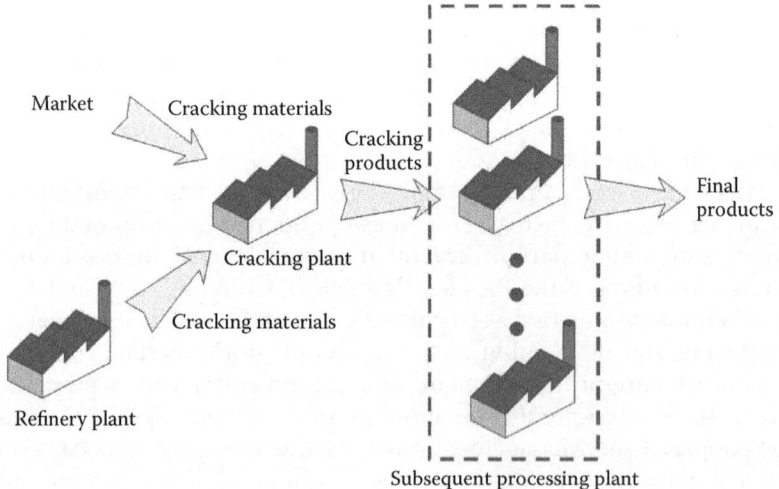

Figure 5.1 Overview of chemical engineering flows.

materials produced by the refinery plant should be basically consumed, so the optimized selection of cracking materials is mainly sourced from those externally purchased. Although oil refining and chemical engineering cannot be integrally optimized, the planner can still compare the advantages and disadvantages of different types of produced naphtha. For the optimization of subsequent processes, our solution will present the flow direction and quantity of materials.

At present, cracking materials mainly include produced naphtha, purchased naphtha, produced light hydrocarbon, purchased light hydrocarbon, purchased tail oil, alkane-rich gas, and olefiant-rich gas. The main cracking products include olefiant, propene, cracked C4, light oil mixture, heavy oil mixture, cracked hydrogen, flare gas, and ethylene bottom oil.

There are many key factors that influence the yield distribution of cracking products—e.g., characteristic of cracking materials, condition of cracking technology, and structure of cracking furnace. These factors are associated and interrelated with one another. However, the properties of cracking materials comprise the most important one. For hydrocarbon cracking material, its components can be classified into alkane, olefin, cycloalkane, and arene, which can be expressed by their PONA (P: paraffin hydrocarbon, O: olefin, N: naphthenic hydrocarbon, A: aromatic hydrocarbon) value. In general, the advantages of raw materials can be measured by the fact that they have more ethylene, propene, and coking resistance; *n*-alkanes > isoparaffin > cycloalkane > arene.

Through several reconstructions and expansions, one subsidiary has now amassed five types of cracking furnaces, 15 units of cracking furnaces, and a large variety of cracking materials. Since 2007, the enterprise's institute has harnessed the "ethylene steam cracking simulation test unit" to perform numerous simulative cracking tests on different materials and furnace types, offering the optimization alternatives of different cracking furnace types.

Its production flows mainly comprise the following: (1) an *olefin plant*, which includes a cracking unit, three gasoline hydrogenation units, aromatics extraction, arene and phthalic anhydride unit; (2) a *plastics plant*, which includes units processing high-density polyethylene, linear polyethylene, polypropylene, polystyrene, styrene, and low-density polyethylene; (3) a *chlor-alkali plant*, which includes units producing chlorine, chloroethylene, polyvinyl chloride (PVC), epoxy chloropropane, and sodium hypochlorite; in this plant, the purpose of producing chlorine is to obtain PVC, and the by-product can be finally processed to sodium hypochlorite; (4) a *rubber plant*, which includes units producing butadiene, butane-1, styrene butadiene rubber, and butadiene rubber (BR); (5) *no. 2 fertilizer plant*, which includes units producing butanol and octanol, and octanol–isobutyraldehyde; (6) an *acrylic plant*, which includes units producing acrylonitrile, acrylon, and acrylic top.

5.2.2 Purchase plan problems of cracking materials in current stage

Two problems appear when preparing a purchase plan for raw materials and the program of subsequent processes.

1. Financial accounting is separate from programming. In general, the planner manually works out the processing program before submitting it to the Finance Department (when he plans the purchase and processing program for the following month). In other words, the planner is not aware of the advantages and drawbacks of the designed purchase and processing program, because financial accounting is done separately from the designed program. This results in a confusing optimization plan.
2. Conventional thinking. At present, cracking materials mainly include produced naphtha, purchased naphtha, produced light hydrocarbon, purchased light hydrocarbon, purchased tail oil, alkane-rich gas, and olefiant-rich gas. On the whole, cracking materials are more varied but some have lower quality than their peers. Thus, to reduce production costs and utilizing the advantage of refining–chemical integration, we can optimize and harness the available resources of ethylene materials to the utmost and improve the yield of diene. Despite these efforts, however, improving the yield of diene has not been thoroughly examined as an optimal choice to improve earnings.

These problems restrict corporate profitability and have led to increased operational costs. The enterprise is in urgent need of a good scientific calculation method and intelligent decision tool to improve its decision quality.

5.2.3 Introduction to the solution to optimizing selection of cracking materials

The optimized selection of cracking materials is an important strategy to enhance the total efficiency of cracking units in the petrochemical industry. Aided by an optimizing tool, material purchase and production programs can be modified accordingly to achieve various economic benefits. The objective of our solution is to present a mathematical model for the entire process of ethylene production that is well tailored for the characteristics (e.g., price change, cracking performance, cracking unit, and product market) of this particular setup. By optimizing the selection of cracking material and subsequent processes, the optimization management software with user-friendly operation interface is therefore developed. To be brief, this proposed solution can be summarized as an "instrumental decision process and intelligent decision tool."

The following functions are enabled in this solution with optimization model and method included in the software, which are well tailored for the production realities:

- Constructing and reinforcing the database of the total flow of chemical engineering, and offering decision to support the optimization of processes.
- Constructing and reinforcing the database of the total flow of chemical engineering, including combining the processes with financial accounting, constructing the detailed database of cracking materials, and linking testing data with actual data.
- Combining the processes with financial accounting. Previously, financial accounting was done separately from the processes. Through the rationalized database, this software combines the processing cost with the unit, and combines the purchase price of raw materials, and the market price of final products with the processes.
- Constructing the detailed database of cracking materials. Previously, there was no available yield database of cracking materials, which makes it more difficult to select the cracking material. Through the rationalized database, the characteristic of the cracking material component is combined with cracking yield, which not only can be applied to optimize the calculation, but can also make it convenient for the planner to control the data of cracking materials (e.g., query or add cracking materials).
- Linking testing data with actual data. The enterprise's institute can run the simulation software to obtain the yield distribution of cracking materials, but it is slightly different from the production reality. As a response, this software will solve it from two aspects. First, transform and align the simulated yield distribution through a chemical equation with the production reality. Second, enabling the function of user adjustment for the benefit of combining the plan and production reality.

Here—based on multiple objectives of optimization—the optimization model of cracking material enables three main functions: the cracking optimization program with multiple objectives; the break-even price calculation of cracking materials; and the analysis of bottleneck resource.

1. Cracking optimization program with multiple objectives (global optimum cracking solution). The optimum cracking solution on the premise of maximized ethylene yield, the optimum cracking

solution on the premise of maximized diene (ethylene + propene) yield, the optimum cracking solution on the premise of maximized triene (ethylene + propene + butadiene) yield, and the optimum cracking solution on the premise of attained high added-value product (triene + benzene + cracked hydrogen) output.

2. Break-even price calculation of cracking materials. The break-even price is important and visualized information of raw material, which should be considered in designing the purchase plan. It means the highest purchase price causing no loss when one material is issued to production. Therefore, a bigger difference between the break-even price and the actual price means a higher profit margin for this material. According to the break-even price distribution of cracking material, the planner may visualize the purchase priority of various materials, which is of great significance to the purchase decision.

3. Analysis of bottleneck resource. According to the available processes, the enterprise is always susceptible to some bottleneck factors. In this concern, its efficiency will be enhanced, provided that these bottleneck factors can be identified. This software has taken into account the bottleneck factors, including the upper and lower limits of cracking materials, the upper and lower limits of the unit processing capacity, and the upper and lower limits of market sales. By analyzing the bottleneck resource, the potential improvement of processes can be identified, thereby optimizing the total processes of chemical engineering.

The selection of cracking material should be examined technologically and economically. Technologically, the yield distribution of cracking products varies from material to material, because they have different components, which affects the product composition. Economically, the cost of different cracking material takes up different proportion in total cost. For example, the expenses constitute about 70–75% of the total cost if naphtha and diesel are selected as the cracking material, whereas it is about 65–70% of the total cost if naphtha and tail oil are selected as the cracking material. In terms of efficiency, cracking products differ in marginal income, so it is required to consider the purchase cost and the distribution of product yield. Therefore, the appropriate selection of cracking material is essential to reducing the cost.

5.3 Solution (implementation method and technological method)

This section is intended to build the theoretical model of oil cracking processes and suggest the solution to this kind of cracking processes, according to the price variation, cracking unit, product market, and other characteristics of cracking material in a petrochemical company.

The established optimization model can be better performed to obtain the following results: (1) The optimum solution of production flows with multiple objectives, including global optimum cracking solution, the optimum cracking solution on the premise of maximized ethylene yield, the optimum cracking solution on the premise of maximized diene (ethylene + propene) yield, the optimum cracking solution on the premise of maximized triene (ethylene + propene + butadiene) yield, and the optimum cracking solution on the premise of high added-value product (triene + benzene + cracked hydrogen) output. This lays the foundation for the optimization of and decision making in oil cracking processes. (2) By running the model, besides the optimum cracking solution with different objectives, all kinds of bottleneck resources can be obtained (e.g., the upper limit of material purchase and unit capacity, offering the decision for the further improvement of oil cracking processes). (3) Through the sensitivity analysis, this model can be performed to obtain the break-even prices of various cracking materials. The break-even price is important and visualized information of raw material, which should be considered in designing the purchase plan. It means the highest purchase price causing no loss when one material is issued to production. Therefore, the bigger the difference between the break-even price and the actual price, the higher the profit margin of this material. The optimization result and computation result can support the oil cracking plan and subsequent processes in all respects.

5.3.1 Hierarchical model

The hierarchical model is the widely used option to solve this kind of problems. To be exact, this model classifies all units throughout the petrochemical processes into several hierarchies, and the optimization of material flow is considered on the basis of each hierarchy. The hierarchical model can be seen as a process of simplification. Fewer hierarchies mean more simplifications, whereas more hierarchies mean more complicated decision variables.

Roughly, the hierarchical procedure can be done in three steps.

Step 1. Combining the units of the same kind in parallel. The units of the same kind can be combined while building the mathematical model.

Step 2. Combining the processing units in serial. The serial processing units mean the one-way linear direction of material flow between the units. To be specific, all material of one kind from one unit flows into the next unit.

Step 3. Simplifying the product unit. To easily understand and compute this model, the main units and materials should be extracted out to simplify the low-value by-product and supporting unit. Furthermore, to ensure the preciseness of this model, all simplified and removed by-products and units can be included into the processing cost.

By these three steps, the original processes can be simplified to obtain the cracking unit and subsequent processing unit in two hierarchy, and cracking material, cracking product, and final product in three hierarchy ("2 + 3" hierarchical structure), thereby greatly simplifying this model on the premise of being precise. Then, rationally hypothesize the process operation, and further obtain the calculation approach that is easily understood and expressed, on the premise of being precise, as shown in Figure 5.2.

As shown in Figure 5.2, the units are classified into two hierarchies: cracking unit and subsequent unit; the materials are classified into three types: cracking material, cracking product, and final product. In this model, decision variables are classified into two types: the processing amount of cracking material in cracking furnace; the processing amount of cracking product in unit. The objective function remains as follows: Total Value of Final Product – Raw Material Cost – Processing Cost. The constraint condition includes the amount constraint of input material (subject to the influence of purchasing capacity), yield constraint and the amount constraint of output material (subject to the influence of policy guidance and market demand), and capacity constraint.

This hierarchical model is widely used in solving the problems of oil refining, which has advantage of making an academic study in macro theory. However, in consideration of the actual production plan, this model

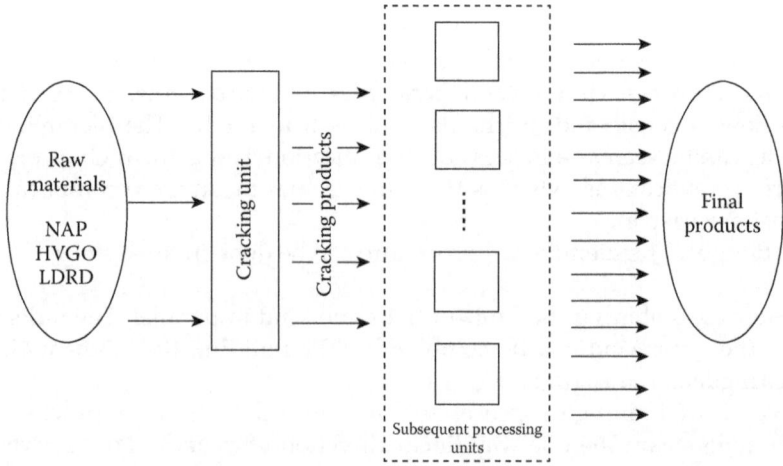

Figure 5.2 Simplified hierarchy of units and processes of chemical engineering.

cannot be oversimplified. This is because the petrochemical industry has a large base—even the slightest difference may greatly affect the profit.

5.3.2 Integrated optimization model

In reality, there are a lot of uncertainties in the processes of oil cracking. For example, the control of temperature parameters varies slightly, thereby affecting the product yield. However, this slight variation (0.1°C) would not lead to a final result beyond tolerance; also, in consideration of the actual decision, a too detailed model does not cater to corporate expectations, which can be attributed to two reasons: (1) the information required is complex because of a too detailed model, which discourages the enterprise to provide such information and further complicate the software use; (2) a large amount of data will greatly diminish the computing efficiency. Based on the needs of actual use, the following hypotheses have been suggested after reaching an agreement with the enterprise.

Hypothesis 1: The cracking yield of cracking material is given, or to be exact, the cracking yield of each alternative cracking material is given, whereas for a new cracking material, its yield function can be recalibrated, or the material yield is similar to its PONA value identified from the materials of calibrated yield, which can be used as the cracking yield of this material. Furthermore, its processing condition is the optimum. As a matter of fact, each cracking material, which corresponds to a cracking furnace, has the optimum processing condition.

Hypothesis 2: Linear output hypothesis. For cracking operation, the material of each cracking furnace should be the same; no mixed material is ever found in the cracking furnace. In this way, the product and material from each cracking furnace are linear output.

Hypothesis 3: Key components of cracking material, intermediate product, and final product can meet the specification. In other words, the constraints of key components including sulfur content and density can be neglected. As a matter of fact, a desulfurization unit can make true of this hypothesis, and this operation has little effect on cost. This hypothesis can preclude the model from the multilinear constraints.

Hypothesis 4: For each unit, its unit processing cost (i.e., processing unit material) is fixed, whereas different units may have different unit processing costs. Processing cost is related with the processing amount in simple linearity.

By these four hypotheses, it can be assured that the yield of each cracking material is given, and this model should be linear.

In this model, decision variables are the material amount of all cracking unit pipelines, where the number of pipelines can determine the number of decision variables. The input and output constraints are considered in each unit. Objective function is the sales profit, which is attained by the equation: Total Value of Final Product – Raw Material Cost – Processing Cost. The following constraint conditions have been taken into account:

1. Input proportional constraint—keeps the input materials to units aligned with the chemical engineering proportion.
2. Input constraint of material amount (subject to the upper limit of purchasing capacity)—keeps the purchase amount of materials within the lower and upper limits.
3. Yield constraint—keeps the yield of various materials aligned with the production reality.
4. Output constraint of material amount (subject to the influence of policy guidance and market demand)—keeps the sales of materials within the limit of state policy and market demand.
5. Capacity constraint—keeps the capacity of each unit aligned with the requirement.
6. Selection of different cracking temperatures.
7. Constraint of input cracking material category. In the cracking processes, the amount of different materials should meet the limit of material category, as subject to the types of unit, upstream production, and market. Cracking materials can be classified into three levels.
8. Cracking materials with the same name and different components should be chosen. In actual production, the cracking materials with the same name should be further classified, after which the enterprise may make the choice according to the needs. In this sense, the constraint of classified material selection is essential.
9. Control of cracking product output. In actual production, the enterprise will control the output of some important cracking products, in order to contrast the benefit of different strategies.
10. Multiple products production in the units. Some units can produce multiple products, but the actual capacity of these products may vary with the physical and chemical properties of different products.
11. Special treatment of aromatics extraction unit. Cracking direct products do not include benzene, but testing data can provide the content of benzene, which makes it possible to calculate the benzene yield. Therefore, a light oil mixture can be divided into benzene and benzene-free light oil mixture, in order to calculate the benzene yield.
12. Separation of cracked C4 and butadiene. The treatment of cracked C4 and butadiene is similar to the treatment of aromatics unit, which classifies the actual product of cracked C4 into butadiene and butadiene-free cracked C4. The difference is that subsequent processes of cracked C4 constitute physical change, which leads to the different treatment of cracked C4 separation unit in subsequent processes. Moreover, the butadiene yield is not a fixed value, which in essence is determined by cracking materials.

Decision variables of integrated optimization model are the material amount of all cracking unit pipelines, where the number of pipelines can determine the number of decision variables. The input and output constraints are considered in each unit. Objective function is the sales profit. Notation and descriptions in this model are listed in Tables 5.1 and 5.2.

Objective function is to maximize the sales profit, which can be obtained as: Product Sales Revenue – Raw Material Cost – Processing Cost, namely,

Table 5.1 Notation of decision variables for theoretical model

Decision variables notation	Description
X_{fFi}	Material amount output from Unit f to Unit F_i
a, b, c	0–1 variable

Table 5.2 Notation of theoretical model parameters

Notation	Description
F	Unit set
f	One unit
f^*	Part-flow unit
f_F	Unit with fixed capacity constraint
f_N	Unit with semifixed capacity constraint
M_L	Cracking materials
M_{L1}	Level 1 cracking materials
M_{L2}	Level 2 cracking materials
M_{L3}	Level 3 cracking materials
M_{LP}	Cracking products
M_P	Other products
M	A large enough and positive number
B	Purchase cost of raw materials
C	Processing cost of unit
P	Sales price of product
A_{fF_i}	Product yield of Unit f output and access to F_i
$A_{M_L,T,o}$	Yield of product output of cracking materials under temperature T
$I_{F_i f}$	Lower limit of materials amount input from Unit F_i to Unit f
$\overline{I_{F_i f}}$	Upper limit of materials amount input from Unit F_i to Unit f
O_{fF_i}	Lower limit of materials amount output from Unit f to Unit F_j
$\overline{O_{fF_i}}$	Upper limit of materials amount output from Unit f to Unit F_j
E_f	Lower limit of Unit f capacity
$\overline{E_f}$	Upper limit of Unit f capacity
R	Proportion of input materials

$$\text{Max} \sum_f \left(\sum_j P_{fF_j} X_{fF_j} - \sum_i B_{F_i f} X_{F_i f} - C_f \sum_i X_{F_i f} \right) \qquad (5.1)$$

1. Input proportional constraint—keeps the input material to Unit f aligned with chemical engineering proportion.

$$\text{For } \forall f, \quad \frac{X_{F_i f}}{R_{fF_j}} = \frac{X_{F_i f}}{R_{fF_i}} \quad i \neq j \qquad (5.2)$$

2. Input constraint of material amount (subject to the purchasing capacity)—keeps the purchase amount of materials within the lower and upper limits.

$$\text{For } \forall f, \ \forall i, \ \underline{I_{F_i f}}, \ \underline{I_{F_i f}} \leq X_{F_i f} \leq \overline{I_{F_i f}} \qquad (5.3)$$

3. Yield constraint—keeps the yield of various materials aligned with the production reality.

$$\text{For } \forall f, \ \forall j, \ X_{fF_j} = A_{fF_i} \sum_i X_{F_i f} \qquad (5.4)$$

4. Output constraint of material amount (subject to the influence of policy guidance and market demand)—keeps the sales of materials within the limit of state policy and market demand.

$$\text{For } \forall f, \ \forall j, \ \underline{O_{fF_i}} \leq X_{fF_j} \leq \overline{O_{fF_i}} \qquad (5.5)$$

5. Capacity constraint—keeps the capacity of each unit aligned with the requirement.

$$\text{For } \forall f_F, \ \underline{E_f} \leq \sum_i X_{F_i f_F} \leq E_f \qquad (5.6)$$

$$\text{For } \forall f_N, \ b * \underline{E_f} \leq \sum_{f_N} X_{Ff_N} \leq b * \overline{E_f}, \ b = 0 \text{ or } 1 \qquad (5.7)$$

The limitation of unit capacity may have two cases—i.e., one is a continuous variable in a fixed interval, which can be expressed by Equation 5.6, whereas another is a discontinuous variable, either meeting the upper and lower limit, or disabling production, where the capacity interval is a discontinuous variable as expressed by Equation 5.7.

6. Selection of different cracking temperatures. Under different cracking temperatures, for cracking material of the same kind, the cracking product yield is differentially distributed, thereby affecting the final returns. In reality, under different temperatures, cracking product yield is a nonlinear function, but nonlinear function has a very low efficiency when solving the large-scale problem; meanwhile, when the temperature changes slightly, it has little effect on product yield. Therefore, the temperature can be discretized, and nonlinear constraint 5.8 can be transformed to integer constraint 5.9.

$$\text{For } \forall A_{M_L,T,o}, \ A_{M_L,T,o} = a_0^{Lm} + a_1^{Lm}T + \cdots a_k^{Lm}T^k \qquad (5.8)$$

$$\text{For } \forall f_L, \ \forall f, \ \forall T, \ X_{f_L,f,T} \leq Ma_{f_L,T}, a_{f_L,T} = 0 \text{ or } 1 \qquad (5.9)$$

Equation 5.9 actually applies if the cracking unit can process cracking under temperature T, which can ensure that cracking products only go to the subsequent unit when the cracking unit processes cracking under temperature T.

7. Constraint of input cracking material category. In the cracking processes, the amount of different materials should meet the limit of material category, as subject to the types of unit, upstream production, and market. Cracking materials can be classified into three levels.

$$\text{For } \forall M_{L2}, \ \underline{M_{L1}} \leq \sum_{M_{L1}} M_{L2} \leq \overline{M_{L1}} \qquad (5.10)$$

$$\text{For } \forall M_{L3}, \ \overline{M_{L2}} \leq \sum_{M_{L1}} M_{L3} \leq \overline{M_{L2}} \qquad (5.11)$$

$$M_{Li} = \sum X_{M_{Li}} \qquad (5.12)$$

8. Cracking materials with the same name and different components should be chosen.

$$\text{For } \forall M_{L*}, \ 0 \leq M_{L*} \leq Mc_{L*}, \ c_{L*} = 0 \text{ or } 1 \qquad (5.13)$$

$$\text{For } \forall M_{L^*}, \quad \underline{M_{L^*}} \le \sum M_{L^*} \le \overline{M_{L^*}} \tag{5.14}$$

$$M_{L^*} = \sum X_{M_{L^*}} \tag{5.15}$$

In actual production, cracking materials with the same name remains to be further divided. The enterprise can select the material according to the needs, so it is essential to add the constraint of selecting classified materials. Equation 5.13 ensures that only one kind of cracking materials with the same name should be chosen, whereas Equation 5.14 ensures the amount limit of cracking materials with the same name.

9. Control of cracking product output. In actual production, the enterprise will control the output of some important cracking products, in order to contrast the benefit of different strategies.

$$\text{For } \forall M_{L^*}, \quad \underline{M_{LP}} \le \sum M_{LP} \le \overline{M_{LP}} \tag{5.16}$$

$$M_{LP} = \sum X_{LP} \tag{5.17}$$

10. Multiple products production in the units. Some units can produce multiple products, but the actual capacity of these products may vary with the physical and chemical properties of different products.

$$\text{For } \forall M_{P^*i}, \quad \underline{E_{P^*F}} \le \frac{M_{P^*i}}{R_{P^*i}} \le \overline{E_{P^*F}}$$

$$b_{P^*N} * \underline{E_{P^*N}} \le \frac{M_{P^*i}}{R_{P^*i}} \le b_{P^*N} * \overline{E_{P^*N}}$$

Here

$$b_{P^*N} = 0 \text{ or } 1 \tag{5.18}$$

$$\text{For } \forall M_{P^*}, \quad b_{P^*N} * \underline{E_{P^*N}} \le \sum_i \frac{M_{P^*i}}{R_{P^*i}} \le b_{P^*N} * \overline{E_{P^*N}}$$

Here

$$b_{P^*N} = 0 \text{ or } 1 \tag{5.19}$$

$$M_{P^*} = \sum X_{M_{P^*}} \tag{5.20}$$

Equation 5.18 ensures each kind of output within the constraint of unit capacity, whereas Equation 5.19 ensures that the output of all types produced from this unit meets the constraint of unit capacity.

11. Special treatment of aromatics extraction unit. Cracking direct products do not include benzene, but testing data can provide the content of benzene, which makes it possible to calculate the benzene yield. Therefore, a light oil mixture can be divided into benzene and benzene-free light oil mixture, in order to calculate the benzene yield.

$$\text{For } \forall F_L,\ M_B = \sum_{F_L} \left(A_B \sum_f X_{F_L f} \right) \tag{5.21}$$

$$\text{For } \forall F_L,\ M_{Q^*} = \sum_{F_L} \left(A_{Q^*} \sum_f X_{F_L f} \right) \tag{5.22}$$

$$M_B + M_{Q^*} = M_Q \tag{5.23}$$

12. Separation of cracked C4 and butadiene. The treatment of cracked C4 and butadiene is similar to the treatment of aromatics unit, which classifies the actual product of cracked C4 into butadiene and butadiene-free cracked C4. The difference is that subsequent processes of cracked C4 constitute physical change, which leads to the different treatment of cracked C4 separation unit in subsequent processes. Moreover, butadiene yield is not a fixed value, which in essence is determined by cracking materials.

$$\text{For } \forall F_L,\ M_{D^*} = \sum_{F_L} \left(A_{D^*} \sum_f X_{F_L f} \right) \tag{5.24}$$

$$\text{For } \forall F_L,\ M_{T^*} = \sum_{F_L} \left(A_{T^*} \sum_f X_{F_L f} \right) \tag{5.25}$$

Table 5.3 Hierarchical model versus integrated optimization model

Metrics	Integrated optimization model	Hierarchical model
Number of decision variables	More, but the number of decision variables can be minimized through designing the database.	Common but unable to be further decreased.
Consideration on constraint	Full consideration on constraint.	The proportional constraint of input materials in single unit is unable to be considered, due to the combination of several units.
Applicability	Stronger, just modify the database to add/delete unit and pipeline, which is not related to this model.	Too weak, hierarchical classification should be reconsidered.
Consideration on special unit	Several special units can be considered in this model.	Unable to do special treatment.
Detailed information	More comprehensive.	Simplified to some degree.

$$M_{D^*} + M_{T^*} = M_T \qquad (5.26)$$

$$M_D = A_D M_{D^*} \qquad (5.27)$$

It should be noted that the simplified model turns to a Network Flow Model (NFM), where each processing unit and material flow can be treated as the node and network flow, respectively. NFM can visually explain the processes of chemical engineering production.

In contrast with the hierarchical model as shown in Table 5.3, the integrated optimization model has more advantages.

Through the integrated optimization model, we can enable the following three main functions: cracking optimization program with multiple objectives; break-even price calculation of cracking materials; and analysis of bottleneck resource.

5.3.2.1 *Generation of cracking optimization program with multiple objectives*

By slightly adjusting the objective and constraint of the above-mentioned model, the cracking optimization program with multiple objectives can be designed according to the needs: (1) the optimum cracking solution on the premise of maximized yield of some material; (2) the optimum cracking solution on the premise of attained high added-value product output.

5.3.2.2 Optimum cracking solution on premise of maximized yield of some material

According to the needs of the actual production plan, the Planning Department usually compares various reference standards when designing the production plan for the following month. Let us take, for example, the optimum cracking solution on the premise of maximized ethylene yield, the optimum cracking solution on the premise of maximized diene (ethylene + propene) yield, and the optimum cracking solution on the premise of maximized triene (ethylene + propene + butadiene) yield. In essence, these optimization objectives are the problems of biobjective optimization, which has obvious priority order. Therefore, this model can enable this function by slightly modifying the objective function.

$$\max \sum_{f}\left(\sum_{j} P_{fF_j}X_{fF_j} - \sum_{i} B_{F_if}X_{F_if} - Ct_f\sum_{i}X_{F_if}\right) \\ + M\sum_{f_L} X_{f_Lf^*} \bigg/ \sum_{F}\sum_{f_L} X_{f_Lf} \tag{5.28}$$

To realize this model, a virtual part-flow unit is manually added in the database, such as the ethylene unit and the propene unit. To be exact, all cracking products of the same kind will first go into the virtual part-flow unit, then go to the subsequent processing unit via the part-flow unit. This chapter will give more details on the advantages of the added part-flow unit in the following parts. Equation 5.28 represents some sort of a part-flow unit. It means that the ratio of cracking product amount to total input of cracking material is the yield of this cracking product. By multiplying the large number M with this factor, this model can ensure that the yield of some cracking product can first be maximized before maximizing the profit.

5.3.2.3 Optimum cracking solution on premise of attained high value-added product output

According to the needs of the actual production plan, the Planning Department may consider that the metrics assigned by the headquarters should be accomplished when designing the production plan for the next month, namely, attaining the high value-added product (triene + benzene + cracked hydrogen) output. To this end, the software should generate the optimum cracking solution on the premise of attained high value-added product (triene + benzene + cracked hydrogen) output.

This problem is a little different from the biobjective optimization described earlier, which adds a standard attained constraint as provided by the user. Certainly, if the standard is set too high, this model will not turn out any solution. According to the needs of this function, this model may be added with a new constraint as shown in Equation 5.29.

$$\sum_{f_L} X_{f_L f^*} \geq \alpha \sum_{F} \sum_{f_L} X_{f_L f} \qquad (5.29)$$

Similarly, to realize this model, the virtual part-flow unit is used in the database. Equation 5.29 represents the set of high-added value product part-flow units. It means the total amount of high-added value cracked products must not be smaller than the total input of cracking materials multiplying the proportionality coefficient, or the yield of high-added value product shall be attained. By adding this constraint condition, the software can ensure that this model considers the cracking optimization on the premise of attaining such standard.

5.3.2.4 Calculation of break-even price of cracking material

According to the needs of the actual production plan, the Planning Department will get the break-even prices of various cracking materials, in order to master the overall direction of decision when selecting the unit cracking material. The break-even price is defined as the available generated value of cracking and subsequent processes subtracting the processing cost, or the critical price that neither generates profit nor causes loss. Then, this break-even price subtracts the actual market price to obtain a price difference, and it can be visually understood that the bigger price difference means better unit profitability of this cracking material. In actual production, the break-even price per unit material will decrease with the increase in material amount, owing to the limit of unit capacity. It can be explained that when using less cracking materials, the optimization model will first process goods by choosing the optimum processing option, with the high profits obtained; when using more cracking materials, the optimization model will choose the suboptimum processing option, owing to the limits of unit capacity, with the second highest profit. Therefore, the break-even price per unit material will decrease with the increase in material amount. Decreasing function is in piecewise linear as shown in Figure 5.3.

The initial break-even price, which is the highest break-even price per unit cracking material, plays an important role in corporate decisions. For calculating the break-even price, this model can only input a few of some cracking materials, after which it calculates the generated return. However, this tool has a big defect in that this software needs to calculate the alternative material one by one; however, as the alternative cracking material in the actual plan may have tens of varieties, this may result in longer computing time. In this concern, this model can indirectly get the

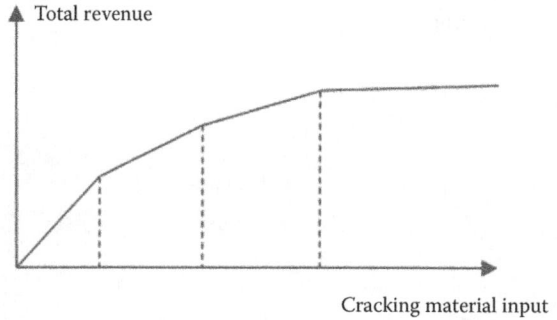

Figure 5.3 Relation between cracking material input and total revenue.

break-even prices of various materials in one time using the sensitivity analysis. Under such sensitivity analysis, this model examines the coefficient of decision variables in objective function, namely, the variation range of the coefficient of decision variables in objective function on the premise of keeping the set of optimum solution intact. The coefficient of decision variables in objective function is determined by processing cost and material price, wherein the processing cost is relatively stable. Therefore, it can be understood that the variation range of the coefficient of decision variables through sensitivity analysis is the variation range of the material price.

5.3.3   Software design

It is too early to solve the actual problem by simply realizing an optimization model; a complete set of application software should be available. Through designing and realizing software modules, such as user interface and database, an effective input/output interface can be supplied for the optimization model. By doing so, our optimization program can be successfully applied. To this end, the design of the software system is introduced in this section.

5.3.3.1   Design of software architecture and function

Software is composed of three parts: underlying database, model construction and algorithm implementation, and program interface, as shown in Figure 5.4.

1. Underlying database. Aspen's Process Industry Modeling System (PIMS) is widely used for the huge database of crude oil contained. The enterprise may process a huge variety of crude oil types (raw materials), different types of crude oil vary greatly, and naphtha, light hydrocarbon, and heavy/hydrotreated vacuum gas oil (HVGO) have complex types. Also, in consideration of the property of

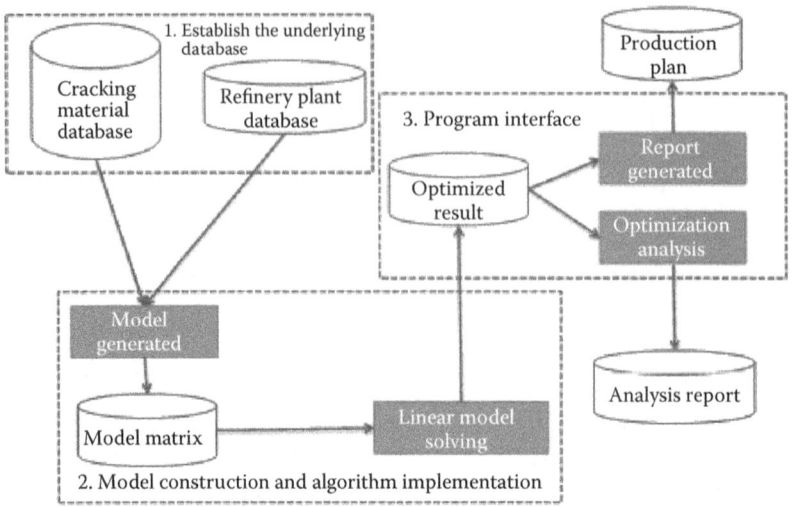

Figure 5.4 Software architecture.

different cracking furnaces and the influence of coil outlet temperature (COT) on cracking yield, cracking material data cannot be input
and maintained manually. By referencing PIMS, the entire cracking
optimization system should establish the database of various cracking materials and plant facilities. By adding the virtual part-flow
unit and "input/output" unit, this database keeps the data access
efficient and complete.
2. Establishment and solution of optimization model (see Section 5.3.2).
3. System interface and interface design. After solving the linear model,
the professional optimization software will generate the report of the
detailed solution; however, some of the information contained in the
solution report is redundant. Therefore, key information should be
extracted from the solution report, and the language of mathematical model should be recovered to production processing plan.

Software function can be classified into three modules: input module, computing module, and output module. The input module allows
users to adjust the input information. The main functions include read,
modify, find, and add. The user first does batch input, or modify and add
individually to obtain the initial input information; then the user selects
various calculation programs, including global revenue maximization,
total revenue maximization on the premise of the maximized yield of
specific cracking product, total revenue maximization on the premise of
the maximized yield of high-added value product, and break-even price
calculation. This software turns out the output results on the basis of the

computing program, including querying the purchase program of the cracking material, subsequent processing program, product sales, break-even price of cracking material, and main bottleneck constraint, which support the user to make a final decision. Here, the user can release the bottleneck constraint that can be optimized in reality, through analyzing the key bottleneck constraint, which can be realized by readjusting the input information. Repeat the procedure until all bottleneck factors can no longer be optimized. This program is not only the optimization program in normal cases, but also allows the user to realize the potential of optimizing the total processes, as shown in Figures 5.5 and 5.6.

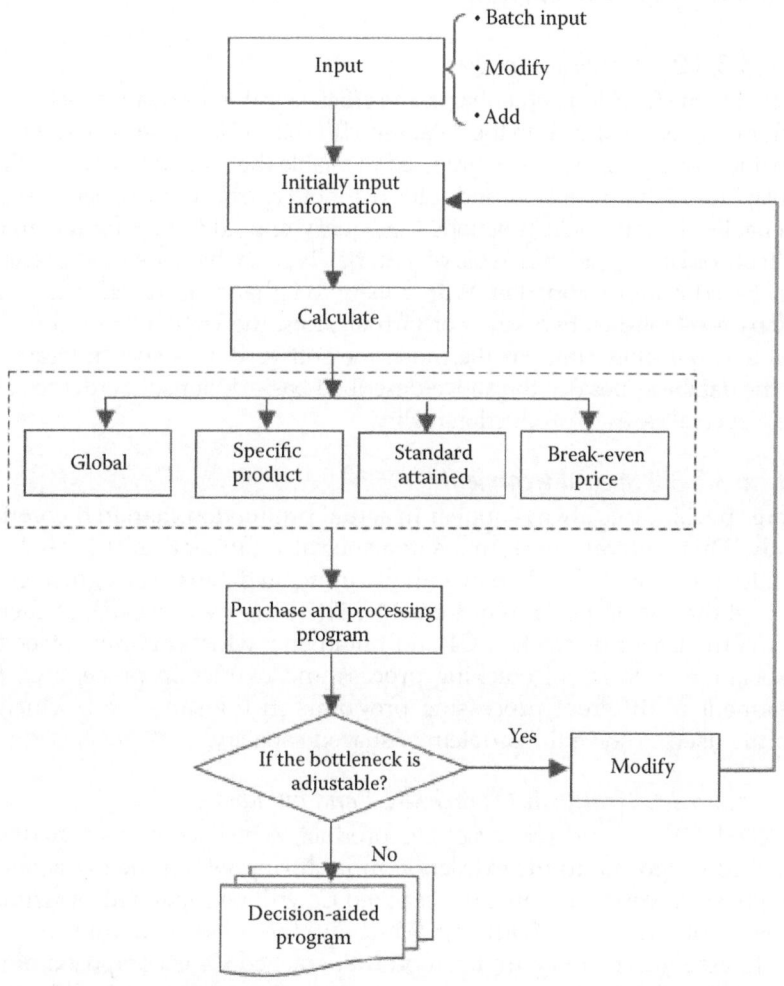

Figure 5.5 Software flow.

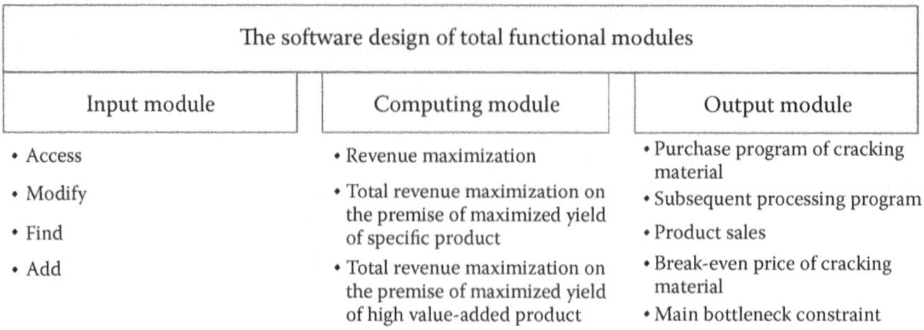

The software design of total functional modules		
Input module	Computing module	Output module
• Access	• Revenue maximization	• Purchase program of cracking material
• Modify	• Total revenue maximization on the premise of maximized yield of specific product	• Subsequent processing program
• Find		• Product sales
• Add	• Total revenue maximization on the premise of maximized yield of high value-added product	• Break-even price of cracking material
		• Main bottleneck constraint

Figure 5.6 Software design of total functional modules.

5.3.3.2   Database design

To enable the function and enhance the efficiency, this software classifies the table design as contained in the database. (1) Table of Units, Table of Materials, and Table of Pipelines, which are used to enable the core functions; (2) Table of Naphtha, Table of Tail Oil, and Table of Light Hydrocarbon, which are used to enable the additional functions (e.g., query and add cracking materials); (3) Table of Unit Types and Table of Material Types, which are used to enable the shared capacity function. With a view to enhancing the efficiency, this database establishes two kinds of virtual units: the virtual "input/output" unit and part-flow unit. Furthermore, the software does special treatment on the database, besides the above-described basic designs, in order to solve some special cases in production reality.

5.3.3.3   Special treatment

More special cases always appear in actual production than in theoretical study. This software performs some special treatment on the database, besides the above-described overall planning on database design, in order to meet the user's special requirements. There are two special treatments: (1) the treatment of cracked C4 and butadiene, which mainly solves the association problem of cracking process and extraction process; (2) the treatment of different processing programs in the same unit, which is mainly used to solve the problem of shared capacity.

5.3.3.4   Treatment of cracked C4 and butadiene

Cracked C4 is the direct cracking product, which contains butadiene. Cracked C4 goes into the extraction unit, during which the extraction is the physical separation process. Cracked C4 will be separated as raffinate C4 and butadiene, which are reprocessed in the subsequent unit.

In essence, it can be understood that cracked C4 is composed of raffinate C4 and butadiene. However, two difficulties will be solved in treating two processes, the output of cracked C4 and further extraction of cracked C4. First, butadiene can be only extracted after actual processing, but the yield of butadiene is determined by cracking material of the previous cracking process and cracking temperature, which means that the yield of extraction unit is not certain. Second, the user pays more attention to the yield of butadiene than the yield of cracked C4, which means that cracking unit and extraction unit should be associated.

Three alternatives can be used to solve these problems.

Alternative 1: Directly combining the extraction unit into cracking unit, when the extracting process is neglected. In this way, raffinate C4 and butadiene can be directly generated. The advantage is that associated units can be combined to lessen the work of the model; however, the disadvantage is that cost accounting of extraction unit may be complicated, which can be attributed to two reasons: (1) the removal of extraction unit may complicate the accounting of extraction cost; (2) the extraction unit may also crack the purchased cracked C4, in addition to the cracked C4 produced from extraction and cracking.

Alternative 2: Associating the cracking unit and extraction unit, when the yield of extraction unit is determined by cracking material. By doing so, the advantage is that the problem meets the production reality, which is more visualized; however, the disadvantage is that uncertainty may appear in the cracking unit and extraction unit. In this concern, the database also needs a lot of virtual extraction units, which may considerably diminish the computing efficiency.

Alternative 3: Establishing two extraction units for butadiene and butadiene-free cracked C4 directly produced from cracking, in which one is used to extract the purchased cracked C4, and the other is input with butadiene and butadiene-free cracked C4. Extracted products are butadiene, raffinate C4, and a few liquefied gases. By doing so, the advantage is that two processes are considered and association is cut down at the same time, which lessens the complication; however, the disadvantage is weak visualization.

By comparison, it can be seen that Alternative 3 is more feasible, which does not affect the real use, although Alternative 3 has weak visualization. Therefore, this software adopts Alternative 3 to solve this problem, as shown in Figure 5.7.

5.3.3.5 *Treatment of different processing programs of the same unit*

In actual production, some units may have multiple processing programs. For example, plastic products have many types, whereas different plates vary in product features and market prices. In this concern, the software should differentiate between the types.

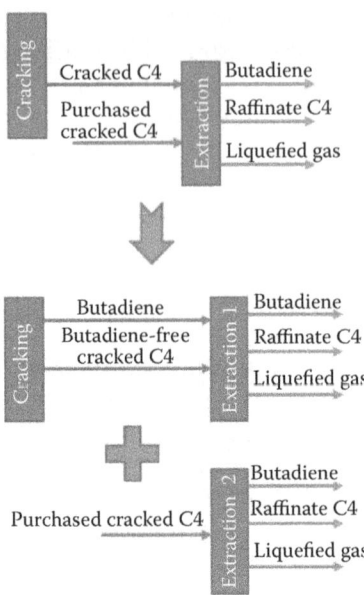

Figure 5.7 Treatment program of cracked C4 and butadiene.

There are three difficulties to solve this problem.

Difficulty 1: Different types of products have different input materials, and different types of products vary in yield, which may disable various types to share one unit in the database.

Difficulty 2: Output of different types of products is a decision variable, meaning that output of different types of products has no direct proportionality, which may also disable various types to share one unit in the database.

Difficulty 3: Different processing programs executed in the same unit. Different materials vary in physical and chemical properties (e.g., viscosity, which results in different processing capacities under different programs), but they share one unit in actual product.

To solve the said problems, the software adopts the program of virtual unit addition and shared capacity. Virtual unit addition means adding the unit in the same quantity with the number of types, and each virtual unit only produces one type. It can solve the difference of input material and product yield. Shared capacity means that these virtual units share one capacity, where the software introduces the coefficient of capacity, because different products vary in processing capacity. The coefficient of capacity is the virtual unit well tailored for shared capacity,

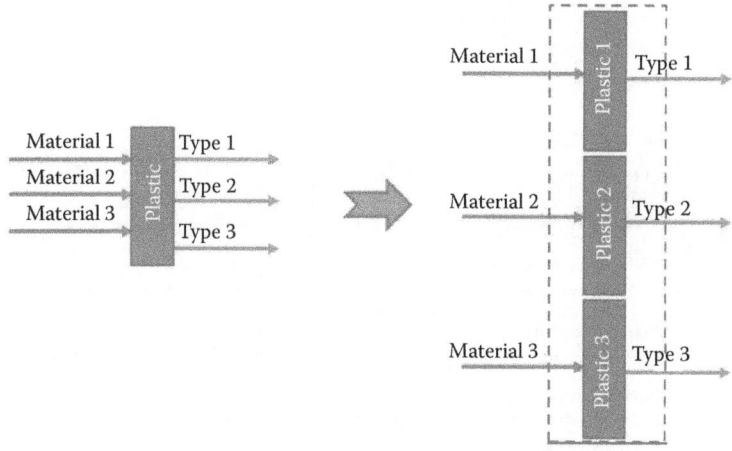

Figure 5.8 Different processing programs in the same unit.

which means that different processing programs are executed in the same unit. Different materials vary in physical and chemical properties (e.g., viscosity, which results in different processing capacities under different programs). To make matters more convenient for the user, the software enables the input of capacity coefficient, as shown in Figure 5.8.

The treatment of shared capacity can be illustrated as follows.

For example, the capacity coefficient of Type 1 is 1; the capacity coefficient of Type 2 is 1.1; the capacity coefficient of Type 3 is 0.9. Furthermore, the total capacity of all plastic products must not exceed the upper limit (100,000 tons). The capacity limit can be calculated as follows:

Capacity coefficient of Type 1/1 + capacity coefficient of Type 2/1.1
+ capacity coefficient of Type 3/0.9 ≤ 100,000 tons

From this example, the software can realize the shared capacity from products with different processing capacities by introducing the coefficient of capacity. In this example, Type 2 has a stronger production capacity, whereas Type 3 has a weaker production capacity, if Type 1 is set as the benchmark.

5.4   Project implementation and application

5.4.1   Data processing

According to production realities, this project should first sort out the flows of all units. For example, the diagram of cracking units is given in Figure 5.9, then import the database information as per this diagram.

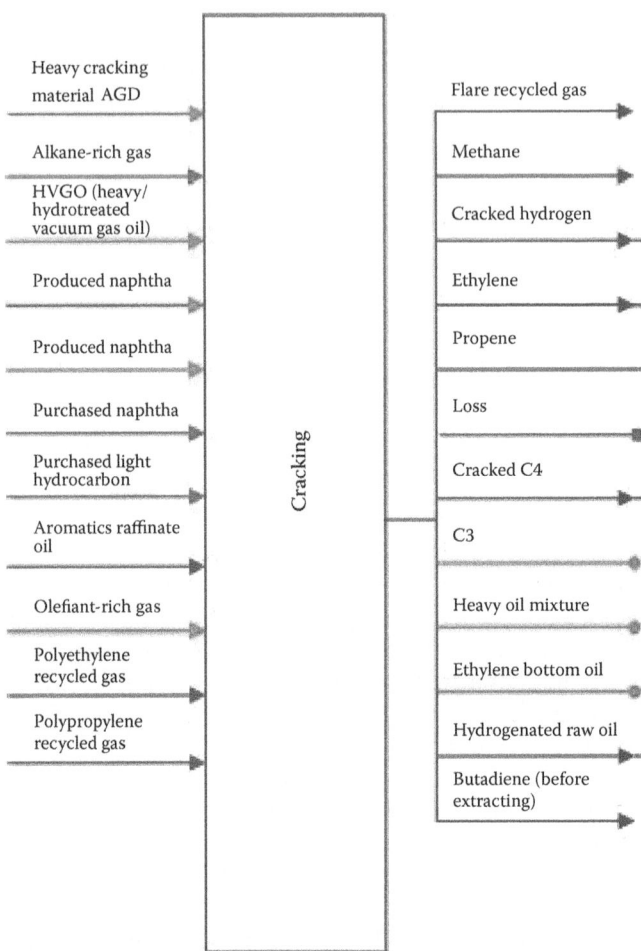

Figure 5.9 Diagram of cracking units.

Basic data required for software computation include the yield distribution under different cracking temperatures of different cracking materials, the purchase price of raw materials, the market price of final product, the processing cost of unit, and the upper and lower limits of materials and capacity.

The yield distribution under different cracking temperatures of different cracking materials and the processing cost of unit cannot be directly provided by the enterprise, which should be pretreated.

The simulated result of yield distribution cannot be directly used in data inputting, because the simulated result of yield distribution by the Research Institute cannot better correspond to the actual processing, so it should be transformed to the actual value of yield first. This procedure is done by two steps. The first step involves initially transforming the original data. In actual production, cracking products do not contain any ethane and propane, which will be completely consumed in the circular reaction of cracking furnace. In the second step, further transform the remainder ethane and propane through Step 1 into other cracking products. In essence, it can be understood that ethane and propane are somewhat cracking materials, and the yield of experimental cracking can be transformed.

Processing cost: When the Finance Department performs accounting for the cost of unit, such items as purchase cost of raw materials, cost of auxiliary material use, cost of energy consumption, and labor cost should be considered. However, as processing cost contains the unit processing cost of the unit, the input amount of raw materials may not need to be considered; meanwhile, labor cost is a fixed value, because the enterprise pays salary to workers, regardless of whether or not the unit is started. In this regard, labor cost neither belongs to variable cost, nor shall it be counted into the unit processing cost of the unit. To sum up, the unit processing cost of the unit, which is imported into the database, should cover the cost of auxiliary material use and the cost of energy consumption.

5.4.2   Result analysis

With the enhanced database of the software, the software realizes the optimized selection of cracking material and subsequent processes by user-friendly operation interface. This part focuses on the decision of planning supported by the software in four aspects, namely, the break-even price calculation of cracking material, the program comparison with multiple objectives, the comparison of optimized result and present program, and the analysis of bottleneck factors.

5.4.2.1   Calculation of break-even price

Cracking materials include produced naphtha, purchased naphtha, produced light hydrocarbon, purchased light hydrocarbon, purchased tail oil, alkane-rich gas, and olefiant-rich gas. By calculating the break-even price, all types of cracking materials can be sequenced according to their advantages and disadvantages.

5.4.2.2   Comparison of programs with multiple objectives

The software optimizes the monthly plan according to the prevailing realities, and compare the following objectives: global optimum cracking solution, optimum cracking solution on the premise of maximized

Table 5.4 Comparison of optimization with different objectives
(in 100 million RMB)

	Global optimization	Ethylene maximization	Diene maximization	Triene maximization	Butadiene maximization
Sales profit	0.56	0.51	0.51	0.52	0.56

ethylene yield, optimum cracking solution on the premise of maximized diene (ethylene + propene) yield, optimum cracking solution on the premise of maximized triene (ethylene + propene + butadiene) yield, and optimum cracking solution on the premise of maximized butadiene yield.

A comparison of multiple programs is presented in Table 5.4. It is evident that conventional thinking has no scientific basis, and the higher yield of diene does not automatically translate to higher total revenue. At present, products using butadiene is more profitable, and the yield distribution of cracking products are contained and associated with one another. Only with the optimum combination can the enterprise maximize its profit. To this end, all kinds of product compositions should be fully considered when designing the production objective.

5.4.2.3 *Comparison of optimized result and present program*

Under the constraints, this software optimizes the production plan dated May 2012 by one petrochemical enterprise. According to the comparison of cracking material input, the program with optimized software has a lower input of cracking material. As shown in Figure 5.10, the optimization program prefers selecting quality materials such as olefiant-rich gas, whereas inferior materials only choose the lower limit of material amount. This basically conforms to reality in recent times, during which some products did not prove profitable. In this regard, material input should be appropriately reduced to avoid any product loss.

As shown in Figure 5.11, by comparing the optimized result and the present program, the profit earned from plastic products is not high, which leads to the decrease in marginal profits of ethylene. A greater output of ethylene cannot generate profit; in fact, the present program indicates an excessive production of ethylene. Meanwhile, the present program is not detailed in the selection of cracking materials, which depends more on past experience. Based on the two main factors described earlier, the sales profit of the optimization program is up around 10%, when compared with the present program.

5.4.2.4 *Analysis of bottleneck factors*

By analyzing the bottleneck factors, the software identifies two evident characteristics: the purchased quantity of produced naphtha is low, and output of plastic products cannot be optimized without taking into

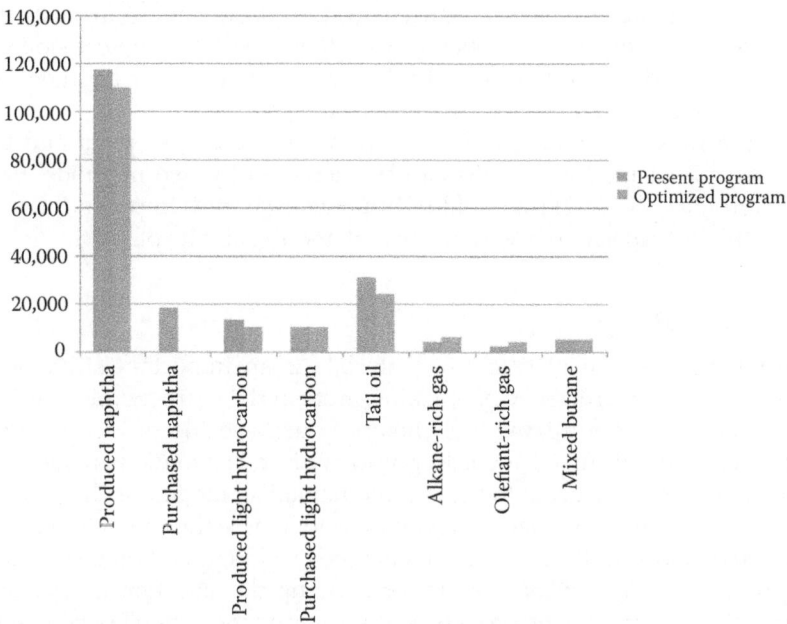

Figure 5.10 Comparison of optimized program and present program in input of cracking material.

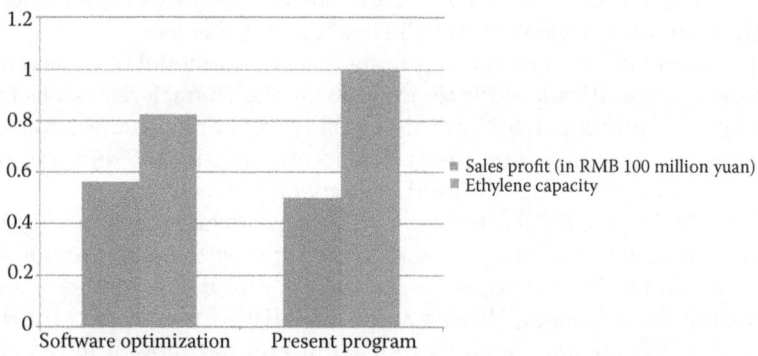

Figure 5.11 Comparison of optimized result and present program.

account the limits of cracking material. The profit margin can be increased by considering the following factors:

1. Cracking materials produced by the refinery plant should be totally consumed. Most of these materials are not very profitable, which constrains the increase in profits of the chemical engineering division to a large extent.

2. Most of the products produced by the plastics plant are causing losses. To solve this problem, two alternatives may be considered: (a) control the ethylene output appropriately; (b) strengthen the management and reduce the processing cost.
3. The most profitable products come from the rubber plant and the acrylic plant. Total profits can be increased by two methods: first, enhance the cracking yield of butadiene; second, improve the production capacity of the rubber plant and the acrylic plant.

5.5 Summary

Optimized selection of cracking materials is an important strategy to enhance the total efficiency of cracking units in the petrochemical industry. Aided by an optimizing tool, material purchase and production programs can be optimized accordingly to achieve economic benefits. The objective of this project is to build a mathematical model for the process of ethylene production that is well tailored to handle the characteristics of this particular situation (e.g., price change, cracking performance, cracking unit, and product market). By optimizing the selection of cracking material and subsequent processes, the optimization management software with user-friendly operation interface is therefore developed.

Many factors were considered in the solution, including model design, database design, program design, and functional process design, in order to (1) enable the basic functions of the software; (2) meet the customer requirement for special use; and (3) ensure its flexible use.

In respect of the model design, the theoretical model of oil cracking processes is specifically built to account for the characteristics of cracking material in this petrochemical company (e.g., price change, cracking performance, cracking unit, product market), in order to study multiple programs for optimizing this kind of cracking processes.

Concerning the database design, three aspects have been mainly considered: (1) enable the basic functions; (2) improve the efficiency of software operation; (3) meet some special user requirements. With a view to enhancing the efficiency, this database established two kinds of virtual units—i.e., the virtual "input/output" unit and part-flow unit, wherein the virtual "input/output" unit keeps the data consistent.

With regard to program design, the software design mainly considered three aspects: (1) enable the basic computing function of the model; (2) combine the unified and special features of the program; (3) improve the flexibility of the program. The main flows of the program can be classified into access, transforming, and calling. This refers to accessing the data contained in the database, and reading into the program structure, which transfers the data in the hard disk to the memory. According to the requirement of reality and model, we transformed the data contained in the program structure into a coefficient matrix, as prepared for calling software and solving problems. In essence, we tapped the available memory data to create the objective function and constraint required by this model.

In terms of flow design, software function can be classified into three modules according to the actual needs: input module, computing module, and output module. The input module allows the user to adjust the input information. The main functions include access, modify, find, and add. In all respects, it provides convenient functional aid for users to adjust the processing condition, and keep the model calculation aligned with reality as much as possible. The computing module is the core module; it allows the user to provide the optimized purchase and production program by calculation, according to the input information. The main computing functions include the optimum solution of production flows with multiple objectives. The output module mainly permits the user to query the optimization program. Its main functions include querying the purchase program of cracking material, subsequent processing program, product sales, break-even price of cracking material, and main bottleneck constraint, which support the user to make the final decision.

chapter six

Optimizing oil production

Yixiao Huang, Lei Zhao, and Simin Huang

Contents

6.1 Introduction

Various optimization problems in oil production are worth studying. In 2012, the Department of Industrial Engineering at Tsinghua University initiated cooperation with the Qilu Branch of the China Petroleum &

Chemical Corporation. This cooperation was intended to optimize crude oil procurement and the oil refining process.

6.1.1   Business background

The China Petroleum & Chemical Corporation (SINOPEC) is a solely funded state-owned company, authorized investment institution, and state-owned holding company. In 2012, its revenue was US$428.167 billion. In 2013, it ranked fourth in Fortune's Global 500. Until 2012, its annual processing of crude oil was 220,000,000 tons, and the number of gas stations owned hit 30,000. In 2011, its domestic sales volume of refined oil was 151,000,000 tons, and the market supply rate was 62%. It has become the largest national supplier of refined oil and petrochemicals and the world's second largest refinery. Its total number of gas stations ranks second worldwide.

The Qilu Branch of the China Petroleum & Chemical Corporation (QLPEC) is a conglomerate of refined oil, chemicals, fertilizers, and chemical fibers, which is engaged with petrochemicals, salt chemicals, coal chemicals, and natural gas chemicals. Its annual processing capacity of refined oil was 10,500,000 tons. The major products of QLPEC include 120 grades of petrochemicals.

6.1.2   Project background

For the oil refinery, the procurement cost of crude oil accounts for 85% of the total production cost (Jiang, 2010), and the percentage is more than 90% for QLPEC. In this respect, the decision of crude oil procurement is essential to achieve profitability. In QLPEC, the procurement of crude oil is vulnerable to the "small scale" and the "big difficulty." Fifty percent of the crude oil is sourced from Shengli Oilfield, and only a small amount of crude oil is procured from overseas because of the economy of scale; QLPEC is located inland, and the cost of ocean and pipeline transport is high. In addition, crude oil with its heavy component, high viscosity, and high condensation point cannot be transported via pipeline.

Contemporarily, QLPEC procures crude oil, including domestic crude oil (Shengli high sulfur) and imported crude oil (Escalantee, Oriente, Zaccoum Super, and 12 other kinds of crude oil). The great variety of crude oil makes it hard to make the decision of crude oil procurement. Further, the market information on crude oil is fluctuating. The planning department is required to forecast market information and quickly respond to any information change, take multiple plans into account, and make comparisons among the plans. Therefore, it requires the optimization problem be solved in a short time.

In the oil processing industry, popular software for optimization and cost calculation are PIMS by Aspen Technology (US), RPMS by Honeywell

(US), and Petro-SIM by KBC. PIMS focuses on the optimization of material processing in production, which has the advantage of production scheduling. But the decision to use PIMS is susceptible to the influence of unimportant details or indicators because it overemphasizes the production details. In contrast, Petro-SIM centers on the optimization of equipment. The QLPEC planning department expects the decision tools to be more focused on the decision of crude oil procurement and its overall influence on the production plan.

Based on the described requirements, QLPEC and the Department of Industrial Engineering at Tsinghua University launched the project to optimize the procurement of crude oil and the refining process. This project is intended to propose an optimization method to solve the procurement problem of crude oil, which is a key to the production planning by QLPEC. In addition, the software will be developed to apply the optimization method subject to the specific oil refining process, policies, and technologies.

6.2   Problem description

The decisions can be classified into strategic, tactical, and operating decisions. In a production system, capacity construction or expansion falls into the category of strategic decisions, which will considerably influence corporate development within a specific period. The planning department procures crude oil on a monthly basis, which is a tactical decision. The design and execution of the production plan is an operating decision. Generally, macro-level (superior) decisions will exert great importance on micro-level (subordinate) decisions, and, in turn, micro-level decisions will influence the evaluation of macro-level decisions.

For QLPEC, the procurement of crude oil and the production plan are closely related. On one hand, the revenue from the refined product should be calculated to evaluate the profit of crude oil procurement. On the other hand, the production plan is decided based on the given procurement decision for crude oil. In this project, we jointly consider the crude oil procurement and production plans. The production plan supports the decision of crude oil procurement, and the procurement plan is decided in order to maximize the profit.

6.2.1   Crude oil refining process

The refining process is closely related to the factory equipment and the type of crude oil. Generally, crude oil first goes through the atmospheric and vacuum distillation unit; then the different separated components flow into the secondary unit for further processing until the final product outputs after a series of processes. Final products usually include gasoline, aviation kerosene, diesel, fuel oil, petroleum wax, asphalt, petroleum coke, and other petrochemicals (Hou, 2012).

The refining process for crude oil is complex, and it involves a lot of physical and chemical processes, equipment, and materials. This project centered on the decision of crude oil procurement, and the production plan is intended for assessing the profit of the crude oil procurement plan. Therefore, we simplified the crude oil refining process for the production plan. The general process of crude oil refining is shown in Figure 6.1.

We divided the crude oil unit into the atmospheric and vacuum distillation unit and the secondary unit. Input material to the atmospheric and vacuum distillation unit is crude oil, which will be separated into sideline products by harnessing different boiling points in the physical process. After that, these sideline products flow into different secondary units and turn out to be the final products after several processes. To highlight the equipment, the equipment name hereafter will be bracketed as, for example, (hydrogenation unit).

Materials involved in the oil refining process can be mainly divided into three categories: crude oil, sideline products, and final products. Wherein crude oil is the main input of the oil refining process, the final product is the output finished product through the oil refining process. A sideline product means the material produced from processing units, which is further processed in other units. It should be noted that some sideline products are named similarly to or even equivalently with final products. To avoid any misinterpretation or ambiguity, materials hereafter mentioned shall be tailed with "product" to mean the final product, for example, "naphtha product," or else default to indicate the sideline product.

In the crude oil refining process, there are many constraints, such as policy and process constraints. Policy constraints are mainly concentrated on the production quantity of some final products. For example, the bottom limit on the production quantity of every gasoline grade is to ensure the overall market supply. Process constraints mean the constraint of mixing ratio when multiple materials are mixed in the same unit due to the specific physical and chemical material property. If the process constraint is not met, it will result

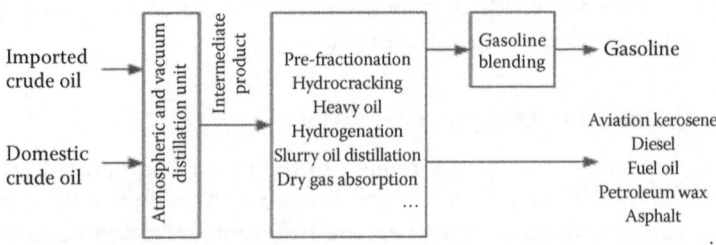

Figure 6.1 Crude oil refining process.

in an infeasible production plan, and if the policy constraint is not met, it will cause potential loss to the oil refinery. In this concern, these production constraints are essential to the optimization of crude oil procurement.

6.2.2   Profit calculation

In QLPEC, the finance department calculates the oil refining profit by the formula herein:

Oil Refining Margin = Sales Revenue − Raw Material Expenditure
− Oil Refining Cost − Fixed Cost

Sales revenue is the output of final products multiplied by the unit price of each product. Raw material expenditure includes the procurement cost of the crude oil, the external procurement cost of the blending material, and the external procurement cost of hydrogen. Oil refining cost includes the processing cost and acid corrosion cost of which the unit processing cost is the cost per unit multiplied by the processed quantity. The acid corrosion cost is the purchased quantity of crude oil multiplied by the acid content of the crude oil with the unit cost of acid value, and the fixed cost means the fixed input arising from procurement and production, which should be deemed as a fixed value herein.

6.2.3   Gasoline blending

Gasoline produced from the oil refining process cannot be sold until it goes through gasoline blending, added with blending materials, such as MTBE, btx aromatics, and aromatic raffinate oil, in order to ensure the final grades of gasoline meeting the quality standard. Standard components have been defined for each gasoline grade, for example, the Beijing Gasoline Standard has been set out in Table 6.1.

Table 6.1 Component standard of motor gasoline

Data		89#	92#	95#
Octane number	≥	89	92	95
Methanol (mass fraction)	≤		0.3%	
Benzene content (volume fraction)	≤		1.0%	
Olefin Content (volume fraction)	≤		25%	
Olefin + arene (volume fraction)	≤		60%	

Source: Beijing Local Standard, DB11/238-2012.

In the model of gasoline blending, we assume the process of gasoline blending as a linear combination of each component, and all blending materials should be externally procured at the market price. So we need to take the procurement cost of blending materials into account.

6.2.4   Break-even price

The "break-even price" or "break-even point" means the price at which revenue offsets cost. The concept of a break-even price has been widely applied in the commodities trade, futures investment, risk management, etc. The break-even price of the crude oil means the procurement price that makes both ends meet in the process of procurement and production, which is referenced to examine if the procurement price of crude oil could be profitable.

In the petrochemical industry, the market price of crude oil is fluctuating, and the break-even price of crude oil is an important guidance for crude oil procurement. Many methods are available for calculating the break-even price of crude oil. Similarly, QLPEC also concluded the experience of calculating the break-even price of crude oil and guiding the procurement of crude oil in which the PIMS Production Plan Income Estimation was performed to calculate the break-even price of crude oil (Niu et al., 2006). In this project, we not only optimize the crude oil procurement and production plan, but also choose the appropriate method to calculate the break-even price of crude oil as provided for each type of crude oil and referenced for crude oil procurement.

6.2.5   Step-function price

According to the relevant policy, when the ratio of total gasoline, diesel, and kerosene output to total procurement of crude oil (percentage) is larger than the specific reward price ratio, the reward price of the excess amount will be obtained based on the ratio of gasoline and kerosene, which is calculated by the formula below.

$$超出部分 = \left(\frac{汽油量 + 柴油量 + 煤油量}{原油采购总量} - 奖励价格比例 \right) \times 原油采购总量,$$

$$超出部分奖励收入 = 超出部分 \times \frac{汽油量}{汽油量 + 柴油量} \times 汽油奖励价格$$

$$+ 超出部分 \times \frac{柴油量}{汽油量 + 柴油量} \times 柴油奖励价格$$

超出部分 Excess amount
汽油量 Gasoline quantity
柴油量 Diesel quantity

煤油量 Kerosene quantity
原油采购总量 Total purchase quantity of crude oil
奖励价格比例 Reward price ratio
超出部分奖励收入 Reward price of excess amount
汽油奖励价格 Gasoline reward price
柴油奖励价格 Diesel reward price

In this project, we should establish the mathematical model for the optimization of crude oil procurement, either considering or ignoring the step-function price, respectively, in order to aid QLPEC to compare different decisions. The next chapter will cover the difference of two situations in the formulation.

6.3   Mathematical modeling

As mentioned, optimizing the procurement decision (tactical decision) about crude oil should consider the subsequent production plan (operating decision). However, the overemphasis on production details will exert little influence on the procurement decision and overcomplicate the model. It is said that PIMS employed by QLPEC spent a lot of time in the calculating process due to the overemphasized description of production details, but it doesn't meet the requirement of a quick solution. Because of detail parameters, PIMS usually results in an infeasible solution, so the planner needs to spend a lot of time to find out the cause. In these concerns, we simplified the production process of crude oil refining in this model and gave assurance to quick solution; as much as possible on the premise of describing the main processes.

6.3.1   Optimization model

In consideration of the flexibility and expansibility of the model, the unit in modeling is regarded as the "functional unit" instead of the physical unit. For example, we combined several hydrogenation units of QLPEC as the (hydrogenation unit), whose capacity is the sum of the capacities of all the hydrogenation units. The "functional unit" has the advantage of flexible adjustment. When the company adds more hydrogenation units as promoted by the production need, the change is the capacity of the (hydrogenation unit). If the company establishes the model in accordance with the physical unit, it is required to add a new production unit and create the relationship between the unit and the input and output materials, which makes the model complicated.

Crude oil is mapped with the atmospheric and vacuum distillation unit, and each atmospheric and vacuum distillation unit has the suitable type of crude oil to be processed. However, when the sideline products output from the atmospheric and vacuum distillation unit flow into the secondary unit, it is not necessary to identify the sideline product from which type of crude oil.

In practice, operators may change the product yield by operating the unit. For example, the atmospheric and vacuum distillation unit performs the distillation, which separates the crude oil into several components at different cutting temperatures. Within a certain range, cutting temperature can be adjusted slightly. Different cutting temperatures lead to varied yields. And the unit yield is relatively fixed in production. In the procurement decision of crude oil, the decision of unit yield is too concrete, so it has little influence on the procurement decision. Therefore, we assumed that each unit could be regarded as a "separation" unit. In other words, some input material (crude oil or sideline product) is separated into several other kinds of output materials (sidelines or final products); for the given input material and production unit, the yield of output materials is assumed to be fixed in this model. Moreover, each unit is able to process multiple input materials, and for the same unit, the yield of output materials from different input materials may be varied (Figure 6.2).

Further, some units may have the same kind of sideline product distributed to several kinds of final products. For example, atmospheric line 1 produced from the (atmospheric and vacuum distillation unit) may be processed as naphtha or diesel. Products from some units may be used as the final product or further processed in other units. For example, vacuum residue produced from the (atmospheric and vacuum distillation unit) may be used as heavy traffic paving asphalt or further processed in the (heavy oil hydrogenation unit). The complex product–product and product–unit distribution will mount the difficulty on the model. Thus, we establish the corresponding "virtual unit" for each product in this model. The product–product and product–unit distribution may be processed to the distribution of sideline products in different units. Atmospheric line 1 and vacuum residue are virtual units as shown in Figure 6.3.

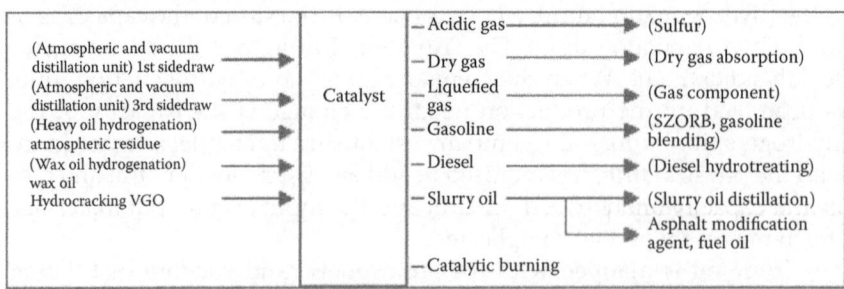

Figure 6.2 Production relationship between unit and material.

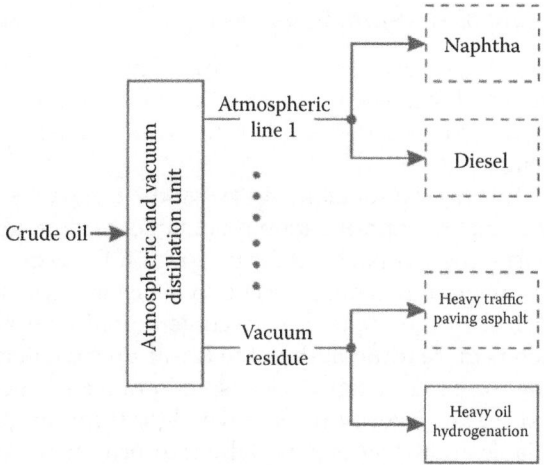

Figure 6.3 Illustration of product nodes.

We also consider gasoline blending as a special secondary unit. However, in contrast with other secondary units, the (gasoline blending) unit contains some external raw materials, and it needs to take the raw material cost into account. Moreover, blended products from the (gasoline blending) unit should reach the required octane number and olefin content in accordance with the national or local gasoline standard.

The crude oil procurement and production problem can be seen as a variant of the network flow problem in operations research. The network flow problem has been widely applied in the transportation or power distribution problem. In this model, we model the simplified production unit (functional unit) and materials (virtual unit) as the nodes of the network diagram, in which the arcs connecting the nodes of the virtual units and functional units represents the inflow and output between the materials and the unit. Each node of the functional unit is limited by processing capacity. However, each arc of the nodes has no limitation of transportation capacity. In other words, it can be understood that in production, there is no capacity constraint of pipelines or transportation equipment.

This model also allows flow furcation in oil refining, for example, vacuum line 1 produced from the (atmospheric and vacuum distillation) unit may flow into (catalyst) or (diesel hydrogenation). This model will optimize the actual flow of each furcation in accordance with factors such as unit capacity and product price.

Ignoring the step-function price, we model the crude oil procurement and production optimization as a linear program. In numerical experiments, the linear program can be quickly solved. In the following paragraph, we will discuss the model in consideration of step-function price.

6.3.2   Modeling of step-function price

In the process of modeling, we may treat the calculation of overall income as a logical process: If the reward price ratio is satisfied, the income shall be added with the excess income; otherwise, the income shall disregard the excess income.

To express this logical relationship, we add a binary (0–1) variable in this model. Note that the optimization model in consideration of the step-function price is a linear mixed-integer program if the ratio of gasoline to diesel output is given. If it is not given but a decision variable of the optimization model, then the model will become a nonlinear mixed-integer program, which is more difficult than the linear mixed-integer program.

In literature, research usually applies an iterative method to find the optimal solution. The majority of iterative algorithms suppose function to be differentiable and harness the gradient information of the function, including the steepest descent method and Newton's method. For the nondifferentiable function, the pattern search method may be used. The search efficiency and convergence property of these iterative methods are dependent on the function property and parameter setting. There may be instability while using these methods.

In managerial practice, the planner intends that the solution time could be controlled steadily within the short time on the premise of ensuring the calculating precision. Considering this, we discretize the ratio of gasoline–diesel output and perform the linear search method by fixed step size, thereby separating the original problem into several linear mixed-integer programs and obtaining the optimal solution, which can be specified as follows:

1. Initialization: Planner sets the range of the ratio "r" of gasoline to gasoline and diesel to [L, U] based on experience, wherein $0 \leq L < U \leq 1$; and set the step size "ε" according to the calculating precision. Therefore, the total search number can be expressed as $N = \lceil (U - L)/\varepsilon \rceil$. Set the initial iteration as $k = 0$.
2. Calculation: For iteration $k \leq N$, set the ratio of gasoline to gasoline and diesel as $r = L + k \times \varepsilon$, solve the linear mixed-integer programming model of fixed "r" value and calculate the ratio "r*" of actual gasoline to gasoline and diesel according to the result.
3. Terminate judgment: If $|r - r^*| \leq \varepsilon$, terminate the search and output the optimal solution and optimal objective value; or else, $k = k + 1$, return to step 2.

By using this linear search method of fixed step size, we need to solve at most N linear mixed-integer programming problems, and the calculating time is approximately linear to N, which ensures the stability. Also, the parameter setting is based on the experience accumulated by the planner, which will be easily controlled and aligned with the quick solution required by the planning department. In practice, this "engineering" method ensuring the quick solution has a good performance.

6.3.3   Objective functions

If the step-function price is considered, the planning department is concerned about the plan of producing more gasoline or diesel as much as possible and comparing the plan of maximizing profit. To this end, we add two extra objective functions, namely maximizing diesel output and maximizing gasoline output under the objective function of maximizing profit. Before solving the model, the user needs to select the corresponding objective function.

6.3.4   Break-even price

In this model, we calculate the two break-even prices of crude oil for one kind of crude oil, which are the full-load break-even price and the low-load break-even price, respectively. When calculating the full-load break-even price of one crude oil, we assume that the system only processes this type of crude oil and that processed quantity reaches the full capacity when the total profit divided by processed quantity is the full-load break-even price of this crude oil. In contrast, when calculating the low-load break-even price of one crude oil, we assume that the system only processes this type of crude oil, and the processed quantity is a unit. The system profit is the low-load break-even price of this crude oil. Two kinds of break-even prices are shown in Figures 6.4 and 6.5. When calculating the

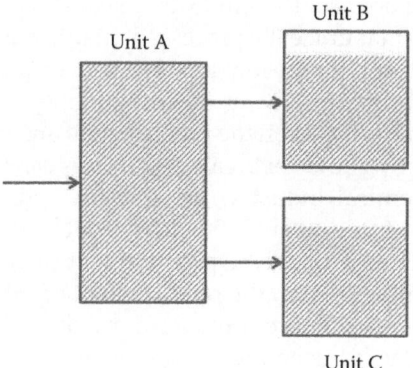

Figure 6.4 Illustration of full-load break-even price (when the crude oil reaches the upper limit of processing at unit A).

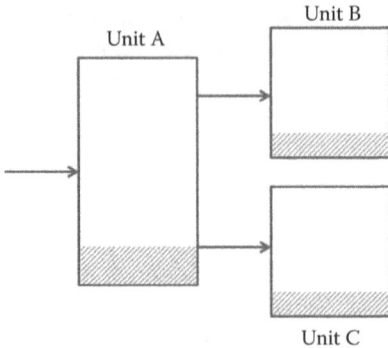

Figure 6.5 Illustration of low-load break-even price.

break-even price, we do not consider the upper or lower limit constraint of the final product output, the procurement quantity of crude oil, or the external procurement quantity or step-function price policy. The capacity constraint of each unit would be considered. It is obvious that two kinds of break-even prices have the following relationship:

Full-load break-even price ≤ Low-load break-even price

The reason is that when the processed quantity of crude oil is a unit, the model will select the production plan with the highest profit; in contrast, when the processed quantity of crude oil reaches the full load, some production plans with lower profit may be selected. Thus the average profit is lower.

Calculating the full-load or low-load break-even price of one crude oil is equivalent to solving the linear programming of crude oil procurement and production optimization. For the full-load break-even price, the objective of linear planning is to maximize the processed quantity of this crude oil, then maximize the profit; and for the low-load break-even price, it is intended to solve this problem by adding the processed quantity of this crude oil as a unit constraint, which is based on the procurement and production of crude oil. In practical use, two kinds of break-even prices can be quickly solved.

In the petrochemical industry, the concept and calculation of two break-even prices of crude oil are innovative, which can be also referenced in practice. For one crude oil, if the procurement price is higher than the low-load break-even price, then the procurement of this crude oil will not be profitable, so the procurement must not be done; if the procurement price is lower than the full-load break-even price, then the procurement of this crude oil will be profitable, so the procurement should be done; and if the procurement price is between the full-load break-even price and low-load break-even price, then the processed quantity, production plan, and further analysis should be done to examine if this crude oil is profitable or not.

6.3.5   Model highlights

In this optimization model, we assure its quickness, convenience, and flexibility. We formulate the optimization of crude oil procurement and production by linear programming. And we add integer variables to formulate the step-function price. Additionally, we transform the nonlinear problems into several linear integer programming problems by fixed step size searching. By doing so, this model can be quickly solved and meet the requirements of the planning department. We emphasize the decision of crude oil procurement and ignore the partial details or factors in production, which adapt to the needs of the planning department. In addition, an introduced concepts of "virtual unit" and "functional unit," guarantee the universality of this model, which is not only limited to QLPEC's existing production flow. When there is any change in production flow and unit capacity, this mathematical model shall remain applicable with some parameters updated.

All the data of this model is saved in data files, including the topology of the refining production network, which is composed of nodes (all units) and arcs (material inflow and outflow among units) and other data (e.g., the capacity and material yield). Therefore, when the network structure of oil refining production changes with the production flow or new crude oil or new product, we can reestablish and solve the model after modifying the corresponding information in the data files without modifying the optimization software. In this way, the flexibility of this model is further enhanced by the process and data, and for other similar software, the modification procedure of the corresponding process is relatively complex and time-consuming, and much detailed data should be input to ensure the running.

6.4   Auxiliary functions

Based on the core function of crude oil procurement and production optimization, we design auxiliary functions for business requirements and operations of the planning department in order to enhance the working efficiency.

6.4.1   Calculation of crude oil price

The procurement of crude oil is calculated in the in-factory price. However, different types of crude oil may be attached to different price systems of crude oil (Li and Huang, 2009; Li and Ma, 2011). For example, international crude oil is usually set in a free-on-board (FOB) price or cost + insurance + freight (CIF); domestic crude oil is usually set in CIF or ex-factory

Table 6.2 Parameters required for converting different price systems to in-factory price

Data	FOB	CIF	Ex-factory
Standard price (USD/barrel)	√	√	√ (Unit: in RMB yuan/ton)
Discount (USD/barrel)	√	√	×
Freight (USD/barrel)	√	×	×
Agent and Insurance (USD/barrel)	√	×	×
Correction (RMB yuan/ton)	√	√	√
Tons/barrel	√	√	×
USD to RMB (RMB/USD)	√	√	×

price. Different price systems of crude oil shall undertake a different scope of responsibilities. For example, the freight and other relevant costs will be borne by the buyer after the delivery point (i.e., the shipment designated by the buyer) if goods are delivered in FOB price. When pricing oil, it will refer to the standard price of oil types and adjust for other factors, such as the quality and freight of crude oil. Moreover, if the freight difference is not considered, the FOB price of Middle East crude oil is sold in about 1–3 USD/barrel to Asia higher than to Europe or America, where the extra cost is termed a "discount."

We enable the function of converting the price of crude oil to the in-factory price, which is well-tailored for different settlement methods of oil price. Data required for converting different price systems to in-factory price is as shown in Table 6.2.

Three kinds of price converted to in-factory price can be presented as below:

$$进厂价 = (离岸价 + 贴水 + 海运运费 + 代理和保险) \times \frac{美元对人民币汇率}{吨桶比} + 修正项$$

$$= (到岸价 + 贴水) \times \frac{美元对人民币汇率}{吨桶比} + 修正项$$

$$= 出厂价 + 修正项.$$

进厂价 In-factory price
美元对人民币汇率 USD against RMB
代理和保险 Agent fee and insurance
贴水 Premium
海运运费 Freight
离岸价 FOB price
吨桶比 Ton/barrel
到岸价 CIF price

修正项 Correction
出厂价 Ex-factory price

In addition, the user may directly input the in-factory price.

6.4.2   Plan presentation

For each production plan, we arranged and combined the results according to the business flow and focused KPIs, which can be presented as comprehensive indicators, such as crude oil procurement, product output, external materials and unit loads, and sideline products. Wherein, comprehensive indicators include the decomposition calculation of each item cost or benefit in profit calculation, and economic indicators such as light oil yield, comprehensive self-use rate, and processing loss rate.

Moreover, we use Excel as the medium of file input and output. The processing plan is not only displayed in the software interface but is also exported to Excel, which makes it convenient for comparing different plans and exporting to other information systems. It is also convenient for communication and coordination among departments and simplifying business operation.

6.4.3   Plan comparison

As the planning department needs multiple solutions under different parameters, we provide the importing and comparison functions of different plans, which makes it easy for planners to analyze the similarity and difference among different plans, and finally select the most appropriate one for crude oil procurement and production.

The presentation of different plans can be compared by comprehensive indicators, such as crude oil procurement, product output, external procurement quantity, unit load, and sideline product. In comprehensive indicators, each plan is aligned in a row, and in other lists, each plan is aligned in two rows as procurement–production–processing quantity, and the corresponding cost or profit. The added plan is inserted from the form rightward for the convenience of horizontal comparison visually as shown in Figure 6.6.

6.4.4   Bottleneck constraint

In linear programming, shadow price and bottleneck constraint are two important concepts. The value of the shadow price is determined as the ratio of the objective function value increment to the resource increment when the resource adds one unit quantity. In particular, only when the resources are fully used, or when this resource is the bottleneck constraint,

Comprehensive indicator	Proposal 1	Proposal 2	Unit load	Proposal 1 processing quantity (in ton)	Proposal 2 cost (in RMB: yuan)	Proposal 2 processing quantity (in ton)	Proposal 2 cost (in RMB: yuan)
Light oil yield	70.00%	75.00%	Unit 1	100.00	3000.00	80.00	2400.00
Comprehensive self-use rate	3.00%	2.00%	Unit 2	20.00	400.00	30.00	600.00
Processing loss rate	0.50%	1.00%	Unit 3	45.00	180.00	50.00	200.00

Figure 6.6 Sample of comprehensive indicators and unit load.

the value of the shadow price equals or is above zero. The shadow price can be construed as the maximum price in order to obtain an extra unit of the certain resource. While solving the linear programming model, bottleneck constraint and the corresponding shadow price are provided to avail more information for the decision.

In the solution plan, we highlight all the units and materials at the upper or lower limit of the bottleneck constraint as instructed for the actual production. Moreover, we display the bottleneck constraint and calculate the shadow price when the units and materials reach the bottleneck. It reveals the contribution of different units and materials to the objective function when the constraint is released. In this way, it will provide favorable information to the future decision, for example, increasing unit capacity and extending the product market. Note that the concept of shadow price exists in the linear planning. Considering the step-function price is the mixed integer program, the shadow price does not show the same meaning. As a result, when considering the step-function price, it will not display the information of shadow price but the bottlenecked units and materials.

6.5   Software design

Based on the mathematical model and auxiliary function design, we implement the "Crude Oil Purchase Management and Decision System." The overall structure and application process is briefed as follows.

6.5.1   Software structure

The software can be divided into three modules: graphic interface, software solution, and data management (in Figure 6.7). The graphic display module directly contacts with users, and users may load and edit the data parameters, solve and display the solved results, and save and export the results. The software solution module solves the model and generates the solution. The data management module connects the graphic interface with the software solution, transforms the input data via user interface into the software solution data. After solving the problem, the data management module transforms the software solution into the graphic interface result. The procedure of the software is described in Figure 6.8.

6.5.2   Software interface

Software interface is mainly used in data management, executing software functions, and graphic display. The hierarchy of interface pages is as shown in Figure 6.9. The interface mainly includes data input and result output. Data input displays the data classified by crude oil, product,

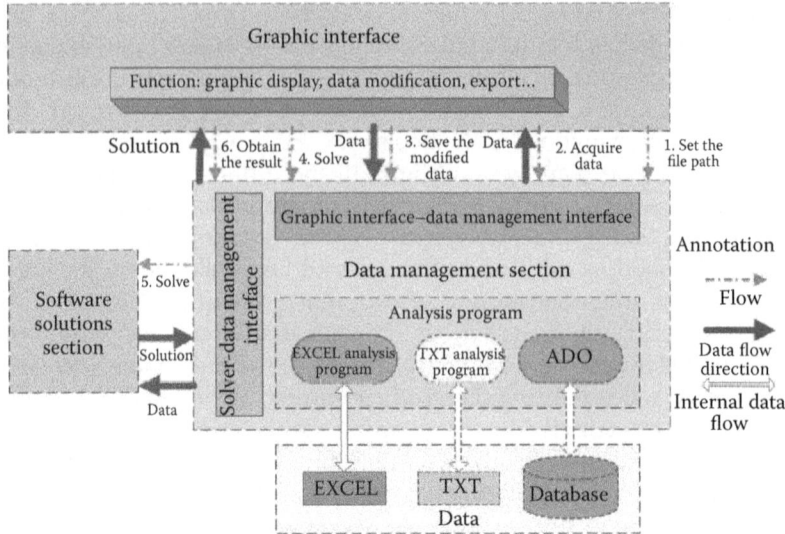

Figure 6.7 Software structure.

blending, and production units. Result output displays the calculation results of the procurement plan, auxiliary analysis, and break-even price and compares the various proposals uploaded.

6.5.3   Data management

For the benefit of business integration and management, input data and calculation results required for the model are saved in Microsoft Excel files. There are 11 Excel files of program data, which can be classified into input and output files.

In practice, input data may be inconsistent with the output data by reasons of any mishandling or mistake, which will intervene against the comparison of plans and mislead the decision. To avoid the foregoing situation, we have the following procedures in respect to data management to ensure the consistency of data and the comparison of plans:

1. Before calculating the procurement plan or break-even price, if there is any modified or inconsistent data, the program will ask "if overwrite the input data file"; after calculating the procurement plan or break-even price, the program will directly save the result as an output file.
2. When the program is about to exit, it will judge if the input file matches the output file. In case of any mismatch, for example, if the user modifies the input data but the user doesn't recalculate the procurement plan or break-even price, or if the user modifies the input

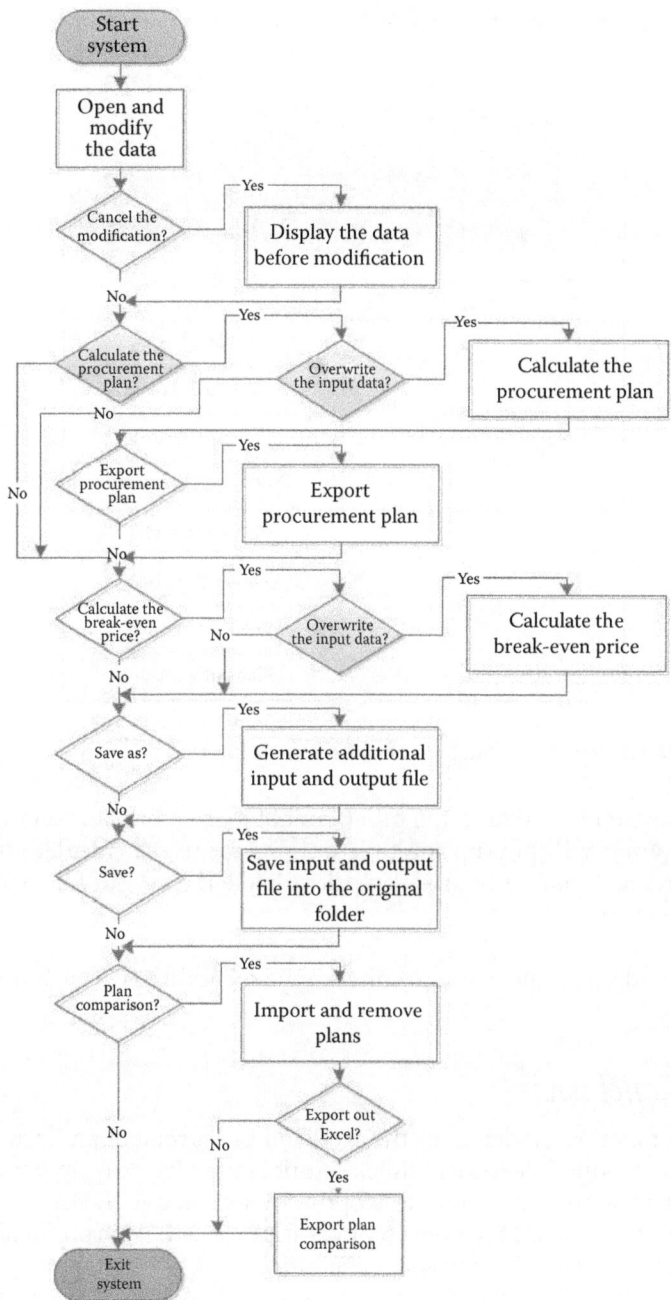

Figure 6.8 Flowchart of the software.

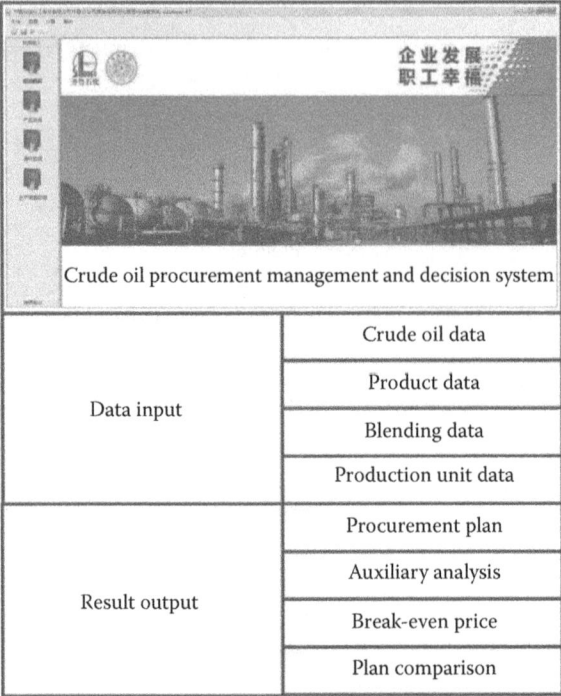

Crude oil procurement management and decision system

Data input	Crude oil data
	Product data
	Blending data
	Production unit data
Result output	Procurement plan
	Auxiliary analysis
	Break-even price
	Plan comparison

Figure 6.9 Hierarchical structure of interface.

data after calculating the procurement plan or break-even price, the program will pop up multiple options for users: calculate the procurement plan or break-even price, delete the output file, or exit the program.

This judgment mechanism enhance the reliability and usability of the software.

6.6   Conclusion

The optimization model takes the crude oil procurement and production plan into account. It features a quick solution, user-friendly operation, and flexible adjustment. Based on this optimization model, the calculation of two types of crude oil break-even prices is proposed. With auxiliary functions embedded in the software, the efficiency of the planning department is further improved. And the procurement plan is also improved via the software.

This software has been in the pilot run in QLPEC since April 2012, and the first official version was released in May 2012. Aided by this software, the planning department has made the decision of the procurement plan, optimized the product structure, and promoted the integration of the oil refining process, which turns out to be satisfying. According to the SINOPEC news report on September 16, 2013, QLPEC's procurement price of imported crude oil was the lowest among 23 corporations of imported crude oil for 20 months consecutively. In 2012, QLPEC procured over 5,590,000 tons of imported crude oil with FOB price 1.2 USD/barrel lower than that of the headquarter, which reduced the procurement cost amounting to RMB 260,000,000.

References

Hou, K. Applying LP model to improve refinery design level (in Chinese). *Chemical Industry and Engineering Progress*, 2012, 31(12): 2811–2814.

Li, H., Huang, L. The composition of price system of international crude oil (in Chinese). *Oil-Gasfield Surface Engineering*, 2009, 28(7): 66–67.

Li, X., Ma, N. Oil price (in Chinese). *CCED working paper series*, 2011, 1–17.

Niu, Y., Bian, S., Kong, L. Study on application of refinery PIMS model in crude oil purchase and production optimization (in Chinese). *Qilu Petrochemical Technology*, 2006, 34(4): 365–371.

chapter seven

Ways of Enterprise T to improve laptop computer quality

Su Wu, Mingxing Xie, and Liang Wen

Contents

7.1   Enterprise status

The PC industry has changed the relationship between the desktop computer and portable computer for a long time. Development of the desktop computer market is almost stalled, and the portable computer market is still growing. Changeable market demand increases the fierce market competition. Limited by marketing strategy and market conditions, brand awareness and quality of enterprises that entered the laptop industry later cannot be compared with domestic top OEM brands, European and American brands, and Japanese brands. Squeezed by these enterprises during the growth and competition, their development becomes very unfavorable.[1,2]

With OEM mode, the marketing mode of the global laptop market is OEM computer manufacturers confirming their channel advantages by their brand advantages. There are very few upper stream suppliers of the global laptop computer industry with a high vertical supply chain and high integrated industrial chain. And these lead to deterioration of the overall competitive environment of the laptop computer market, fiercer price wars, lower gross profits, and low controllability degree of resources. The controllability of the upper supply chain is basically mastered by the top 10 of the most famous European and American and Japanese brands.

Laptop computer OEM manufacturers basically choose upper suppliers according to the occupied market share and brand influence.[3] OEM manufacturers with large scale, high level, and strong competitiveness will choose suppliers such as Quanta, Compal, and Flextronics to achieve powerful combinations. Accordingly, second-line OEM manufacturers will choose those second-line suppliers such as First International Computer, Clevo, and Elite. Naturally, upstream factors of the industry chain account for a great proportion of product quality of OEM computer manufacturers, and a low industry-access barrier is also an important reason for the variable quality of laptop computers. If an enterprise wants to obtain a favorable position in the laptop computer industry of the marginal profit era, it is very urgent and necessary for the enterprise to keep improving quality. Because each key components of the product are is a purchased items, the related functional index and reliability index have to be guaranteed by tests during the purchasing. These purchased components will be assembled, sold with the enterprise brand, and represent the product quality of the enterprise. OEM enterprises need to be responsible for product after-sales maintenance with complete after-sales service policies. Thus, enterprises need to improve the quality level of the whole process by quality improvement activities so as to meet user demand, to improve customer satisfaction, to reduce after-sales service costs, to improve competitive advantages of the laptop computer enterprise, and to make the laptop computer quality rise to a new level.

XX company is a well-known domestic listed company with total assets of more than 30 billion RMB in 2011 and annual revenue of more than 20 billion RMB. Since the company was established in 1997, the computer system headquarters has achieved extraordinary development after four years of struggle. During 1997 to 2001, sales of X brand computer have increased with 100% progressive growth for four consecutive years. In the year 2000, X computer was ranked top three in sales rankings of PC computers in mainland China. In 2001–2003, due to persistent efforts, the sales surpassed a million, which not only consolidated its position in the top three best sellers list in the Chinese market, and has been ranked in the top eight PC selling sales list in the Asia-Pacific market by IDC since the third quarter of 2002. Moreover, in the third quarter of 2003, the company was ranked in the top six of in the Asia-Pacific region. Since 2004, according to CCID, its selling of home computers has jumped to second place in China. In the first quarter of 2006, according to CCID, its desktop computer was stably ranked in the top 10 of the best selling list in the global market.

At present, the product lines of this company have covered desktop PCs, laptop computers, workstations, servers, digital and computer peripheral products, and industrial application solutions, etc.

As to home computers, the company has launched the most complete solution in the industry currently, "DIGI HOME Infinite Digital Home," and has perfected the product line segmentation as the Z Series Entertainment Computer, HY Series Game Computer, LY Series Film and TV Computer and Home Computer, etc. As for commercial computers, the company has launched comprehensive products and related solutions, such as the CX Series mid–high end industry-specific computer, CY1 Series small and medium enterprises (SME) Applied Computer, and CY2 Series education-specific computer, etc., and has been unanimously approved by various industries such as education, finance, securities, military, taxation, Internet bar, government, small and medium-sized enterprises, etc.

The company maintains a leading position in aspects of thinness and lightness, economy, fashion, and cost performance. A complete family of server products, such as super brand tower type, rack type, and stunning graphic workstation, etc. launched by the company is among the leading technology in China. As a good supplement for PC peripherals in IT application solutions, the company has also launched digital products, such as mobile HDD, mobile flash disk, and digital photo frame, etc.

The same as other small and medium-sized computer enterprises in China, the company encountered the same problems during the quality control of enterprise laptop computers. The company still has insufficient awareness from application of dead on arrival (DOA) and mean time between failures (MTBF) to quality improvement. DOA and MTBF are considered as key indexes that need to be controlled during the quality control of laptop computers under such production and sales modes. Therefore, the company conducted quality improvements according to problems reflected by statistic data and these two key quality indexes.

7.2   Problems and influence

7.2.1   About DOA

DOA refers to conformance and consistency of product performance during the unpacking and acceptiance when the product is finally sent to the customers after the transportation and storage. It is a kind of evaluation index of product quality level and also the customers' first impression of product quality. The daily quality control of enterprises is usually based and focused on this index, and a good DOA value will improve profit space and competitiveness of the enterprise. The quality improvement analysis that is oriented by DOA is a necessary work of quality control of enterprises and usually is taken as quality assessment index of the inner organization of enterprises.

7.2.2   About MTBF

MTBF is another important index to measure product quality (especially electronic products) and also can be said to be the most important index to measure reliability and stability of computer products.[7] Whether product functions can be performed or not is greatly dependent on the reliability of the product.

Reliability improvement technology, which is introduced and centered on MTBF can obviously improve the quality level of products. It has made up the weaknesses of a traditional quality index and can indicate time and quality characteristics of products. It indicates the ability of products to maintain functions in the specified time and is more able to fit the actual needs of customers. The degree of MTBF will obviously influence customer satisfaction. Continuous and repeated quality problems will cause a psychological effect on customers and also will cause great influence on brand image of the enterprise. Such an associated effect is very unfavorable for sustainable development of the enterprise.

MTBF degree is reflected by after-sales service cost variance. Most after-sales service policies in the laptop industry are the same: commitment of free maintenance and repair on key main-board components for three years. A laptop computer company has 2.2% gross profit of computer products with an average MTBF of about 330, an annual maintenance and repair quantity of 250 thousand sets, a maintenance and repair rate of 9–10%, and its annual maintenance cost is about 5.6 million RMB. Compared to the benchmarking enterprises of IBM, Toshiba, HP, Dell, and Lenovo (shown as Table 7.1), MTBF of this company is much lower with

Table 7.1 DOA and MTBF of benchmarking enterprises[4-6]

Computer manufacturer	IBM (China)	Lenovo	Dell (China)	HP (China)	Toshiba (China)
DOA (%)	0.5	0.8–1.0	1.0	0.8	0.75
MTBF (Day)	570	410	/	440	480

almost double the maintenance and repair costs of a single set of computers. The stubbornly high after-sales maintenance and repair cost has transferred to great pressure on the enterprises' operating costs.

Therefore, DOA and MTBF are extremely important for enterprises. X company, as a new entrant in laptop computer manufacturing, both its two indexes are lower than enterprises listed on Table 7.1. On the basis of importance and necessity to improve laptop computer quality according to DOA and MTBF, this chapter analyzes the influence factors of DOA and MTBF and improvement measures and concludes with a relatively comprehensive analysis and improvement measures.

7.3 Reason analysis on laptop computer deficiency based on DOA

7.3.1 Definition of DOA

Products produced by enterprises are generally through steps from R&D, manufacturing, operating, maintenance and repair, discard, etc., and this is the whole life cycle of products. Among which, product problems occurring during the process after outgoing quality control (OQC) to packaging, handling, transportation and storage, etc., to unpacking and accepting when customers receive the products are defined as DOA.

As shown in Figure 7.1, disqualification found by users during the accepting is called DOA disqualification, which is usually called first

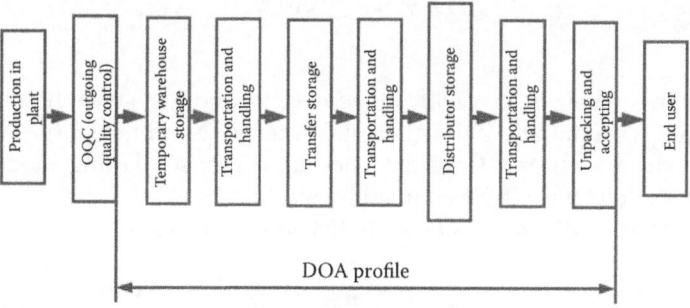

Figure 7.1 DOA process profile.

unpacking disqualification, and it is the disqualification of $t = 0$ moment in quality and time characteristics. DOA focuses on conformance and consistency of product performance. It usually shows appearance defects, component damage, configuration inconformity, abnormal boot, abnormal display, and abnormal use, etc.

From statistical analysis on DOA data of a laptop computer, we can see that the product nonoperating failure rate, such as in storage and transportation, is much lower than the failure rate in the working state. However, influenced by various conditions, such as handling, transportation, and storage, etc., the failure rate accumulated by accepting should not be ignored. Thus, related analysis is necessary.

7.3.2   *Statistical approach of DOA*

The DOA unqualified rate is also called first unpacking unqualified rate. It is the percentage that total unqualified Qty n are divided by total selling Qty N, which is

$$DOA = \frac{n}{N} 100\%$$

When actually calculating the DOA unqualified rate, time factors have been considered. The feedback time of after-sales information of each agent and exclusive shop has direct influence on DOA statistics. Monthly DOA data can usually be counted after several months, and such information transmission log will bring enterprises certain difficulty in quality control. Consider the influence of marketing mode, enterprise information system, and the level of information; the calculation method is different from one enterprise to another. In general, there are several formulas as follows:

The way to calculate DOA with direct marketing mode:

$$DOA = \frac{n_w}{N_w} * 100\%$$

Among which, n_w is weekly defective Qty, and N_w is weekly sales Qty.

Foreign capital and Taiwan capital enterprises on the Chinese mainland, such as Dell and Gateway, conduct direct marketing mode; thus their DOA defective Qty is counted by week.

The way to calculate DOA with distribution mode:

$$DOA = \frac{n_m}{N_m} * 100\%$$

Among which, n_m is monthly defective Qty, and N_m is monthly sales Qty.

Main laptop computer enterprises with distribution mode are Lenovo, Hasee, Founder, and Company T, etc. Their DOA defects are counted by month.

According to different information and marketing modes, presentation of DOA indications also have some differences. During actual operation, many enterprises will use the proportional relationship of production Qty to approximately replace actual sales Qty. Other common calculation methods are as follows:

$$DOA_1 = n_m/(N_{lp} * 40\% + N_{dP} * 60\%)$$

$$DOA_2 = n_m/(N_{lp} * 50\% + N_{dP} * 50\%)$$

$$DOA_3 = n_m/(N_{lp} * 55\% + N_{dP} * 45\%)$$

Among which, N_{lp} is production Qty of last month, and N_{dp} is production Qty of this month.

Although the above DOA calculations are different in time, all of them reflect the quality-control level of the enterprises within a certain allowed error rage.

7.3.3   Current status of DOA

According to the definition and calculation methods of DOA, we can learn that enterprises' DOA description includes the following main contents: monthly DOA defective Qty and monthly sales Qty or production Qty, monthly DOA defective Qty and monthly sales Qty or Production Qty of each model, DOA or first unpacking qualified rate, annual DOA cost, and classified statistics on defects.

1. Monthly DOA Defective Qty and Monthly Sales Qty

   According to statistics from the laptop computer after-sales service department of X company, total defective Qty of this company is 2314 sets in 2007. Weekly DOA defective Qty is shown as Table 7.2.

Table 7.2 Monthly DOA defective Qty of X laptop computer

M	1	2	3	4	5	6	7	8	9	10	11	12
DOA Qty	168	132	194	157	148	156	223	282	259	256	191	148
Sales Qty	8842	6111	10,486	8771	10,137	9176	12,120	14,315	14,469	14,713	12,993	9367

2. Monthly DOA Defective Qty and Sales Qty or Production Qty of Each Model

DOA defective Qty and sales Qty or production Qty of each model of this company in 2007 are shown as Table 7.3.

3. DOA and Qualified Rate of the First Unpacking

According to statistics on the date of after-sales DOA data and sales data of the laptop computer department of X company, the DOA and qualified rate of the first unpacking of this company in 2007 (shown as Table 7.4), the average DOA is 1.76%, and the qualified rate of the first unpacking is 98.24%.

According to the above monthly DOA data, use theory of statistical process control to analyze whether the process is controllable. Conclude the following by control-limit expression of P control chart[8]:

$$\text{UCL} = \bar{p} + 3\sqrt{(1-\bar{p})\frac{\bar{p}}{n}}; \text{CL} = \bar{p}; \text{LCL} = \bar{p} - 3\sqrt{(1-\bar{p})\frac{\bar{p}}{n}}$$

Control limit of monthly DOA is calculated as follows:

$$\text{UCL} = \bar{p} + 3\sqrt{(1-\bar{p})\frac{\bar{p}}{n}} = 0.0176 + 3\sqrt{0.0176 * 0.9824/n}$$

$$= 0.0176 + 0.3945/\sqrt{n_i}$$

Table 7.3 DOA defective Qty and Sales Qty or production Qty (sets) of each models in 2007

Suppliers	CELEVO	FIC	MSI	TOPSTAR	AMOI	Total
DOA defects	460	918	468	285	183	2314
Sales Qty	25,078	51,506	27,067	17,038	10,811	131,500

Table 7.4 Monthly DOA and first unpacking qualified rate (%) of laptop computer products of X company in 2007

M	1	2	3	4	5	6	7	8	9	10	11	12
DOA (%)	1.90	2.16	1.85	1.79	1.46	1.70	1.84	1.97	1.79	1.74	1.47	1.58
FU	98.1	97.84	98.15	98.21	98.54	98.3	98.16	98.03	98.21	98.26	98.53	98.42

Note: FU: first unpacking.

$$CL = \bar{p} = 0.0176$$

$$LCL = \bar{p} - 3\sqrt{(1-\bar{p})\frac{\bar{p}}{n}} = 0.0176 - 3\sqrt{0.0176 * 0.9824/n}$$

$$= 0.0176 - 0.3945/\sqrt{n_i}$$

Analysis results are shown as Table 7.5. We can see that all DOA unqualified rates are under the control line, which means the process is under control. It means variation of the unqualified rate is not influenced by special factors, but the mean value is abnormal.

4. Approximate Calculation Method of Average DOA

In 2007, total DOA defective Qty of a laptop computer company for a whole year n_y = 2314 sets, with actual production Qty N_y = 139,426 sets, DOA = $\frac{n_y}{N_y}$ * 100% = 1.66%, and the first unpacking qualified rate is 98.34%.

Compare to annual DOA of 1.76%, the DOA of 1.66% of this approximate calculation is 0.1% less, which is within the allowed error range. Therefore, during the analysis on average DOA, the sales Qty can be replaced by production Qty.

Table 7.5 Monthly DOA data and computation of diagram *P* of X laptop computer company in 2007

Month	Sales Qty n	Defects Qty D	Defective rate p%	UCL of figure p	LCL of figure p
1	8842	168	1.9	0.0218	0.0134
2	6111	132	2.16	0.0227	0.0126
3	10,486	194	1.85	0.0215	0.0138
4	8771	157	1.79	0.0218	0.0134
5	10,137	148	1.46	0.0215	0.0137
6	9176	156	1.7	0.0217	0.0135
7	12,120	223	1.84	0.0212	0.0140
8	14,315	282	1.97	0.0209	0.0143
9	14,469	259	1.79	0.0209	0.0143
10	14,713	256	1.74	0.0209	0.0144
11	12,993	191	1.47	0.0211	0.0142
12	9367	148	1.58	0.0217	0.0135
Total (average)	131,500	2314	1.76		

5. Cost of DOA Defects

   In 2007, the annual DOA value of laptop computers of this company was 1.76%. Due to DOA defects, the annual cost was about 700,000 Yuan.

6. Classified Statistics on DOA Defects

   As to DOA defective data through the year 2007, conduct data classification according to the acceptance process and logic order of laptop computer use: an appearance check on the packing box, on the attached materials after unpacking, on the appearance of the computer (scratch and structural damage), boot and inspection with power on (boot or not, boot with nothing displayed), boot but cannot enter the system, abnormal display after entering the system, operation inspection after entering the system, inspection on software drive, and system installation.

Data statistical results are shown as Table 7.6.

Table 7.6 Statistics on DOA defective Qty of a laptop computer company in 2007

Defective items	Defective Qty	Accumulated defective Qty	Percentage of total defective Qty	Cumulated percentage
Operation inspection after entering the system	868	868	37.51%	0.00%
Boot and inspection with power on	596	1464	25.76%	37.51%
Abnormal display after entering the system	377	1841	16.29%	63.27%
Check appearance of the computer	170	2011	7.35%	79.56%
Check software installation	143	2154	6.18%	86.91%
Boot but cannot enter the system	104	2258	4.49%	93.09%
Check packing box	21	2279	0.91%	97.58%
Check attached materials	19	2298	0.82%	98.49%
Others	16	2314	0.69%	99.31%
Total	2314		100.00%	100.00%

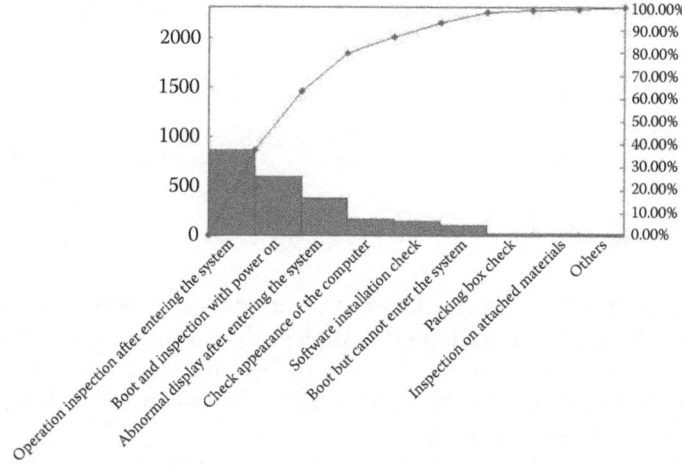

Figure 7.2 DOA defects pareto chart of a laptop computer company in 2007.

The statistical results show that defects during operation inspection after entering the system, boot and inspection with power on, abnormal display after entering the system, check appearance of the computer, and check software installation account for 86.91% of total defects. However, according to specific defects, the phenomenon is complicated, especially the defects during boot with power on are much higher than the qualified rate of components and batches of OQC (Figure 7.2).

According to above description of the current status of DOA, analysis on P control and quality level of benchmarking of the comprehensive laptop computer industry confirms the goal of quality improvement of laptop computers to reduce the defective rate to 1% from 1.76% on the basis of DOA.

7.3.4 Reason analysis on DOA problems

First, get to know whether there is any correlation between DOA defects and the above five kinds of models by data. Compare the defective rate of Celevo and Topstar by hypothesis testing.[9]

Set production Qty of Celevo is m_1, defective Qty is n_1, production Qty of Topstar is m_2, defective Qty is n_2. Learn according to the data of Table 7.2 DOA defective Qty and total production Qty of each model in 2007:

$$m_1 = 25{,}078, n_1 = 460, \text{defective rate } p_1 = 0.018343;$$

$$m_2 = 17{,}038, n_2 = 285 \text{ defective rate } p_2 = 0.016727;$$

$n_1 + n_2 = 745$, $m_1 + m_2 = 42{,}116$, total defective rate $p = 745/42{,}116 = 0.017689$

$$z = \frac{p_2 - p_1}{\sqrt{p \times (1-p)\left(\dfrac{1}{m_2} + \dfrac{1}{m_1}\right)}} = \frac{0.018343 - 0.016727}{\sqrt{0.017689 \times 0.982311 \times \left(\dfrac{1}{25{,}078} + \dfrac{1}{17{,}038}\right)}}$$

$$= 1.234 < 1.96;$$

Also, hypothesis test results of other models are shown as Table 7.7. All Z statistic values of each model are less than 1.96. It shows that there are no significant difference after-sale defects for different models ($\alpha = 0.05$), which shows no obvious relationships between DOA defects and models. It is a general problem.

Influence of models has been excluded as a high unqualified rate of DOA, conduct cause–effect map analysis on aspects such as materials, machines, personnel, methods, and environment as shown in Figure 7.3. By exclusive method, three main influence factors have been confirmed as followed:

a. Insufficient protection provided by buffer to mainframe.
b. Unreasonable design of inner interface structure of mainboard of laptop computer.
c. Inadequate measurement methods.

As to the above three main factors, determine the research content and conduct investigation and analysis, shown as Table 7.8.

A. Investigation No. 1

Laptop computer will enter links of storage and transportation after passing the OQC and finally be sent to the users. The whole transportation process is shown as Figure 7.4.

To verify the protection of buffer for the mainframe, a tracking test on the transportation environment has been conducted and simulated the whole transportation process.

Table 7.7 Z Value of various models

Models/z	CELEVO	FIC	TOPSTAR	MSI	AMOI
CELEVO		0.508	1.234	0.908	0.928
FIC	0.508		0.944	0.539	0.643
TOPSTAR	1.234	0.944		0.445	0.246
MSI	0.908	0.539	0.445		0.126
AMOI	0.928	0.643	0.126	0.246	

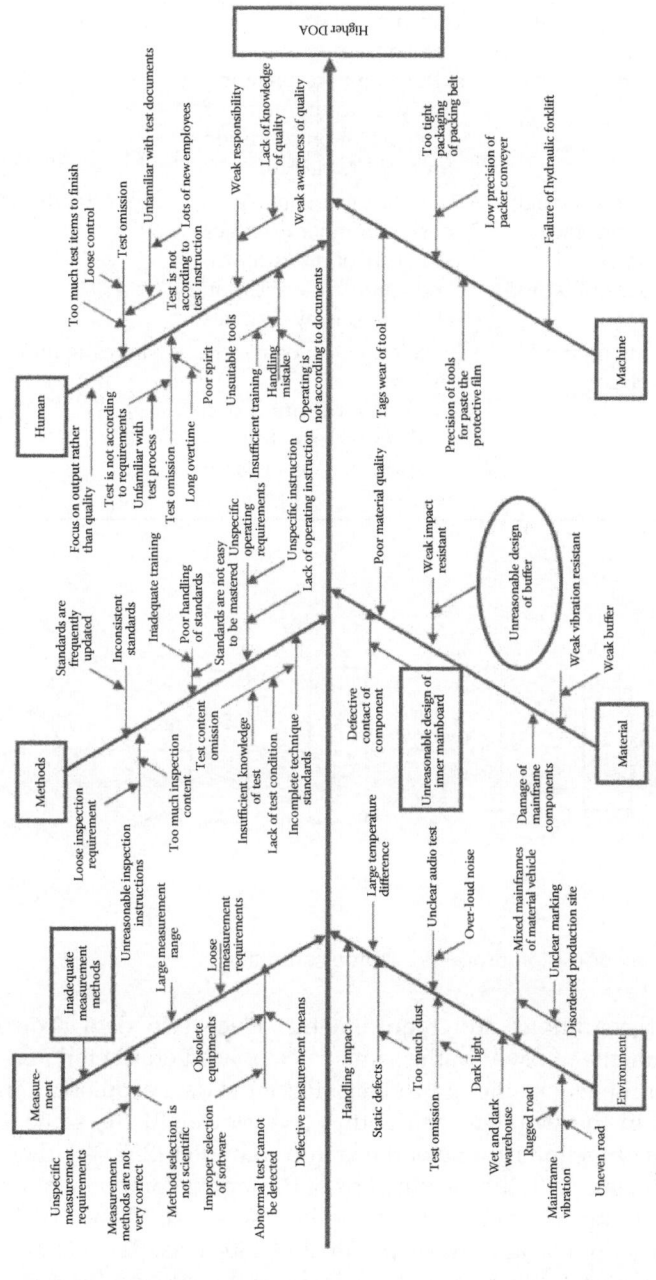

Figure 7.3 Fishbone diagram of unqualified DOA.

Table 7.8 Questionnaire on main factors of DOA

No.	Main factors	Investigation items	Investigation no.
1	Insufficient protection provided by buffer to mainframe	Tracking test on transportation test: According to test results, analyze requirements of transportation conditions for mainframe buffer	Investigation No. 1
2	Unreasonable design of inner interface structure of mainboard of laptop computer	Analyze unreasonable design of inner interface structure of main board, and conduct verification and analysis by drop test	Investigation No. 2
3	Measuring method is not in place	Installation test of system driver Conduct verification and analysis by drop test according to experimental results	Investigation No. 3

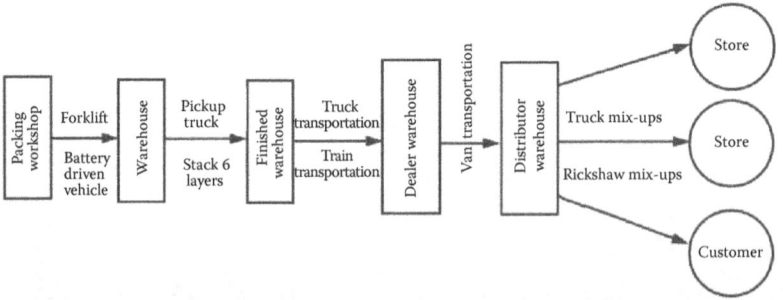

Figure 7.4 Transportation process of laptop computer.

Test conditions and requirements: Place two data acquisition instruments of Saver in the same truck, placed on the tail and front part of the truck among products. Place the data acquisition instruments in the wooden case with a counterweight the same as the weight of the mainframe of the laptop computer (2.35 kg), then pack instruments with the packing box of the laptop computer, and make the marking. The vehicle used for the test is a 9.6 M van truck (size: 57 m^3) with a total number per load of 1800 sets; 325 sets of laptop computers are in storage in the Wuxi warehouse, and the remaining 1475 sets of computers are sent to a dealer in Zhejiang. DOA defects during the whole process of transportation, unpacking, and accepting of this batch of laptop computers will be tested and tracked. The track record of the transportation environment is shown as Table 7.9.

Collected data of vibration, impact, and dropping during the test process is shown as Table 7.10. Total distance of motor transport is

Table 7.9 Track record of transportation environment

Date	Starting	Terminal	Departure	Destination	Transport and storage
2007-5-16	10:45	18:25	Workshop	Transfer Warehouse	Forklift, battery driven vehicle
2007-5-16	13:20	19:45	Transfer Warehouse	Xin Da Warehouse	Freight (pickup truck)
2007-5-17	9:00	(5–18) 22:25	Xin Da Warehouse	Wuxi Warehouse	China Railway Express van truck
2007-5-19	15:00	22:30	Wuxi Warehouse	Distributor Warehouse	Motor transport
2007-5-20	9:30	10:30	Distributor Warehouse	PC Mall	Rickshaw, pickup truck

Table 7.10 Tacking test results of transportation

Recorder A	Recorder B
Total Events 1196 (vibration, impact, and dropping)	Total Events 1280 (vibration, impact, and dropping)
Time-triggered 1097 times	Time-triggered 1193 times
Signal-triggered 99 times	Signal-triggered 87 times
During the collecting process, temperature: 31.67°C–16.95°C, relative humidity: 37.9–52.6%	During the collecting process, temperature: 31.46°C–18.35°C, relative humidity: 34.2–54.7%
Dropping: 12 times	Dropping: 10 times
One time of throwing with distance of 1693 mm; five times of drop heights are: 451 mm, 391 mm, 384 mm, 334 mm, 200 mm, 286.2 mm; all other drop heights are lower than 200 mm	Four times of drop heights are 439 mm, 383 mm, 378 mm, 335 mm; all other six drop heights are lower than 200 mm
Impact: 72 times	Impact: 73 times
The biggest impact acceleration is 38 g. Times of impact more than 20 g: three times. All other impact accelerations are less than 200 mm	The biggest impact acceleration is 32 g. Times of impact more than 20 g: five times. All other impact accelerations are less than 200 mm

2150 km from a plant warehouse in Beijing to the Wuxi warehouse to a distributor warehouse in Zhejiang and finally to arrive at the mall.

Vibration Results

Total power spectral density of time-triggered vibration, average value: 0.074 Grms, max value: 0.773 Grms. Total power spectral density of signal-triggered vibration, average value: 0.47 Grms, max value: 1.295 Grms. DOA defect: 1475 sets of laptop computers are loaded by the vehicle, and total DOA defects according to tracking and statistics are 29 sets. Sixteen sets have packing box damage and buffer with six sets of DOA defects. Among the packing boxes without damage, there are 23 sets of DOA defects. The hypothesis test shows an obvious relationship between DOA defects and package protection. Improvement of protective performance of packing materials should be taken into consideration.

B. Investigation No. 2

Then, conduct statistics on DOA defects of laptop computers of X company in 2007. It was found that each column "defects removal" of the after-sales DOA report is filled out as "defective contact with main board." Wherein "defective contact with main board" is 523 PCs, and total "defective main board" is 1228 PCs. The "defective contact with main board" accounts for 42.56 of the total "defective main board" and accounts for 22.6% of the total DOA defects. The DOA defects statistics are shown as Table 7.11. And "defective contact with main board" is shown as Table 7.12.

The main board of the laptop computer is the largest piece of circuit board in the computer system. It is the platform for stable operation of each component, the link between the inner component interface and the external device interface, and ensures coordinated and orderly operation of each component of the computer. The main board of the laptop computer has various and numerous components, a complicated technique and manufacturing, and lots of process links. The quality of components and process control will influence performance of the main board, and performance of the main board has a direct relationship with the performance of the whole computer. Thus, it is the key point that needs special attention.

From a statistical table of "defective contact with main board," we can learn that defective contacts with the memory slot, LCD screen cable, wireless card, keyboard, and touch board account for 82.22% of total defective contacts. Wherein, defective contact with the memory slot accounts for 41.30% of total defective contacts. By further analysis, we can learn the main memory slots with defective contact are Concraft memory slots. Thus, analysis and research should be conducted on defective contact of Concraft memory slots.

Table 7.11 DOA defect statistics of laptop computers of a company in 2007

Defective items	Defective Qty	Accumulated defective Qty	Accounts for total defective Qty	Accumulated percentage
Main board	1228	1228	53.07%	53.07%
LCD screen	177	1405	7.65%	60.72%
Hard disk	96	1501	4.15%	64.87%
CD drive	87	1588	3.76%	68.63%
Horn	79	1667	3.41%	72.04%
Memory	75	1742	3.24%	75.28%
Screen cable	73	1815	3.15%	78.44%
Webcam	72	1887	3.11%	81.55%
Battery	67	1954	2.90%	84.44%
Booster board	63	2017	2.72%	87.17%
Plastic parts	54	2071	2.33%	89.50%
Others	51	2122	2.20%	91.70%
Keyboard	45	2167	1.94%	93.65%
Touch board	39	2206	1.69%	95.33%
CPU	26	2232	1.12%	96.46%
Switch plate	22	2254	0.95%	97.41%
Fan	21	2275	0.91%	98.31%
Wireless card	21	2296	0.91%	99.22%
Power adapter	18	2314	0.78%	100.00%
Total	2314		100.00%	

By detailed analysis on the main board, which is returned due to defective contact, lots of defective computers cannot have the phenomenon reappear that is listed on the after-sale reports. The off memory card is the only one that can be confirmed and will return to normal when extracted and plugged in again. According to the main board with phenomena that is difficult to have reappear, related tests need to be conducted to simulate and verify. Thus the drop test has been designed.

Drop test: 20 sets of DOA unqualified computers, which have been returned, were randomly selected, and two drop tests were conducted on every mainframe so as to ensure the randomness of the test. Before the drop test, the structure and performance should be checked as normal and then packed the mainframe. Use four sensors to separately record maximum dropping acceleration of two points of memory slots and two points where the card is plugged in. Dropping should be conducted according to international

Table 7.12 Statistics on defective contact with main board of a laptop computer company in 2007

"Defective contact with main board" types	Defective Qty	Accumulated defective Qty	Accounts for total defects Qty	Accumulated percentage
Memory slot	216	216	41.30%	41.30%
LCD screen cable	113	329	21.61%	62.91%
Wireless card	41	370	7.84%	70.75%
Keyboard cable	32	402	6.12%	76.86%
Touch board cable	28	430	5.35%	82.22%
CD driver	27	457	5.16%	87.38%
Hard disk	21	478	4.02%	91.40%
CPU	17	495	3.25%	94.65%
Radiator module	12	507	2.29%	96.94%
Webcam	12	519	2.29%	99.24%
Horn cable	4	523	0.76%	100.00%
Total	523		100.00%	

requirements: one angle three ridges, six sides with drop heights of 1 m, 80 cm, and 60 cm.

Twenty sets defective mainframes have been dropped 40 times. In six computers, nine times the memory fell off the slots, and three sets of mainframes had memory loss, serious deformation of snaps, copper leakage between memory and metal, and slight copper leakage and deformation of snaps when memory is normal each time. The maximum accelerations of memory with normal or abnormal dripping of each test face are shown as Table 7.13.

Table 7.13 Dropping acceleration (g) of normal and abnormal dropping

Drop faces	Max acceleration with normal memory slot	Max acceleration with abnormal memory slot
Front right base angle	17.21	18.13
Front right ridge	23.69	22.83
Front base ridge	28.31	28.11
Front right top ridge	27.55	27.87
Front side	31.55	32.05
Back side	31.95	31.70
Left side	49.85	58.16
Right side	44.51	53.43
Top side	35.27	31.95
Base side	36.15	32.35

The drop test shows that when dropped on the left side, the max impact acceleration due to dropping will be up to 58.16 g. With the max acceleration, the memory will fall down from the slot. Meanwhile, we conducted the contact inspection on memory after the drop test, and we can found severe abrasion and copper leakage between the metal snap and memory in the memory slot.

The analysis suggests that the intensity of the metal snap is not strong enough, and this makes the memory snap against the dropping impact. And this is the main reason for defective contact of memory and memory loss.

C. Investigation No. 3

According to after-sales statistics on DOA, we can see 143 cases of defective installation of the system driver, shown as Table 7.6. But no similar phenomena can be found in the maintenance and test records in the year 2007. By analysis on the above differences, the reason should be the difference of the operation system.

By comparison and analysis, we found that the mainframe with the Freedos operation system accounts for 70% of the total mainframe with 121 sets with defective installation of the system driver. This system is the Freedos operation system, which can be used only after the Windows or Linux operation system software and a driver that has been installed by the customers after they bought the computer. As to the mainframe with an operation system when leaving the factory, the customers can directly boot without reinstalling the system and driver. The mainframe without the system when leaving the factory obviously has higher DOA than the mainframe with the system. This suggests that the installation will influence the DOA.

Installation processes have certain differences among users. Compare users' installation of system and driver and online installation of system and driver; users install the system and software by CD driver, or the online system and driver are copied by the mother disc. They are quite different.

Simulated and tested according to system and process differences, the production process of the current laptop computer is as follows. After the final inspection, the system and driver were installed by CD before packaging with a month of test time: April 26, 2008–May 25, 2008. In the same time, check the after-sales DOA defects of this month. The results are shown as Table 7.14.

Defective phenomenon of the test: Seven cases of defective system installation, two cases of defective driver, five cases of automatic power off during system installation, five cases of blue screen during system installation, two cases of CD driver without loading, three cases of crash during system installation, and one case of reboot

Table 7.14 Records of defective installation of system and driver

Production date	Installation Qty	Defective installation	Defective rate %	After-sales DOA defects	After-sales defective rate %
2008-4-26	820	1	0.1220	1	0.122
2008-4-28	870	3	0.3448	1	0.115
2008-4-29	353	1	0.2833	0	0.000
2008-4-30	303	0	0.0000	0	0.000
2008-5-6	388	1	0.2577	1	0.258
2008-5-7	415	2	0.4819	0	0.000
2008-5-8	337	1	0.2967	0	0.000
2008-5-9	768	3	0.3906	0	0.000
2008-5-12	804	1	0.1244	1	0.124
2008-5-13	394	2	0.5076	0	0.000
2008-5-14	402	1	0.2427	0	0.000
2008-5-15	205	0	0.0000	0	0.488
2008-5-16	316	1	0.3165	0	0.000
2008-5-19	428	1	0.2336	1	0.234
2008-5-20	534	2	0.3745	0	0.000
2008-5-21	908	3	0.3304	1	0.110
2008-5-22	785	1	0.1274	0	0.000
2008-5-23	385	1	0.2597	1	0.260
Total	9425	25	0.2653	7	0.074

during system installation. The phenomenon of testing is basically accordant with the phenomenon presented in the after-sales DOA.

The test results show that installation without the online system driver will greatly influence the after-sales DOA. It is also shows that the previous test is not entirely reasonable.

Based on DOA indicator analysis, control chart analysis, hypothesis testing, and cause and effect analysis, the main reasons for the undesirable level of DOA is identified. For the main reason of DOA, transport simulation, drop test, and comparison are introduced. The detailed methods of DOA analysis related to the quality problem is formed.

7.4   Defects analysis based on MTBF of laptop computers

MTBF refers to the interval between two failures of the products, or it can be referred to as the durability that is reflected by the failure during use after the product was checked by the customers. It is an important evaluation index to measure the quality level of a laptop computer. Conducting MTBF-oriented analysis of quality problems of laptop computers is an important part of quality control of enterprises. We started from after-sales lifetime data of laptop computers of X company, built a lifetime distribution model upon laptop computers with relevant failure characteristics, and estimated model parameters. Then we found the main reasons that influence the MTBF.

7.4.1 Reason analysis on laptop computer quality defects that influence MTBF

The reasons for variation of MTBF also line up with six aspects of personnel, materials, methods, machines, measurement, and environment. We conducted cause–effect graphing analysis on these six aspects as shown in Figure 7.5.

In conducting the above MTBF reason analysis, we learned that lots of factors influence the MTBF, but further analysis was still needed to analyze the main reasons. First, we conducted analysis, selecting, and statistics on failure of after-sales laptop computers that were sent for repair. The statistical results are shown as Table 7.15.

Statistical results show that disqualifications of main board, hard disk, LCD screen, audio speaker, and optical storage are much higher than for other components and account for 80% of total disqualification. The main board defect accounts for 55.12% of the total maintenance defects, which means that the main board will greatly influence the MTBF. Therefore, we take the main board failure as the key quality characteristic and conduct the analysis.

The framework of the laptop computer main board is according to the design code of IBM PC/AT, and the main components of the main board system are as follows: north bridge chip, south bridge chip, display chip, embedded controller (EC), BIOS chip, network card and sound card chip, power management chip, and various interface chips, etc. (shown as Figure 7.5) so as to realize functions such as mobile computing, mobile network, storage, display, audio, and multimedia, etc.

Analysis of the manufacturing process of the main board, from the main board schematic diagram designed for printed circuit board assembly (PCBA), two major manufacturing processes: printed circuit board (PCB) and PCBA are needed, and each of these two major manufacturing processes on average conclude 30–40 working procedures. Technological processes are various and complicated with sophisticated testing content. With OEM mode, these two processes are manufactured by different manufacturers. An increase of supply point has made the process control become more complicated, and this very easily causes the hidden trouble of main board quality.

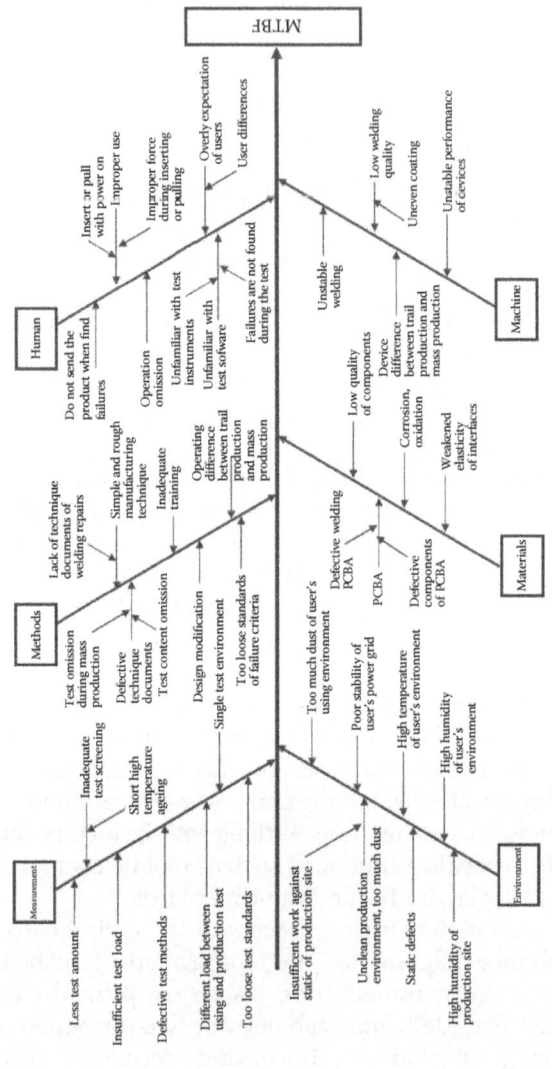

Figure 7.5 Fishbone diagram of low MTBF.

Table 7.15 Classification of after-sales failure of a laptop computer company in 2007

Type	Defective Qty	Accumulated defective Qty	Accounts for total defective Qty	Accumulated percentage
Main board	16,078	16,078	55.12%	0.00%
Hard disk	3027	19,105	10.38%	55.12%
LCD screen	2238	21,343	7.67%	65.50%
Audio speakers	1223	22,566	4.19%	73.18%
Optical storage	1201	23,767	4.12%	77.37%
CMOS battery	1086	24,853	3.72%	81.49%
Adapter	1045	25,898	3.58%	85.21%
Memory	879	26,777	3.01%	88.79%
Keyboard	836	27,613	2.87%	91.81%
Radiator	358	27,971	1.23%	94.67%
Battery	352	28,323	1.21%	95.90%
Touch board	224	28,547	0.77%	97.11%
Switch board	171	28,718	0.59%	97.87%
Webcam	150	28,868	0.51%	98.46%
CPU	135	29,003	0.46%	98.97%
Network card	91	29,094	0.31%	99.44%
Others	73	29,167	0.25%	99.75%
Total	29,167		100.00%	100.00%

With analysis of industrial characteristics after the completion of the laptop computer main board (PCBA), manufacturers of laptop computer main boards will assemble them into bare bones according to the requirements of OEM manufacturers, and transport bare bones to OEM manufacturers. The final assembly of products and OEM sales will be conducted by OEM manufacturers. The product chain is very complicated, and this has greatly increased the process uncontrollability, which has led to huge hidden trouble.

We conducted further analysis on defective performance of main boards as shown in Table 7.16. From the failure classification table, we can learn that failures, such as "unable to boot," "abnormal interfaces," "boot but nothing displayed," and "boot but frequent crashes," account for about 80% of mainboard defects. Therefore, we need to conduct some analysis on these main unqualified items.

According to principle analysis, main board failure of laptop computers can be divided into power failure, bus failure, and component failure. Power failure includes power failure and PWRGD signal failure, such as rechargeable batteries, principal voltage transformed by external power adapter, +3VAL, +5VAL, +V3.3S, +5VS, +V1.8S, +V0.9S, +V1.2S, and +V1.5S

Table 7.16 Classification of defective performance of main board

Items	Defective Qty	Accumulated defective Qty	Percentage	Accumulated percentage
Unable to boot	5721	5721	35.58%	35.58%
Abnormal interfaces	3626	9347	22.55%	58.14%
Boot but nothing displayed	1795	11,142	11.16%	69.30%
Boot but frequent crashes	1192	12,334	7.41%	76.71%
Boot, entering system with screen mass up	914	13,248	5.68%	82.40%
Abnormal audio during use	779	14,027	4.85%	87.24%
Power down/off during use	527	14,554	3.28%	90.52%
Blue screen during use	412	14,966	2.56%	93.08%
Abnormal installation of system and software	275	15,241	1.65%	94.79%
Auto reboot during use	266	15,507	1.51%	96.45%
Blank screen during use	242	15,749	1.23%	97.95%
Automatic shutdown	198	15,947	1.71%	99.19%
Others	131	16,078	0.81%	100.00%
Total	16,078		100.00%	

of the main board. Bus failures include failure of itself and failures caused by bus control. Component failures include failures of resistance, capacitance, inductance, and integrated circuit chip as well as various interface components in the main board.

7.4.1.1 *Analysis of reasons why main board is unable to boot*
According to principles of main board failure analysis, the main reasons why the main board is unable to boot (black screen occurred and power indicator is not on after pressing the power button) are as follows:

1. Circuit
   RTC circuit is used to keep the operation internal lock of the main board and ensure CMOS configuration information will not be lost

when power is off. ALWAYS circuit is mainly used to ensure the normal operation of EC. The adapter and battery power input to isolation circuit on the main board mainly supply the power to the whole computer system. The boot circuit boots the south bridge peripheral voltage, boots the CPU, conducts the power on self-test process. Any abnormities, such as open circuit or short circuit, occurring in these main board circuit systems will make the main board unable to boot.

To check whether the corresponding voltage measurement point and test signal of each circuit are normal or not, plug in the power to see whether voltage +V3.3AL, +V5AL of PWM circuit of power management is normal or not. Press the power button to see whether a low pulse will be sent to the EC controller, whether the voltage of system power and DDR power, +V3.3S, +V5S, +V1.8S, +V0.9S, is normal or not, and whether PWRBTN# between the EC controller and south bridge, RSMRST, PWROK, CPU Voltage +VCC_CORE, North Bridge Reset PLT_RST#, and CPU reset H_CPURST# are normal or not.

2. Component Damage and Defective Welding

Component damage includes power management chip damage or defective welding, BIOS chip damage, defective welding and BIOS internal procedure damage, MOSFET component damage, defective welding of power input and isolation circuit; gate circuit damage of the boot circuit, damage of triode and diode where power goes through, PWRBIN# diode damage, damage of voltage stabilizer of south bridge supply circuit or defective welding, south bridge damage or defective welding, north bridge damage or defective welding, and defective welding of CPU socket, etc.

7.4.1.2 Defective interface on main board

According to principles of main board failure analysis, reasons for defective interface on mainboard are as follows:

1. Circuit

The interface circuit is the fulcrum of each main board circuit. By principle, defects of an interface circuit make the interface function of this circuit unable to be normally used but will not influence normal use of other components. From the circuit diagram and PCBA, we can clearly see voltage and signal measuring points of the corresponding main board of each interface circuit. Conduct measurement according to the characteristics of each interface, and see whether the corresponding voltage and signal are normal or not.

2. Damage of Interface Components and Defective Welding

Each interface on the main board is the link between the main board and external devices. Abnormal functions are mainly reflected by two aspects: interface component damage and defective welding

Table 7.17 Failure statistics on each interface of mainboard

Interface type	Defective Qty	Accumulated Qty	Percentage	Accumulated percentage
USB	1044	1044	28.79%	28.79%
Power	865	1909	23.86%	52.65%
Network card	567	2476	15.64%	68.28%
Audio	299	2775	8.25%	76.53%
Keyboard slot	211	2986	5.82%	82.35%
Wireless card	198	3184	5.46%	87.81%
Battery	134	3318	3.70%	91.51%
CD Driver	108	3426	2.98%	94.48%
PCMCIA slot	67	3493	1.85%	96.33%
Memory slot	45	3538	1.24%	97.57%
LCD cable slot	36	3574	0.99%	98.57%
Hard disk slot	33	3607	0.91%	99.48%
VGA slot	19	3626	0.52%	100.00%
Total	3626		100.00%	

between the interface component and the main board. From further statistics on the failure of each mainboard component (Table 7.17), we learn that failures of USB interface, DC battery socket, network card interface, and audio interface are the top four failures of interface failures. On one hand, it is related to the quality of the interface component of the main board; on the other hand, it may be related to frequent use, times of insertion and extraction, and wrong insertion and extraction.

7.4.1.3 Boot with nothing displayed

According to the principles of main board failure analysis, reasons for "boot with nothing displayed" (black screen and power indicator is on after pressing the power button) are as follows:

1. Circuit

   After boot with nothing displayed, the power indicator is on, which shows the boot circuit of the main board is normal, then verify whether VGA interface and external display is normal or not. If it is normal, then the LCD display circuit is abnormal. If both the LCD and VGA interfaces with nothing displayed, then the join point between the north bridge and LCD and VGA or peripheral circuit of display card chip is abnormal.

   Measure weather the voltage of main board LCD display circuit, the LCD booster board, and the datum point of the mainframe interface circuit is normal or not. Measure memory voltage of the datum point; measure voltage jump of BIOS address cable and data cables such as A0~A15, D20~D24; measure whether signals of AC_Reset#, PCI_RST #, and CPU_RST # are normal or not; conduct debug card auxiliary test to see if there are reset signal, clock signal, chip select signal, and various voltage, etc.

2. Component Damage and Defective Welding

This includes defective welding of main board VGA interface or damage; defective welding or damage of mainboard LCD slot, defective welding of north bridge chip or damage, damage of display card or defective welding, component damage of peripheral circuit of display card; defective welding of memory slot, etc.

7.4.1.4   Boot with crash

According to principles of main board failure analysis, reasons for frequent crashes are as follows:

1. Circuit

Abnormal control circuit of memory chip; wrong input of delayer; unstable supply circuit of main board memory; increased impedance corrosion of circuit due to dust on main board.

2. Component Damage or Defective Welding

False welding of PIN memory slot, false welding of CPU socket, defective interface or false welding between keyboard and main board, improper or poor program of BIOS chip of main board, defects or defective welding of south bridge chip.

7.4.1.5   Frequent blurred screen

According to principles of main board failure analysis, reasons for frequent blurred screen are as follows:

1. Circuit

Abnormity of LCD display circuit of main board and abnormity of interface circuit, which is related to VGA.

2. Component defect and defective interface welding

Defective memory slot and damage of memory exclusion; defective welding of north bridge chip; defective contact or false welding between main board and LCD socket pin.

7.4.1.6   Other failure cause analysis

According to principles of main board failure analysis, detailed analysis on 29,629 pieces of main boards, which have poor performance and phenomena, such as unable to boot, abnormal interface, boot with nothing displayed, boot with frequent crash, and blurred screen. There are total 308 defects, and the results are shown as Table 7.18.

Table 7.18 Reason analysis on defective performance
of main board

Name	Qty	Defective Rate
Component damage (resistance, inductance, capacitance, diode, triode, field-effect tube, and IC, etc.)	121	39.29%
Defective welding of components	69	22.40%
Interface slot/socket defects	54	17.53%
Defective PCB board	38	12.34%
Trimming socket/slot	19	6.17%
Others	7	2.27%
Total	308	100%

Meanwhile, by further analysis, we found average MTBF of main board with component damage and defective component welding is 283 days, which is different from 330 days of average value of average MTBF by total after-sales maintenance statistics.

Above analysis shows that defective components and defective component welding are main reasons for defective performance of the main board. And these two defects have great influence on MTBF and need to be improved by corresponding measures.

7.4.2 *MTBF-oriented quality improvement and application of laptop computers*

MTBF-oriented quality improvement is quality improvement that is based on establishment of the laptop computer lifetime model, analysis of after-sales computer maintenance data, characteristics of OEM manufacturing mode, and practical application of enterprise manufacturing.

Weibull distribution can represent most lifetime distribution types, so it is selected for laptop lifetime analysis. The parameter is estimated by the original failure data of X company (omitted here), and approximate lifetime distribution of X company laptops is as follows:

$$F(t) = 1 - \exp\{-(t/\eta)^\beta\} = 1 - \exp\{-(t/230)^{0.89}\}, \beta > 0, \eta > 0$$

$\beta = 0.89 < 1$ indicates the failure function is decreasing, which means that the laptops are within the early failure stage during operation after sales. Another important conclusion is that defective components and defective component welding are two main reasons for defective performance of the main board. Such defects have the greatest influence on MTBF. As to the above analysis results, there are two ways that can be considered to improve MTBF. One is to extend the time of high temperature aging to eliminate products with initial failure so as to improve the MTBF of the whole machine; another way is to decrease the failure rate of components so as to improve MTBF of the whole machine. Here are evaluations of these two measures.

7.4.2.1 Evaluation of measures of extending time of high-temperature aging

The way of extending time of high-temperature aging needs to consider aspects, such as current manufacturing technique, time of high temperature, and cost, etc.

1. Technological Evaluation on Extending High-Temperature Aging

   To conduct the technological evaluation needed to consider the actual conditions of the production line, such as filed, the capacity of the high-temperature aging room, quantity of debugging, and feasibility of manufacturing techniques, etc.

   Related data on the current production of laptop computers at the company: The time of high-temperature aging screening is two hours, and the average production capacity of laptop computer is 85 sets/ hour. The max capacity of the high-temperature room is 352 sets of laptop computers, and the high-temperature heating equipment is designed according to the max capacity of the high-temperature room. The quantity of debugging cars is 51 units, and each car is with 18 sets. The vacancy rate of current high-temperature aging space is a bit higher.

   Consider the limitations of the production workshop; the high-temperature room cannot be expanded. Calculated according to current capacity, the longest time of high-temeprature aging when fully filled is four hours and no high-temperature devices need to be increased, but need to add 25 units of debugging cars, connecting devices of debugging cars, wiring board of power line, and transformation of debugging cars.

2. Comparison of Different Aging Screening Time

   Conduct test design for main models, K40A, T9, K400, and K411 with the same configuration for the same model, and accumulatively test 1000 sets under normal conditions. Count quantity of main boards that didn't pass the high-temperature aging. The results are shown as Table 7.19.

   According to annual production quantity, 131,500 PCs, screened defects after three hours aging is shown as Table 7.19:

   $$42 - 32 = 10 \text{ PCs}; 10/1000 = 0.010; 131{,}500 * 0.010 = 1315 \text{ PCs}$$

   Similarly, after four hours of aging, screened defective main boards are 2499 PCs.

Table 7.19 Quantity of unqualified main boards of each model after high-temperature aging (unit: PCS)

Time	K40A	T9	K400	K411	Average screening Qty
2 hours	34	31	32	29	32
3 hours	40	39	43	44	42
4 hours	52	50	53	49	51

3. Analysis on Cost of Quality

Calculation of costs that need to be increased.

Cost of debugging car device TC_{cost}: 25 sets of debugging car, with each cost ¥1500 (include costs such as power line and packaging, etc.)

Cost of debugging car $TC_{cost} = 25 * 15 = 37,500$ (¥)

Cost of electricity D_{cost}: two hours will be increased for adding the system driver installation. The power of a laptop computer is 65 w, calculated according to total production capacity in 2007; cost of industrial electricity is ¥1/per KWH, and 1 KWH = 1000 w/h.

Electricity Cost $D_{cost} = 131,500 * 2 * 65 * 1/1000 * 1 = 17,095$ (¥)

Human Cost R_{cost}: Increase of online obsolete mainframe, two maintenance personnel and one person for high-temperature room need to be increased. The human cost is ¥1500/month.

Human cost $R_{cost} = 12 * 1500 * 3 = 54,000$ (¥)

Increased cost: total costs of debugging car, electricity, and human cost are ¥108,595.

Decreased cost: Maintenance cost, annual cost 5.6 million/29,167 * 2499 = 479,803 (¥)

By comprehensive comparison, in terms of cost, this proposal is economical.

7.4.2.2   Evaluation of decreasing failure rate of components

Average types of current components of computer main board are about 200 kinds, and average quantity of components is 1200 PCs. Assuming that a reliability model of laptop computers is series model, evaluate λ_{Gi} of each component by the component counting method according to system failure rate λ_S.

$$\lambda_S = \sum_{i=1}^{n} N_i (\lambda_G \pi_\theta)_i$$

In formula

λ_{Gi}: general failure rate of the i-component:

$\pi_{\theta i}$: quality coefficient of the i-component:

N_i: quantity of i-component:
n: types of all components.

Assume that all quality coefficients of each component are 1.

The average MTBF of laptop computers of this company is 330 days. The average MTBF of benchmarking enterprises is 475 days as shown in Table 7.1. Calculate formula according to failure rate:

$$\lambda(t) = (\beta/\eta)(t/\eta)^{\beta-1} = 0.00387\left(\frac{t}{230}\right)^{-0.11}$$

Failure rate $\lambda_{S1} = 0.00372$, $\lambda_{S2} = 0.00211$, as shown in Figure 7.6.

Assume the quantity is the same, and obtain $\lambda_{G1i} = 0.000003099$, $\lambda_{G2i} = 0.000001754$. This means it improves the quality of each component to at least 1.8 times as high as its primary quality. The MTBF of the product can be improved by decreasing the failure rate of components.

The above two evaluation methods, considering the OEM mode of laptop computer manufacturing and involved quality improvement of multistage suppliers, current testing equipment of enterprises is not perfect, and their collaboration level with upstream suppliers is quite low. Implementation of a decreasing failure rate of components is uncontrollable. The way to improve MTBF of the whole machine by extending time of high-temperature aging is more practicable. At present, this improvement measure has been applied to T9, the new model that is developed in August, and specific improvement of MTBF should be verified further.

7.4.2.3 Quality improvements of component welding defects

The section of MTBF-oriented analysis on quality improvement of laptop computers has elaborated the important influence of defective welding of components on the reliability of the main board while defective welding of components is directly related to the PCBA manufacturing process. As to MTBF-oriented quality improvement, which is dominated by OEM

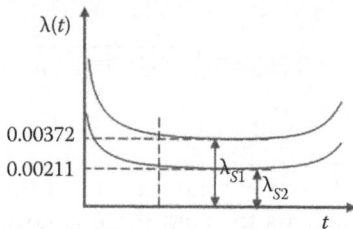

Figure 7.6 Comparison of defective rate of laptop computer.

manufacturers, it is necessary to conduct deep analysis of the PCBA welding process and then provide operable quality improvement measures, which are accepted by suppliers, and also can play a supervisory role.

PCBA manufacturing processes of laptop computer: Solder paste printing, surface mounting, reflow soldering, solidification, turnover, solder paste printing, surface mounting, reflow soldering, solidification, reversing, manual insertion, wave-soldering, cleaning, and inspection, etc. Solder paste printing, surface mounting, and wave-soldering are three key techniques of the PCBA process and have a key influence on the reliability of main board.

According to analysis of three key technique factors of the PCBA manufacturing process, to improve the quality of solder paste printing, to optimize technological parameters of reflow soldering and wave soldering is the direction of quality improvement of reliability of component welding. Considering the characteristics of production mode OEM of laptop computers, and no fixtures and test equipment for inspection and testing on PCBA welding, thus the problems of defective welding of components have to be reported to upstream suppliers, and ask them to conduct corresponding quality improvement. According to a regular supplier audit and SMT process control and requirements of comparative suppliers, strengthen control on the welding process and build a control chart. On-site audit, understand control of the welding technology by analysis of quality problems so as to ensure the stability of quality during welding.

Collect production data of T9 of this company, and calculate corresponding control limits of p chart as shown in Table 7.20. Prevent by control chart, so as to ensure stability of welding of the PCBA process.

$$\text{UCL} = \bar{p} + 3\sqrt{(1-\bar{p})\frac{\bar{p}}{n}} = 0.055 + 3\sqrt{0.055 * 0.945/n} = 0.055 + 0.6839/\sqrt{n_i}$$

$$\text{CL} = \bar{p} = 0.055$$

$$\text{LCL} = \bar{p} - 3\sqrt{(1-\bar{p})\frac{\bar{p}}{n}} = 0.055 - 3\sqrt{0.055 * 0.945/n} = 0.055 - 0.6839/\sqrt{n_i}$$

According to the situation of the sample (produced on day 15), the process is out of control. Analyze the reasons for this, find out the abnormal factors, and take actions to avoid these abnormal factors, and then repeat the above steps until the process is stable. At this moment, the P chart can be applied to daily production management as a control chart.

Table 7.20 Defective welding data of model T9 and calculation of *P* chart in August

Date	Sales Qty	Defective Qty	Defective rate	CL	UCL of *P* diagram	LCL of *P* diagram
1	1748	73	0.042	0.055	0.071	0.039
4	717	39	0.054	0.055	0.081	0.030
5	1662	110	0.066	0.055	0.072	0.038
6	1282	70	0.055	0.055	0.074	0.036
7	1622	89	0.055	0.055	0.072	0.038
8	718	41	0.057	0.055	0.081	0.030
11	1647	88	0.053	0.055	0.072	0.038
12	1463	72	0.049	0.055	0.073	0.037
13	3225	214	0.066	0.055	0.067	0.043
14	3149	145	0.046	0.055	0.067	0.043
15	6958	462	0.066	0.055	0.063	0.047
18	4418	233	0.053	0.055	0.065	0.045
19	2347	111	0.047	0.055	0.069	0.041
20	3065	165	0.054	0.055	0.067	0.043
21	3090	132	0.043	0.055	0.067	0.043
22	2962	177	0.060	0.055	0.068	0.043
23	3076	196	0.064	0.055	0.067	0.043
25	3088	175	0.057	0.055	0.067	0.043
26	2560	142	0.055	0.055	0.069	0.042
27	2874	162	0.056	0.055	0.068	0.042
28	4179	201	0.048	0.055	0.066	0.045
29	2032	93	0.046	0.055	0.070	0.040
Total	57,882	3190	5.511%			

7.5   Experience and lessons

This case is from lower product quality and high after-sales service costs, which exist in operation of an OEM laptop computer enterprise. Analyze the influence of brand awareness, negative factors of the upstream environment, price war, and marketing strategy on quality of product. On the basis of the summary and analysis of DOA and MTBF data, be customer-oriented and take benchmarking enterprises as targets. The DOA- and MTBF-oriented improvement measures have been established and been applied to practical production and achieved the target of quality improvement and operation cost optimization. As to DOA, with comprehensive cost and operability, we have put forward the quality improvement measures and corresponding code requirements and pre-liminarily formed specific improvement measures of DOA. Finally, apply these quality improvement measures to actual operation, and DOA of the enterprise has been obviously decreased. As to MTBF, we found defective components and defective welding of components are two major reasons that influence the MTBF. Specific measures have been put forward and improved the reliability.

By statistics and analysis of DOA and MTBF data, we have confirmed the lifetime distribution law of laptop computers, got a deeper understanding of the production, system, and supply chain as well as quality improvement and expanded the idea of quality analysis and improvement. By implementation of measures that are oriented in DOA and MTBF, we have confirmed the cost analysis and comprehensive consideration of measures. The quality improvement measures of this chapter have been preliminarily applied in actual production of laptop computers and have obtained a good effect on quality improvement.

References

1. Yang, X. Y. Vertical Integration Research for Taiwan Industrial PC Market: Master thesis. TaiWan: Chung-Hua University, 2007.
2. Nicotin. Consumer Survey of Laptop Computer In Spring, 2007. http://www.sina.com.cn http://www.pconline.com.cn, 2007.2.15.
3. L. Xianping et al., OEM King. Economy of Shenzhen Special Zone. 2007.5 28–43.
4. Internet Weekly. Lenovo VS MTBF. http://www.sina.com.cn 2005.04.14.
5. Nassar, S. M.; Barnett, R. IBM Personal Systems Group. Applications and results of reliability and quality programs: Reliability and Maintainability Symposium, 2000. Proceedings. Annual Volume, Issue 2000, Page(s): 35–43.
6. R. Meyers; M. M. Roffman. PRAISE: A PC-based reliability evaluator: Reliability and Maintainability Symposium, 1989. Proceedings, Annual 24–26 Jan 1989, 491–494.
7. State Bureau of Technical Supervision. GB9813-88. General Technical Specification of Microcomputer. Beijing: China Standard Press, 1988.
8. H. Furong. Modern Quality Management. Beijing: China Machine Press, 2007.8, 393–401.
9. L. Shi, Applied Statistics, Beijing: Tsinghua University Press, 2005.

chapter eight

China's implementation of manufacturing management reform

Li Mo and Xiguang Hu

Contents

China South Industries Group Corporation (CSGC) is the giant SOE incorporated with the approval from the State Council, the investment institution, and assets management entity and one of the most vigorous civil–military giant industries. In 1999, when CSGC was incorporated, its operating income was only RMB 21 billion yuan with close to RMB 2 billion yuan, and it was known as one of the most struggling enterprises in the national defense industry. In 2013, CSGC annual sales revenue hit RMB 36 billion yuan with more than 10 billion in profits, ranking fifth among China's Top 500 manufacturers, 22nd among China's Top 500, and 209th among the World's Top 500, marking its elevation to the most influential civil–military giant group worldwide. In this amazing change, a package of management innovation and reform, as represented by manufacturing management reform, has played an irreplaceable role.

To date, CSGC has owned more than 40 industrial manufacturers, such as Changan Automobile, Changan Industry, Tianwei Baobian Electric, Tianwei New Energy, China Jialing, and Jianshe Motorcycle, and more than 10 R&D institutions and other entities, which cover special equipment, automobiles, motorcycles, new energy, electric transmission and transformation equipment, photoelectricity, pharmaceutical chemicals, etc.

The annual sales of automobiles and engines reaches 2.6 million sets, ranking among the biggest in China's automobile industry, which is known as the largest producer with the most extensive production bases among the self-owned brand of automobile groups. According to the statistics released by FOURIN in October 2013, one researcher specialized in the world automobile industry, Changan Automobile under CSGC sold 1.22 billion cars of self-owned brands in 2012, ranking first nationwide and 14th worldwide, making it the first Chinese self-owned brand for seven years in a row. Also, Changan Automobile is the only Chinese automobile manufacturer with the self-owned brand to hit one million. Its motorcycle can be represented by famous brands, such as Jialing, Jianshe, Dayang, and Qingqi, with annual sales reaching 6 million sets, ranking among the world's best for its industrial scale. With a view to power transmission and transformation, CSGC occupies more than 10% market share of large transformers, with ultra-voltage transformer technology among the world's best, making it well known as one of the transformer suppliers with the largest coverage of voltage classes, the most product categories, and the largest output of transformers per factory in the world.

On the one hand, CSGC sticks to technological innovation, pushes innovation progress, which is awarded as the national-level innovative enterprise, with six research institutes, six national-level R&D centers, and 37 provincial-level technological centers; on the other hand, CSGC performs lean manufacturing management to push forward management innovation and transformation. Technological innovation and management innovation are complementary and promote one another, which leads to good and rapid development.

8.1 Background of manufacturing management reform of a lean production (LP)-based diversified large group

8.1.1 Enhancing technological development capacity and modern management

Since CSGC was incorporated in 1999, its rapid growth and enhanced technological innovation never stops. However, these achievements don't mean improved internal management as shown by no standard integrated management system and assessment criteria, chaotic management, differential management level, nonstandard management flow, low efficiency of total management; no ideology and methodology of modern management, inattention of process optimization and management improvement, extensive internal management, weak management, uncontrolled extravagance, over-consumption of irrational factors. These drawbacks threaten

CSGC's survival and development. CSGC as specified in manufacturing should feel more stress, sense of responsibility and mission, and quicken and push the lean manufacturing management despite the long process of management reform and innovation. With due effort, CSGC should have its production method and management method transformed from extensive production to lean production, from traditional mode to modern mode, and from too much dependence on factor input to management innovation and laborer quality improvement. In that regard, LP-based manufacturing management reform is required for enhancing the capacity of technological development, securing the national defense, and fulfilling the economic, political, and social responsibility. As a civil–military enterprise, CSGC has a good management system as a military enterprise with consolidated and accumulated advanced managerial experience, so it should be able to push for the integration and fusion of its manufacturing management system.

8.1.2   Adapting to the intensified market competition, cost reduction, and efficiency improvement

The cost of the majority took a large proportion of sales revenue, some hit 100%, and some hit 110% with low profitability, and some even suffered loss, which could be attributed to bad internal management, resulting in cost that were too high and irrational expenditure. In contrast with those advanced enterprises, they had underperformed in the effect of lean production management, including efficiency improvement, quality improvement, logistics optimization, management information construction, and quick response to the market. Particularly in response to the negative impact of the external environment, for example, the financial crisis, domestic and foreign enterprises act to enhance lean manufacturing management. For CSGC, which specializes in manufacturing, it has been more squeezed by cost stress and price competition stress than has ever been seen. The cost stress is indicated by the drastic volatility of raw material prices and high-level operation, for example, steel, bronze, coal, crude oil, and rubber. Although price competition stress features the domestic capacity expansion and intensified peer-to-peer competition, periodic excess capacity can be found in automobiles, motorcycles, power transmission and transformation, and new energy. In this concern, LP-based manufacturing management reform is also required for improving quality and efficiency, responding to various crises, and promoting sustainable development.

8.1.3   Development to be among the world's best and facilitating soft power

World famous enterprises should further strengthen the construction and implementation of internal management systems. In addition, according to common experience, world famous enterprises should perform well in technological innovation, brand building, market share, team building, information construction, and risk control. National policy also provides for the stimulation of technological progress of economic development, laborer quality improvement, and management innovation and transformation with a push for thrifty intensive development; reducing product cost and price; enhancing customer satisfaction; and shouldering economic, political, and social responsibility. CSGC is committed to fulfilling the mission to "safeguard the national defense, develop to be strong and enrich people." CSGC sets optimizing group control, promoting specified management, and implementing international development as strategic policy to attain the goal of establishing itself as a modern group with international competitiveness. To this end, excellent internal management should be the fundamental requirement, and a managerial level among the world's best should be secured. Lean manufacturing management, which is a tailored approach for manufacturers with the most appropriateness and best efficiency, should be the only choice for CSGC to exercise management innovation, according to the realities. It has always been evidenced that LP-based manufacturing management reform with characteristics can increase the strength, improve the operating efficiency, and enhance international competitiveness.

8.2 Content and approach of manufacturing management reform of lean production (LP)-based diversified large groups

Lean production (LP), invented after World War II, is a kind of brand-new management ideology and methodology, which is seen as the advancement of bulk and stacking production, as represented by U.S. enterprises. By now, LP has been widely recognized and practiced by world industries, from which Japan, the United States, Brazil, and China, either in civil industry or military industry, have been greatly rewarded with big success, which, in turn, injects vitality into LP.

Based on LP, CSGC borrows managerial experience from domestic and foreign enterprises with proven success, by which LP is enabled in every respect or on every level, and thrift, intensity, standards, and high efficiency are in LP DNA. Further, the manufacturing model and management model can be extended in two directions of the internal value chain and industrial chain.

In LP-based CSGC manufacturing management reform, the LP idea and awareness are essential; a modern manufacturing management system is a top-level design. The brains behind the development of LP management is the guarantee; waste elimination, cost reduction, and efficiency improvement are the essences; the enhanced basic management is the foundation; Kaizen is the core; 3-D synergy advancement is the approach; the enhanced competitiveness to the maximum is the objective; a harmonious win–win result among interested parties is the highest hierarchy of the management system.

From 2005 onward, CSGC has vigorously implemented LP in Changan Automobile, Chongqing Tsingshan Industrial, and other key industries. Since 2008, 40 industrial enterprises have exercised lean manufacturing management as per the PDCA approach, enhanced the top-level design, tailored the action plan and program, deployed careful arrangement, reinforced the implementation mechanism (e.g., evaluation and incentives), enhanced the lean manufacturing management, continuously improving the LP-based manufacturing management system, with evident achievements obtained and economic, political, and social responsibility fulfilled. By doing so, CSGC's core competitiveness has been enhanced as shown in Figure 8.1.

8.2.1   Guidelines, principles, and objectives of LP-based manufacturing management reform

CSGC has suggested the guidelines, principles, and objectives of LP-based manufacturing management reform as prompted by the evident contradictions and problems of manufacturing management in all businesses, according to the overall survey, summary, and study of more than 40 industrial enterprises.

8.2.1.1   Guidelines
With LP fully implemented, the transformation and upgrading should be made from traditional to modern, from extensive to lean management, and accelerate these transformations to the lean manufacturing management model. With this model extended to all nodes of the internal value chain, such as design development, planning, finance, HR, sales, after-sales, and purchase, further expanded to its partners, either upstream or downstream, the ideological reform of lean manufacturing management has been promoted and lean enterprise alliances with the core of major businesses and key enterprises can be formed, hence developing CSGC into a modern group with technological development capacity improved and international competitiveness enhanced.

8.2.1.2   Principles
1. Centering on high-end manufacturing, that is, transformed from low end to middle or high end, do our best to ensure manufacturing with high added value and high efficiency.

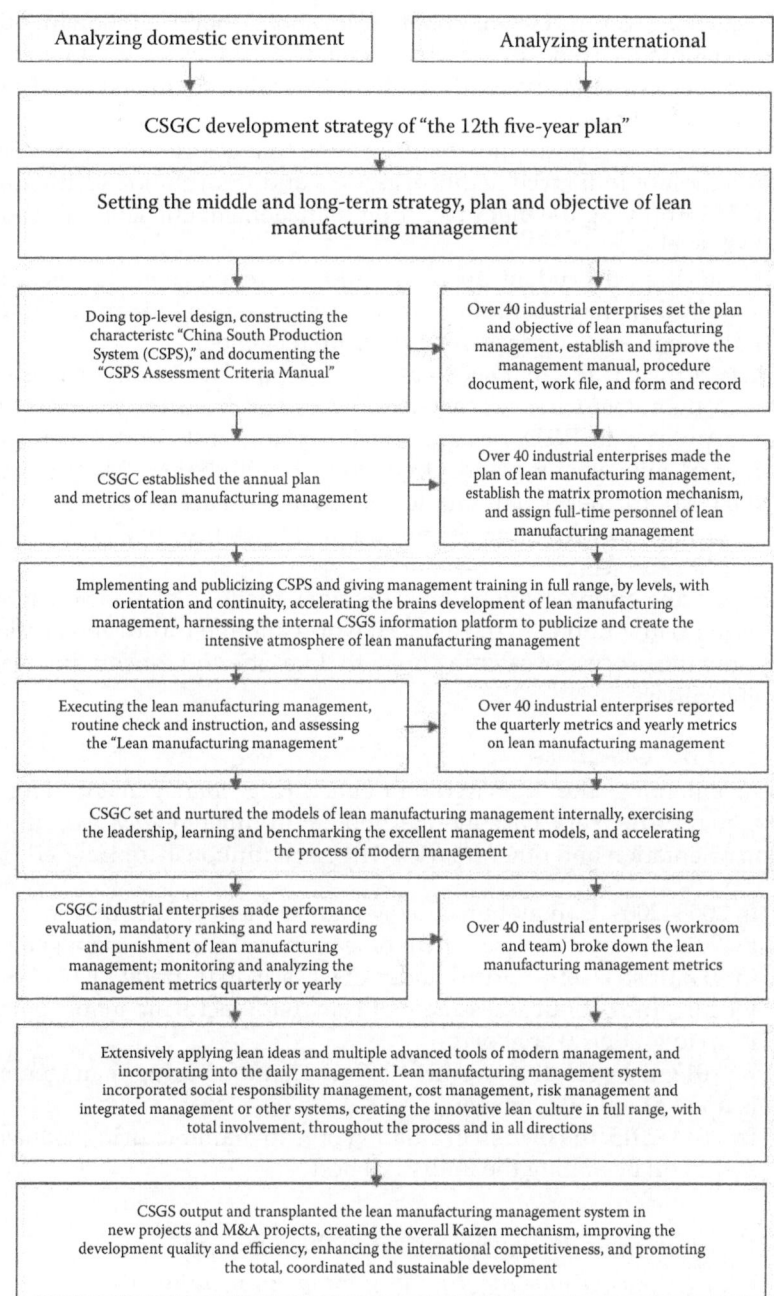

Figure 8.1 The mechanism and workflow of the CSGC manufacturing management system.

2. Centering on service-oriented innovation, that is, following the development trend of service-oriented manufacturing, enhancing the service management throughout the value chain and satisfying customer demand.
3. Centering on information management, that is, promoting the combination of industrial engineering (IE) and information technology (IT), improving the manufacturing management efficiency and control level.
4. Centering on prudent operation and risk management, executing standard management and process optimization, enhancing internal control and strengthening basic management.
5. Centering on the management of the total value chain, extending lean manufacturing management gradually from the manufacturing process to R&D, design, logistics, sales, cost control, environmental management, and Occupation Health Safety Management (OHSM), further expanding to the front and back of the industrial chain and applying it to the management practices in a new project or M&A project.
6. Centering on international management, that is, global vision, managing with outlook to the modern world and the future and accelerating the process of catching up with domestic and foreign first-class enterprises.

8.2.1.3 Objectives

By implementing the LP-based manufacturing management reform, safety, quality, delivery, cost, morale, and environment (SQDCME) information orientation and other metrics will be continuously optimized.

1. In 2008–2009, lean manufacturing management is fully driven.
2. By 2010, the overall assessment of lean manufacturing management in 40 industrial enterprises under CSGC are up to grade 1.5 or above.
3. By 2011, the overall assessment of lean manufacturing management is up to grade 2.0 or above.
4. By 2012, the overall assessment of lean manufacturing management is up to grade 3.0 or above.
5. By 2013–2015, the overall evaluation of lean manufacturing management will be among the industrial best.

8.2.2 Strengthening the organizational leadership of LP-based manufacturing management reform

CSGC has suggested three functions of headquarters, namely, strategic decision, organizational command, and management service. To this end, group control, professional management, and international development should be implemented as soon as possible. In response to these require-ments, insufficient management, weak basic management, uncontrolled extravagance, and high operating costs, the CSGC party group has made a package of deployments, urging all industrial enterprises to further implement the LP-based manufacturing management system, by releas-ing annual reports, giving instruction, checking and evaluating, and demanding the continuous optimization of SQDCME and other metrics. Further, CSGC established a lean manufacturing management office at headquarters, which functions as the management institution for daily work and implementation.

At the CSGC headquarters, each department should be highly aware of the necessity and urgency of promoting and implementing the LP-based manufacturing management system along with the implementation of lean manufacturing management, in an attempt to transplant LP ideas into the day-to-day work. The CSGC lean manufacturing management office should reinforce the organizational implementation, gradually establishing regulations and standards and improving, executing, man-aging, and evaluating the mechanism. The department of development and planning should inject lean ideas into project construction, ensur-ing lean planning. Finance is required to implement the combination of LP and cost leadership as well as lean manufacturing management and budget management. Human resources should do their best to change the management awareness of leaders and implement the combination of lean manufacturing management and training. Science technology and information is required to promote the combination of IE and IT as well LP and lean R&D design. The department of the west-south area should give more routine instructions to 16 enterprises under CSGC, which are distributed in southwest China.

Leadership focus is essential in promoting and implementing the manufacturing management system. Lean manufacturing management is a reform starting from the leaders. The confidence, determination, perseverance, and driving force of leaders, particularly of the "leader of leaders" are of vital importance. In other words, it will make no dif-ference, provided the "leader of leaders" doesn't organize, involve, or effectively allocate various resources, duly shooting all kinds of troubles. As per CSGC's request, all the leaders of more than 40 enterprises have paid much attention to the management reform with lean manufactur-ing management–leading team, the LP office/department, the task force/cross-functional force, and the coordinator or other promoting institu-tions built. By doing so, the matrix-structured organization is well estab-lished with clear responsibilities and is effectively run.

As per CSGC's request, more than 40 enterprises have made clear the responsibilities, which also implement the policies of lean manufactur-ing management accordingly; prepared an outline of lean manufacturing management in a Gantt chart every year; and specified the input, process, and output, including content, objectives, departments concerned, and work node. Further, each industrial enterprise has enforced the work of lean manufacturing management into the performance assessment system accordingly, with incentives and restraint mechanisms. With the PDCA method adopted, thanks to the evaluation, assessment, regular meeting, and case release and other mechanisms established, each enterprise creates a virtual competition in personnel management. For instance, some vigorously award the outstanding employee with the "Horse Prize" and motivate the underachievers with the "Snail Prize" in an attempt to motivate others and create a virtual competition among them in respect to corporate management.

8.2.3 Building a LP-based manufacturing management system

A well-established manufacturing management system is the key to promoting lean manufacturing management, and a system is the support of total implementation of lean manufacturing management. CSGC borrows managerial experience from international giants. In this way, CSGC is oriented toward lean management on the total value chain, which always emphasizes the construction and upgrading of the LP-based manufacturing management system. The construction and implementation of CSGC's LP-based manufacturing management system goes through the process from copying foreign standards to tailoring the one with China's enterprise realities considered, from the trial in the pilot enterprise to the full implementation in the group, and from the single automobile manufacturer to diversified industries.

8.2.3.1 Constructing and executing CPS
With the effort of CSGC, Chongqing Changan Automobile constructed "Changan Production System" (CPS) in 2005, which contains 12 management modules in relation to manufacturing production, including logistics and quality. This system was well tailored for the management realities of Changan Automobile, which borrowed the managerial experience of a joint venture. In early 2006, CPS was applied in the No. 3 Plant, No. 4 Plant, No. 5 Plant, and other manufacturing plants with good effects obtained.

8.2.3.2 Constructing and executing SPS
In December 2005, China South Industries Motors Company (CSIMC) was incorporated with the assets of Changan Automobile vehicle and parts integrated. To suit the management needs, CSIMC constructed "South Production System" (SPS). This system contains 10 management modules, which is an enhanced version of CPS, as tailored for the vehicle, parts, and

machinery manufacturer. Later, SPS was pilot run in seven enterprises, including Chongqing Tsingshan Industrial, with good effect obtained. Seven pilot enterprises have been set as models of CSGC LP.

In 2008–2010, CSGC urged more than 40 members to implement LP management as per SPS.

8.2.3.3 Constructing and executing CSPS

In 2010, according to the examined, accumulated, exercised, and strengthened internal manufacturing management system in a few years, CSGC expected to be among the domestic or world's best in the manufacturing industry, based on management theories such as LP ideas, system theory, industrial engineering, and internal control. CSGC borrowed the expertise of advanced manufacturing systems and managerial experience from Japanese, U.S., German, and other world-famous companies. CSGC combined its realities with management tradition and tailored a LP-based manufacturing management system with CSGC characteristics for the management needs of automobile, motorcycle, power transmission and transformation, new energy, special products, and other industries or products.

Earlier in 2014, CSGC established the task force and office to work out the system proposal. For almost one year, more than 30 chief experts on manufacturing management examined and discussed this proposal several times, ending up with the CSGC manufacturing management system and standards studied, integrated, and constructed, known as the latest version of China South Production System (CSPS). CSPS is composed of assessment criteria and utility software, which can be further divided into 12 management modules as shown in Table 8.1, written in 250,000 Chinese characters.

As per CSPS, all industrial enterprises under CSGC, including special products, automobile, motorcycle, power transmission and transformation, and new energy, have processed, constructed, and reinforced the multilevel system documents of LP management (e.g., management manual, procedure document, work file, form, and record as shown in Figure 8.2), and initially merged and continuously optimized multiple management system standards. More than 40 packages of manufacturing management standards and procedures have been worked out as valuable management wealth for industrial enterprises.

CSGC also launched the standard procedure of management assessment. After releasing and executing CSPS, CSGC was committed to finishing new tasks, for example, how to effectively implement and execute CSPS as per PDCA, how to measure it in the evaluation process, and how to ensure the impartial and honest disclosure of the results of management assessment. Expecting to stand in the forefront, CSGC launched the standard procedure of management evaluation, organized a panel of internal

Table 8.1 The content and structure of CSGC Industrial Manufacturing Management System (CSPS)

Seq.	"12 modules"	Basic objectives	Remark
	General	Suggest accelerating the process of modern management	Each module basically includes: Note, management procedure, main content, assessment content, management tool and management technology, cost control and Kaizen, etc.
1	Leadership	Management at all levels understands, emphasizes, and effectively organizes the execution of lean manufacturing management, and promote the continuous improvement of management metrics	
2	Brains development	Develop a team of experts and a team of improvement masters and multi-skill operators	
3	Team building and management	Solidify the basic management and effectively execute the management	
4	Image building and code of conduct	Develop the standard image language and efficient behavioral model	
5	Manufacturing engineering and process control	Standardize the project management and improve the manufacturing process	
6	Equipment and fixture management	Improve the equipment perfectiveness ratio, availability ratio, enhance the overall equipment efficiency and production security	

(Continued)

Table 8.1 (Continued) The content and structure of CSGC Industrial Manufacturing Management System (CSPS)

Seq.	"12 modules"	Basic objectives	Remark
7	Quality management	Effectively run quality management system and improve the quality qualified ratio	
8	Logistics and plan management	Scientifically plan, optimize logistics, and reduce the logistics cost	
9	Cost control	Decrease cost and increase benefit in full range, with total involvement, throughout the process and in all directions	
10	Information construction and application	Synchronize the industry and information and improve the management efficiency	
11	Occupation Health Safety Management (OHSM)	Stick to the human-centered principle, set "zero accident" as the objective, and ensure safety	
12	Environment management	Apply the low-carbon technology, execute clean production, realize green development, and establish the "resource-economical and environment-friendly society"	
	Definitions and practical tools	Offer the method and tools to push for the lean manufacturing management	

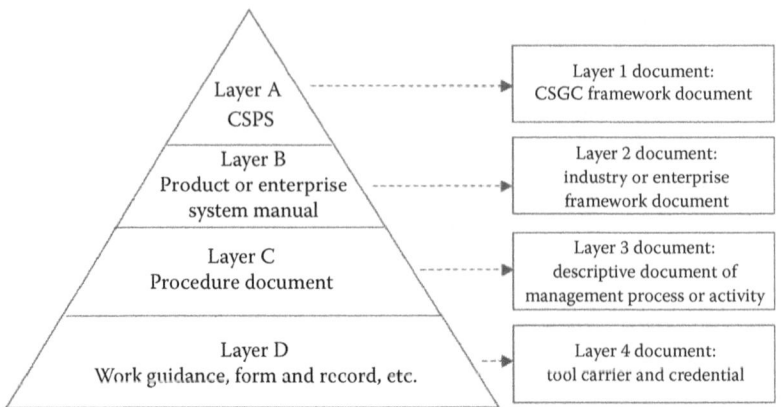

Figure 8.2 CSPS-centered document structure.

and external experts, and documented work manuals of management assessment (e.g., evaluation template, notes, and quantitative evaluation metrics). Further, with the evaluation panel trained, these experts understood the method and procedure of this evaluation, which led to the implementation and application of advanced managerial ideas and methods.

8.2.4 Driving the LP-based manufacturing management system

8.2.4.1 Establishing and reinforcing the plan, executing the strategy, and orderly promotion

As per PDCA, a complete action program, including ideology and objectives, shall be tailored before implementing and executing the manufacturing management system. CSGC made and released the "12th five-year plan of management innovation" about the implementation and execution of the manufacturing management system, including *Promotion Program, Evaluation Method, Annual Assignment, Management Instruction Opinion of Star Lean Production Team, Management Method of Experts Panel*, etc. In this effect, the management objectives, action measurements, and supporting measurements were suggested. Each industrial enterprise established the blueprint, plan, and essentials for implementing and executing the manufacturing management system, and they were gradually extended and promoted to the team and workroom.

8.2.4.2 Developing and setting the example of management, publicizing, and exchanging experience

Further, CSGC executed the reform and implementation of LP-based manufacturing management and set some models of internal management in order to exercise the modeling effect and typical guidance. In this effect, some models of lean manufacturing management also appeared, for example, Chongqing Changan Automobile Corporation, Chongqing Tsingshan Industrial, China Jialing Industrial, Chongqing Jianshe Motorcycle, Sichuan Jianan Rear Axle Automobile, Baoding Tianwei Baobian Electric, etc. For example, Changan Automobile pushed forward the merging and information construction of the manufacturing management system. With the basic system showing output and good effect, Changan Automobile has developed into the largest automobile manufacturer with the most self-owned brands. Tsingshan Industrial consistently executed lean manufacturing management with quick process and thorough implementation, creating the strong culture of lean manufacturing management, particularly of pull production, one piece flow, standard operation, and standard management (stock turnover ranks the first in this industry), making it the largest transmission manufacturer. Enterprises including Baoding Tianwei and Jianan Rear Axle Automobile had a well-established and characteristic management team with good effect (5S, value flow analysis, and other management tools were outperformed), supporting the rapid growth and making them known as domestic giants.

CSGC released the typical cases and exchanged experience in multiple forms, including teleconferencing, on-the-spot meeting, information platform, and print media, in a vigorous attempt to explicitly share the implicit knowledge, share the explicit knowledge, effect the shared knowledge, and construct the learning-based enterprise. Led by the management models, other industrial enterprises actively promoted lean manufacturing management by site visits and group discussions, unveiling a lot of optimum operating methods.

To promote the exchange of management experience among members, CSGC prepared the well-documented Experience and Case Study in CSPS Promotion, which expounds on more than 40 cases and the experiences of more than 30 enterprises (written in more than 300,000 characters). This document totally summarized the typical management cases and optimum operating methods in recent years, which is not only a material of experience exchange but also a training material. CSGC lean production management has proven to be effective as indicated by the overwhelming popularity among all industrial enterprises.

8.2.4.3 *Enhancing and effectively promoting the construction and management of the "Star Lean Production Team"*

Upon the request of the State-owned Assets Supervision and Administration Commission (SASAC) under the State Council, CSGC kept focused on team building and management as indicated in the compiled "Management Guideline of Star Lean Team" and "8 tasks" of team management in respect to lean manufacturing management, including basic team management, safety, health and environmental protection management, team quality management, team standard operation management, total productive maintenance management (TPM), team cost management, team training management, team rationalized proposal, and Kaizen management, thereby promoting the total employee involvement, continuous consolidation, and improvement of management. Further, the enhanced team training developed a lot of LP experts, improvement masters, and multiskill operators. In Chongqing, three terms of "Training Class of Lean Manufacturing Management and Star Lean Production Team" had more than 430 attendances. Under the efforts of SASAC and Tsinghua University, "Remote Training of Team Leader Management Capability Qualification" was given with 232 attendances accumulated. By reviewing and applying, 14 red flag classes and 13 outstanding employees were awarded by SASAC. With effective organization, CSGC enterprises had established and improved some 20 pieces of "Team Management Manuals," promoting the construction of Star 1–5 Lean Production Classes. Each enterprise continuously applied standard operation, value stream mapping (VSM), total production maintenance (TPM), single minute exchange of die (SMED), Pokavoke, visual management, and other tools of lean manufacturing management, improving management efficiency.

The sustainable development of LP cannot be possible without the total involvement. In this concern, CSGC urges every employee or organization to exercise enthusiasm, initiative, and creativity, which are the cornerstones of successful LP. Some enterprises practiced the total involvement idea—"Lean Production is a careful work, while improvement should be made by everyone"—enhancing the positive incentive and exercising the creativity and motivating the enthusiasm of every employee. By rationalized suggestion or improvement proposal, aided by advanced management tools and management technology, management optimization is enabled, solidifying the construction of the Star Team, promoting the on-site Kaizen management, and continuously improving the managerial level.

8.2.4.4   Conducting a wide range of training, publicity, and learning, enhancing the cultural influence

The implementation of a LP-based manufacturing management system involves complex systems engineering, which consists of corporate management, technological correction, and total employees. A lot of training, publicity, and education should be conducted before establishing the employee's awareness of LP, mastering the management technology of LP. To this end, the implementation of lean manufacturing management technology should be enabled by sticking to the ideology of talent development, emphasizing training, and highlighting training as the basic assurance of lean manufacturing management. Also, training is essential to the employee's career development, which demonstrates the important role of brains.

Every year, CSGC conducted a wide range of oriented training in multiple levels, including classroom training, on-site simulation, Doujou training and consulting training. By now, 130,000 attendances have been accumulated. Some members under CSGC promoted managers internally from those who have working experience in the lean manufacturing management department (office). By doing so, the lean manufacturing management department has cradled the intermediate and senior management brains. Further, CSGC has made public the concept, action, experience, and case study in relation to lean manufacturing management via China Ordinance News, the group website, internal office website, and its members' website, creating a strong atmosphere.

CSGC also strengthened the management communication and cooperation with Tsinghua University, Beijing University of Aeronautics and Astronautics (BUAA), social management and consulting institution, taking in ideas and manpower, in a joint effort to implement lean manufacturing management through multiple effective ways, such as International Forum of Manufacturing Management, Entrepreneur Summit, Project Improvement Management Consulting and Graduate Internship.

8.2.4.5 *Aligning with domestic and foreign famous enterprises and proceeding on benchmarking management*

CSGC strengthened the benchmarking management with typical manufacturers in "World Top 500" and "China Top 500," benchmarking sales revenue, profit, manufacturing cost, and overall labor productivity.

CSGC also urged core members to proceed on benchmarking management, aligning with the world-class manufacturing management. As per CSGC's request, Chongqing Changan Automobile earnestly conducted the benchmarking enhancement in alignment with Toyota's value chain, achieving a systematic benchmarking outcome and ensuring continued improvement in an attempt to stand out as a "world-class carmaker." Tianwei Group, which specializes in power transmission and transformation, launched the benchmarking management with some world-famous enterprises, such as ABB. Some industrial enterprises took in the managerial experience from the joint venture, enhancing the management level. Further, some enterprises conducted cross-disciplinary benchmarking management. Through the annual output of management innovation, more than 20 members under CSGC constantly summed up experience, aligned with or even surpassing the giants.

Moreover, CSGC learned from other enterprises and exchanged experiences with them. For instance, in order to benchmark the excellent management modes of several domestic and foreign giants specializing in vehicle, engine, and major equipment, CSGC learned the approaches of

lean manufacturing management from more than 30 domestic and foreign famous enterprises within two years, including the construction of
the manufacturing management system and on-site management, benefit-
ing and developing more than 300 intermediate and senior management
brains.

8.2.5   Enhancing the evaluation and assessment of the LP-based manufacturing management system

To implement CSPS to the maximum, CSGC established the *Evaluation
Method* of the LP-based manufacturing management system, after which
the evaluation outcome was considered in the performance evaluation of
industrial enterprise leaders.

CSGC built the panel of experts to make diagnosis, inspection, and
on-site instruction from time to time. Special training for the managers,
workroom directors, and team leaders were ready to make management
analysis. In each quarter, CSGC convened the video meeting featuring
the analysis of lean manufacturing management, summed up experience,
analyzed problems, and urged correction.

Further, CSGC makes a total check of lean manufacturing management every half year and evaluation every year. Particularly every fourth
quarter, 4–6 evaluation teams are built to conduct the annual management
evaluation, when more than 30 experts of lean manufacturing management
are assigned to make systematic management evaluation of all industrial
enterprises in 12 respects of CSPS as shown in Table 8.2, after which the
evaluation outcomes were referenced to examine, award, or punish industrial enterprise. Further, the punishment system was enhanced with the
reinforcement of routine mechanisms and on-site inspection mechanism,
thereby creating the longstanding mechanism and strong atmosphere in
order to promote the lean manufacturing management.

CSGC assessed CSPS ranging from Grade 1 to Grade 5, after which
the management-assessed outcome was released and included in the
evaluation system. Total CSPS grade for Industrial Enterprise = (CSPS
Assessment Scores * 65%) + (Outcome Confirmation Scores * 15%) + (Lean
Behavior Grade Scores * 10%) + (Self-evaluation Grade Scores * 10%). With
a review of the punishment system, any enterprise with top ranking will
be praised and awarded and vice versa.

CSGC was urged to continuously improve the manufacturing method
and management method until it is perfected. In the process of establishing the manufacturing system, executing system, assessing and evaluating, confirming effect, and reinforcing system, the Kaizen of business
performance was improved and competition strength in the industry was
further enhanced with PDCA circulated.

Table 8.2 The outcome of 2010 CSGC CSPS management evaluation

Seq.	CSPS management evaluation outcome	Number of enterprises	Remark
1	Grade 3.3	1	(1) Grade assessment result can reflect the enterprise progress and effect of lean manufacturing management, which can be used as the total metrics to assess the capability of enterprise manufacturing management.
2	Grade 3.0	1	
3	Grade 2.8	1	
4	Grade 2.7	1	
5	Grade 2.6	2	
6	Grade 2.5	1	
7	Grade 2.4	1	
8	Grade 2.3	2	(2) The higher assessment result means more reinforced enterprise management system, more significant management effect, and higher degree of modernized management.
9	Grade 2.1	1	
10	Grade 2.0	5	
11	Grade 1.9	3	
12	Grade 1.8	3	
13	Grade 1.7	11	
14	Grade 1.6	7	
15	Grade 1.5	6	
16	Grade 1.4	1	

CSGC urged each member, each department, and all employees to actualize LP. In order to promote on-site LP, the wage and evaluation system should be established for the lean manufacturing management, HR training, remuneration, and punishment when equipment management, information construction, and corporate culture should be enhanced before realizing the organic reform and growth.

With the instruction of CSGC's evaluation, industrial enterprises emphasized more on the progress and performance of lean manufacturing management, competed with each other for the full promotion, passed the evaluation stress down to the employees and operators, creating the improvement mechanism of total involvement. Some even mapped the lean manufacturing management into the middle and long-term development strategy and "the 12th five-year plan," vigorously pooling the efforts to implement the top-down lean manufacturing management. Further, with the considerable investment in manpower, wealth, and material resources, CSGC accelerated the process of modern management.

8.3 Effect of manufacturing management system on LP-based diversified group

The past years witnessed the implementation of the manufacturing management system, which has strongly supported CSGC to improve quality and efficiency. In this effect, enterprise efficiency, total development quality, and capability have been further enhanced.

8.3.1   Improved development quality

CSPS is well tailored for automobile, motorcycle, power transmission and transformation, new energy, and other business units, which is also applicable to all manufacturers under CSGC. With CSPS established, a manufacturing management standard has been developed in alignment with CSGC characteristics, which promoted the management regulation, process, and standardization; revealed the CSGC manufacturing management philosophy—"To be a pioneer, to be best, to ensure high-end manufacturing, and to serve the innovation"; accelerated the transformation, upgrading, and growth of CSGC manufacturing; and guided the total balanced enhancement of the industrial management level.

CSGC established CSPS, integrated manufacturing management, and pushed for the corporate integrated management and supply chain management, which were oriented at the standardization, information, and horizontal levels. Further, CSGC pushed for the integration and merging of the management system, from the top to down, from upstream to downstream, and vigorously output management to the outside, enhancing the total management level of CSGC stock resources and new resources.

In 2008–2013, it is roughly estimated that CSGC accumulated more than 3 billion yuan in cost decreases and benefit increases, and saved more than 200,000 m^2 accumulated in the production area, after implementing the LP-based manufacturing management system. As a result, a good situation that features overall deployment, linkage mechanism among all departments at headquarters, industrial sector driving, active promotion, LP highlights presentation, self-evident management effects, and Kaizen of development quality. In 2013, CSGC annual sales hit RMB 360 billion yuan with its profits over RMB 10 billion yuan, ranking the fifth in "China Top 500 Manufacturers," 22nd in "China Top 500 Enterprises," and the 209th in the "World Top 500." It is a surprising change from the most struggling enterprise of national defense technologies in early 1999 to the giant group with the world influence.

For example, in 2010, more than 40 industrial enterprises saved the cost of RMB 840 million yuan, saved the production area of 60,000 m^2, increased 0.63 times the stock turnover rate, organized employees to suggest 310,000 pieces of rationalized advice, gave training of LP with 60,000 attendances, and prepared more than 40 documents of the manufacturing management system (each set contains the management manual, procedure document, work file, form, and record). By doing so, the enterprise management process has been optimized, LP improvement cases of 400 systems/Model LP Line have been presented, enterprise production efficiency, quality acceptance rate, and other management operation metrics have been opti-mized, manufacturing cycle time has been shortened, employee morale has been stimulated, team building has been enhanced, on-site management has been improved, and total management capability has been reinforced as shown in Table 8.3. By assessing the management, which was organized by CSGC, it is expected that CSGC will reach Grade 1.5. Sixteen enterprises will reach Grade 2.0 or above and two enterprises will reach Grade 3.0 or above, reflecting the reinforced industrial management system, enhanced management effect and improved management modernization, with CSGC development quality and efficiency further improved.

In CSGC, LP has been deeply rooted in everyday thinking and the behavior of many employees, and a lot of lean manufacturing management

Table 8.3 CSGC Main Metrics of Economic Operation in 2009–2010

Metrics	Unit	(2010) 2010	Actual works completed in 2009	Variation amount	Variation ratio (%)
			versus 2009		
Sales (business) revenue	In RMB 10,000 yuan	25,706,981	19,262,968	6,444,014	33.45
Total profit	In RMB 10,000 yuan	746,769	567,893	178,876	31.50
Return on equity (ROE)	%	10.35	9.70	0.65	–
Total return on assets (ROA)	%	4.77	4.42	0.35	–
Rate of sales (business) profit	%	2.41	2.12	0.29	–
Capital maintenance and increment ratio	%	105.28	108.58	–3.30	–
Overall labor productivity	In RMB 10,000 yuan	15.2	14.83	0.37	2.5
Current assets turnover ratio	%	2.07	1.90	0.17	–
Asset-liability ratio	%	72.24	72.75	–0.51	–
Energy consumption per RMB 10,000 output	t/standard coal	0.0462	0.0564	–0.0102	–18.09

professionals and technicians have stood out as the endogenous genes of management reform. With lean manufacturing management implemented, CSGC has improved the product quality, reduced the energy consumption, cut off the product cost and price, benefited a lot of consumers, effectively fulfilled the social responsibility, and enhanced the social benefit.

8.3.2 Enhanced core competitiveness

With the LP-based manufacturing management system continuously implemented and executed, CSGC avoided the consumption and non-value-added activity and enhanced the core competitiveness.

To be exact, with the LP management reform further promoted, LP has affected every aspect of enterprise management, enhancing the core competitiveness of product, technology, market, and team. First, lean manufacturing management can push for the optimization of development strategy. "Intensive operation promotes development, innovative management improves quality, and lean manufacturing boosts efficiency" has been found throughout the execution of development strategy and operation activity. Second, lean manufacturing management can push for the adjustment of the technological innovation system. The enterprise lean R&D, lean design, and promotion can be conducive to innovative lean manufacturing management tools, and enhancing the research and application of the newest manufacturing management, such as agile manufacturing (AM) and service-oriented manufacturing. Third, LP management can push for market restructuring, adapt to the market change, optimize the market layout, rationalize the allocation of management resources, and implement just-in-time production (JIT). Fourth, lean manufacturing management can push for personnel restructuring. Rearrange and reinforce the enterprise's lean manufacturing system, optimize management flow, drive the optimization of enterprise structure, optimize the HR allocation, and improve the labor productivity. As per the principle of "anyone who leans will be benefited," employee enthusiasm will be stimulated.

As shown in Figure 8.3, according to the contrasted business revenue in 2004–2013, during the period of "the 11th five-year plan," CSGC saw an annual increase rate of 27.49%, 3.42-fold the end of "the 11th five-year plan"; in 2010, CSGC business revenue hit RMB 200 billion yuan, and in 2012, it hit RMB 300 billion yuan, marking the change of company scale and strength.

As shown in Figure 8.4, according to the contrasted business revenue in 2004–2013, during the period of "the 11th five-year plan," CSGC saw an annual increase rate of 49.31%, 7.24-fold the end of "the 11th five-year plan"; in 2013, CSGC total profit hit the record high of one million yuan. To

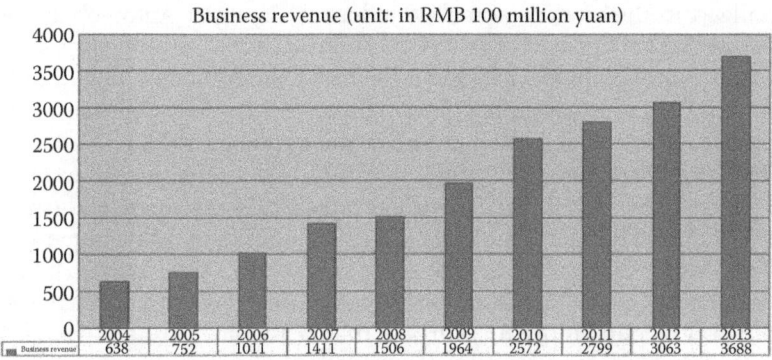

Figure 8.3 2004–2013 CSGC business revenue barchart.

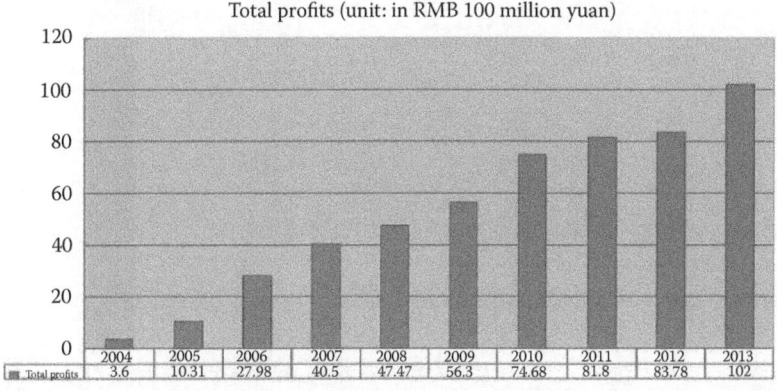

Figure 8.4 2004–2013 CSGC total profits barchart.

be specific, it is the implemented LP-based manufacturing management reform that boosts the quality and efficiency of corporate development.

8.3.2.1 Case study: "Chongqing Changan Automobile" under CSGC

Through manufacturing management reform, Chongqing Changan Automobile under CSGC has implemented and executed the Changan Production System (CPS) since 2005, marking it among the best of China automobile manufacturers successfully. In 2013, Changan Automobile hit two million in annual sales, making it the giant state-owned enterprise with seven domestic bases, 23 vehicle and engine plants, and output close to RMB 150 billion yuan. Its brand has climbed to RMB 38.202 billion yuan, ranking second among Chinese automobile manufacturers. In October 2013, according to the statistics released by Fourin, one researcher specialized in the world automobile industry, Changan Automobile under CSGC sold 1.22 billion cars of self-owned brands in 2012, ranking first nationwide and 14th worldwide, making it the first Chinese self-owned brand for seven years in a row. Also, Changan Automobile is the only Chinese automobile manufacturer with the self-owned brand to hit one million.

In 2004–2010, Changan Automobile annual sales hit more than 1.9 million cars, up from less than 0.58 million cars in 2014 as shown in Figure 8.5.

Moreover, Changan's multiple metrics of operation management have ranked among the best in China, according to 2010–2011 industrial data. For instance, its capacity utilization rate (as given in Figure 8.6) and one-line output (as given in Figure 8.7) surpass that of Toyota. Through promoting

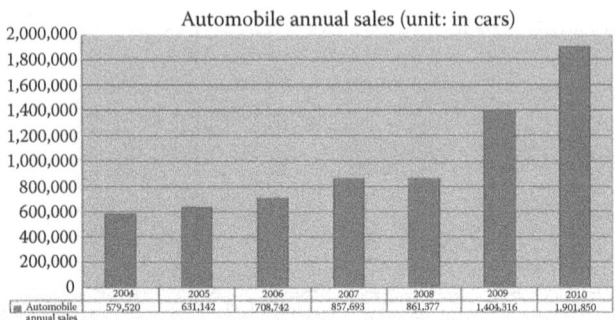

Figure 8.5 2004–2010 Changan Automobile annual sales barchart.

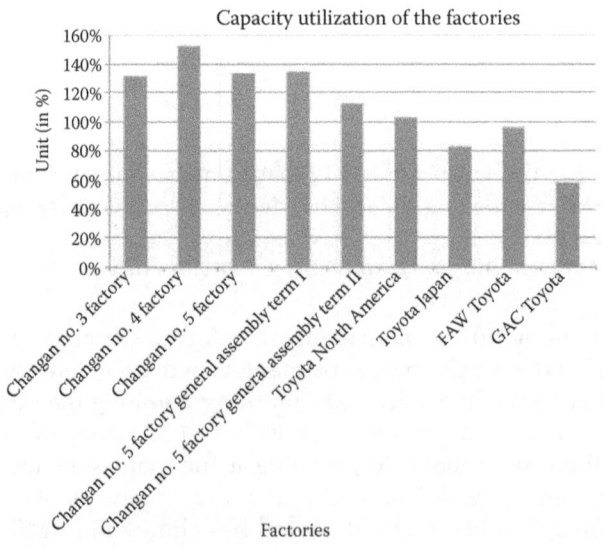

Figure 8.6 The rate of capacity utilization barchart (Changan vs. Toyota).

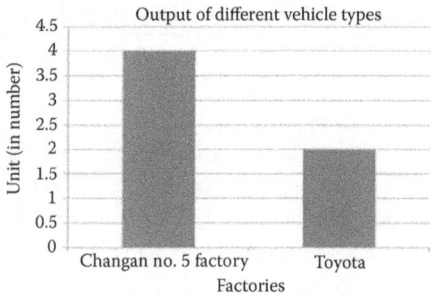

Figure 8.7 One-line vehicle output (Changan vs. Toyota).

and applying the Changan Automobile Production Development System (CA-PDS), its total product development system has altered remarkably. In this effect, the development cycle time of new vehicle type is shortened to 36 months from the previous 42 months, with costs off 25%–30%. "Changan's own development process is peerless among the China self-owned brands," Dr. Dong Xianquan commented, the former president of BMW China.

8.3.2.2 Case study: "Chongqing Tsingshan Industrial" under CSGC

Through the construction and execution of the manufacturing management system, Chongqing Tsingshan Automobile under CSGC has been well known as the passenger car transmission manufacturer with the largest production and marketing scale and most diversified product categories. Further, it worked out and released many state automobile standards, such as the assembly technology conditions of minicar mechanical transmission, and the stock turnover rate and other operating management metrics have ranked first in this industry as shown in Table 8.4.

8.3.2.3 One Changsha-based special product manufacturer under CSGC

With the instruction and push by CSGC, this manufacturer initiated the LP-based manufacturing management reform in manufacturing special product since 2012, which was executed by the following:

1. Reinforcing the management system. As per the standards of the lean manufacturing system, establishing a four-level architecture of system documents, that is, management manual, procedure document, work guidance, form and sheet; by analyzing and evaluating the value stream, reviewing and optimizing eight core business

Table 8.4 2008–2010 Tsingshan Industrial Lean Metrics

Lean Metrics	Unit	Year 2008	Year 2009	Year 2010
Sales revenue	In RMB 100 million yuan	10	15.45	21.92
Stock Turnover	Times	15.89	22.37	23.65
Overall equipment efficiency (OEE)	%	75.5	81.45	85.1
Occupation of production funds	In RMB 10,000 yuan	408	386	381
The ratio of management, finance, and business cost to business revenue	%	10.86	8.06	4.37
Overall labor productivity	In RMB 10,000 yuan/ person	11.73	13.01	17.9
The ratio of cost rxpense to business revenue	%	98.24	99.08	89.98

flows and 20 key business flows; as per the method of "one business flow, one procedure document, and one set of management forms and sheets," checking if there is any omission, creating and revising 11 management procedure documents and 31 work guidelines; through "three-form auditing" by managers and technicians and "four-form auditing" by operators, promoting the standard management and operation.

2. Optimizing the production organization. Centering on planning management; fulfilling planning preparation, execution, change, and examination; applying management tools, such as capacity assessment or manufacturing BOM; reinforcing monthly, weekly, and daily work planning and management system, and innovating the examination method of production planning. The monthly planned fulfillment rate of the Changsha New Area hit 90%, up from 60% in early 2014 as per the "general assembly and partial assembly," lean restructuring one assembly production line, and establishing the "one piece flow" pulsing assembly line.

3. Enhancing the on-site management. Strengthening the on-site 5S management, standardizing employee's conduct, and conducting TPM.

4. Executing the Kaizen. Revising the Kaizen management procedure, evaluation criteria and incentive method; breaking down the Kaizen metrics into each department, cell, and individual, and including

these metrics in the performance evaluation; convening presentation of improvement outcome and releasing excellent cases in every quarter; rewarding employee improvement and releasing the improvement cases in every month.

5. Standardizing data management. Enhancing the norm management, executing the special action of the norm system establishment, working out the promotion program, building six special task forces, sticking to the synchronous progress of the reinforced standards and mechanism construction, and cut through six norm standards, namely main material, auxiliary material, fixture, work hour, tool, and action; executing metrics management, arranging the KPI metrics system in three levels, namely company, department and team, and making clear the logic between metrics; through company and department scorecards and an employee performance plan, breaking down the metrics from the top down, while, in turn, performance is supported from down to the top; standardizing reporting management, reinforcing the reporting management system, unifying the template and data structure, standardizing reporting time and frequency, and creating the down-top effective data collection, statistics, analysis, and application mechanism.

According to the contrasted data in 2013, this enterprise has achieved a significant effect in the wake of implementing the LP-based manufacturing management reform with substantial improvements in the plan fulfillment rate of special products, equipment utilization rates, and production efficiency.

1. Through optimizing the plan management flow and strictly controlling the process, the plan fulfillment rate of some special products is approximately 90%, up from around 60% with stable tendencies as shown in Figure 8.8.

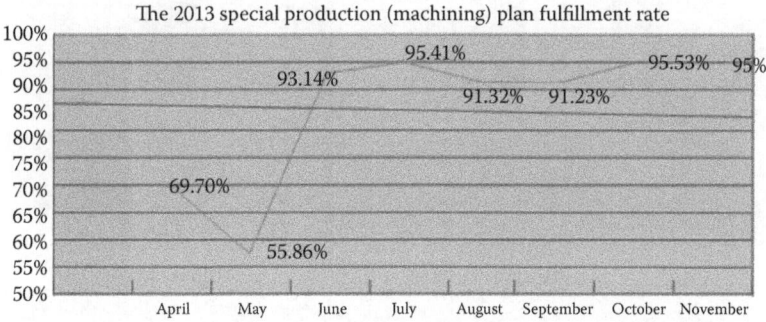

Figure 8.8 The broken-line chart of plan fulfillment rate change trend.

2. Equipment utilization rate. In 2012, its equipment utilization rate was no more than 50%, but a substantial rise was seen in the second half of 2013, up from no more than 50% before 6 months to 70% or above with stable tendencies as shown in Figure 8.9.

3. In 2013, the special products workroom saw a substantial rise in the ratio of the average hour for work completed to the highest work hour in a single month. Wherein, the average work hour completed rose to 10,057.57, up from 7411.98 in 2012, increasing by 35.9%; the highest work hour in single month rose to 12,719.28, up from 10,563.76 in 2012, increasing by 20.4% as shown in Figure 8.10.

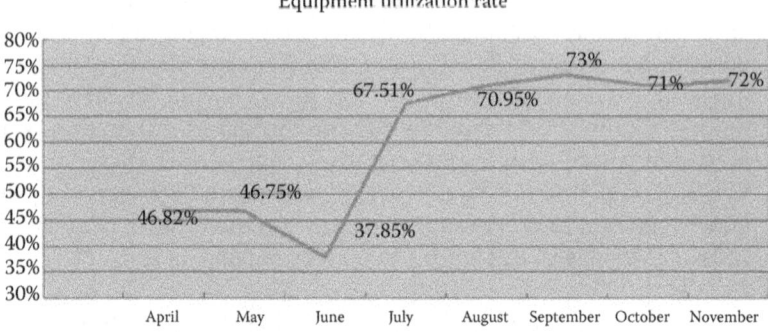

Figure 8.9 The broken-line change trend of equipment utilization rate change trend.

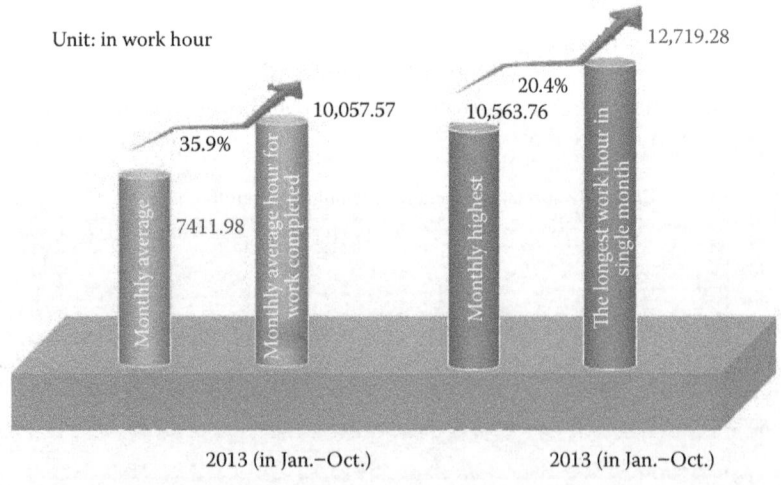

Figure 8.10 The average hour for work completed to the highest work hour in single month (special products workroom).

8.3.2.4   *Case study: One Chongqing-based special products manufacturer under CSGC*

One Chongqing-based special products manufacturer under CSGC actively examined the different application models of lean tools in the production process of special products, such as "standard operation," "one worker multiple machine (OWMM)," and "SMED," which were well tailored for the production characteristics of special products. With the trial run, the application method of lean tools of special products is determined, which proves to be effective as shown in Figure 8.11.

1. By focusing on the key components of military and civilian products with stable technological conditions, the mass loss rate of main products decreased from 0.3% to 0.15%, down by 50%.
2. By making SMED for processes, such as bottleneck equipment processing, long time period parts processing and parts assembly, "one product gear grinding process" in pilot workroom decreased from 2747 seconds to 450 seconds.
3. By executing the OWMM on numerical control equipment, the coverage rate of numerical control reached 60%, up by 40%, of production efficiency.

By managing the change point, systematically doing abnormal identification, abnormal treatment (quick response mechanism), and closed-loop management of problems and innovating the control mode of production process. On-time handling rate of production anomaly rose 31.3% and the repeated rate of abnormal problems cut by 18.75%.

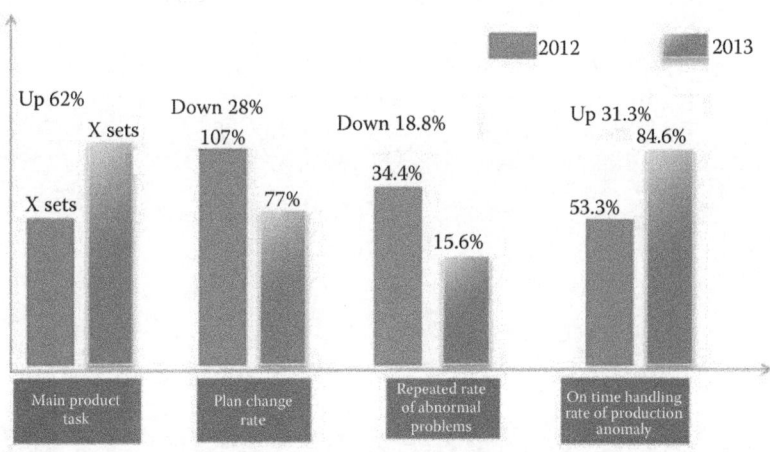

Figure 8.11 The completion of process control metrics.

8.3.3   Enhanced CSGC international competitiveness

In the period of "the 11th five-year plan," particularly of the end of "the 11th five-year plan," CSGC executed the construction and implementation of lean manufacturing management with total involvement, in full range, in all processes and with all aspects, optimizing the allocation of all factors, including person, machine, material, law, and environment, increasing return on invested capital (ROIC), accelerating the growth of scale and efficiency and enhancing the international competitiveness. In 2009, CSGC was elevated into the "World Top 500," ranking 428th; in 2010, CSGC business revenue and total assets hit RMB 200 billion yuan, up from approximately RMB 100 billion yuan, reaching RMB 257 billion yuan and 219.3 billion yuan, respectively. In 2010, CSGC jumped to 275th in the "World Top 500," up 153 places within one year; in 2011, CSGC ranked 225th in the "World Top 500" with further enhanced standing; in 2013, CSGC ranked 209th in the "World Top 500," growing to be in the world-class group.

chapter nine

Restructure of the supply chain

Linning Cai

Contents

9.1 Background

Yuchai Machinery or YC Diesel, known as a large-sized SOE, is one of China's top 100 machine manufacturers and China's largest production base of internal combustion engines. For the complex processes, many varieties, and more than 20,000 types of parts, Yuchai purchases most parts from external suppliers, which are located in Shandong, Zhejiang, Hunan, and other provinces.

Purchasing logistics is the forefront of the total supply chain, which plays a significant role in logistics operation, prompting modern manufacturers to realize the operating efficiency of a purchasing network as it can ensure the rapid and steady development of the total supply chain.

With a growing number of advanced enterprises being equipped with the advanced supply chain, the high efficiency and economic efficiency of a purchasing network has captured more public attention. By analyzing the supply mode of about 400 suppliers for Yuchai Machinery (abbreviated as "YC Diesel"), one third of the purchasing logistics has been assigned to Yuchai Logistics, known as a professional third-party logistics provider, and the other two thirds are transported by suppliers or other third-party logistics enterprises. However, YC Diesel's scattered supply system of parts would reduce the scale of the economic effect of parts supply. In this system, centralized management and strict control of delivery time cannot be possible, which is considered to be its biggest drawback.

In this concern, Yuchai Logistics cooperated with Department of Industrial Engineering of Tsinghua University to tailor a "milk run," which is one of the supply chain management models. This model establishes the parts supply network, integrates all parts of Yucai suppliers under unified distribution management, and reduces the cost of parts logistics through centralized purchasing logistics.

9.2   Model analysis of parts supply logistics

It is understood that three modes of purchasing logistics were used previously, namely supplier–production line, supplier–parts warehouse, and third-party logistics–parts warehouse (e.g., Yuchai Logistics), from the investigation of the transporting operation of YC Diesel purchased parts. Purchased parts in the supplier–production line are mainly concentrated in the vicinity of Yulin Guangxi, and the majority is inappropriate to be stocked in the parts warehouse due to the large size. Despite that, the majority of suppliers choose the road transportation on the way to the warehouse, which is followed by production assembly and consignment settlement. At present, Yuchai Logistics is assigned to transport diesel engines by YC Diesel, with a lot of trucks at 30–36 tons waiting to be loaded and heading for major sales places nationwide. Yulin is situated on the south border, which causes a lot of trucks to transport diesel engine in large batches northward to the geographical location; further, a lot of parts are demanded in the assembly process, hence generating a lot of parts transports. Many suppliers are located in Wuxi, Tianjin, Chengdu, and other places, and engines transported from Yulin to all places and backhaul of parts can be the main instrument to improve the efficiency of truck transporting and reducing the transporting cost.

As shown in Figure 9.1, the small circles stand for regular suppliers of YC Diesel, and the triangles stand for major clients. As contrasted in the figure, the great majority of areas are distributed with an automobile factory in need of diesel engines and suppliers of parts, including pistons, cylinder covers, and shock absorbers. Therefore, by combining engine and

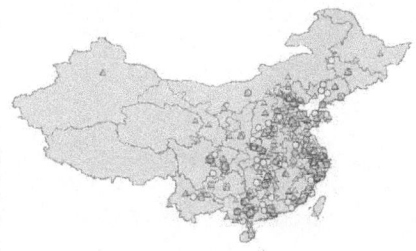

Figure 9.1 Geographical distribution of YC Diesel clients and suppliers.

parts transporting into a closed ring, high efficiency and low cost can be ensured.

9.2.1   Demand analysis of parts logistics

The final parts demand of YC Diesel is to satisfy the production assembly demand of its no. 1 engine factory, no. 2 engine factory, no. 3 engine factory, and power plant. To this end, YC Diesel assigns purchasing to the purchase department, which is authorized to purchase parts in the form of a purchase order and supply contract, and inventory control is enabled through consignment mode. To date, YC Diesel exercises the production model of great varieties and small batches, prompting higher requirement for a material supply system. To ensure smooth production, YC Diesel reserves one area for the parts warehouse, in order to buffer the stress, but out-of-stock (OOS) remains one of the urgent problems.

Concretely, YC inventory is audited twice per day, and purchasing is managed based on the market forecast by the sales company. The supplied quantity may slightly vary with the purchased quantity due to the consignment mode. By analyzing the data given in YC Diesel purchase orders, materials distributed in the full amount by suppliers is 66% according to the purchase plan, and about 10% of purchased materials have not been entered. Fill rate in weighted average turns out to be 46.2% (Figure 9.2).

9.2.2   Supply quantity and location distribution characteristics of YC Diesel suppliers

By now, Yucai Machinery has owned more than 20,000 kinds of parts. By analyzing one year's annual data, the number of regular suppliers is around 400 (excluding one-time suppliers and unknown suppliers as a result of an unavailable summary of supplier records), up to 74% of the total number with 47 kinds supplied on average. The supplier can provide

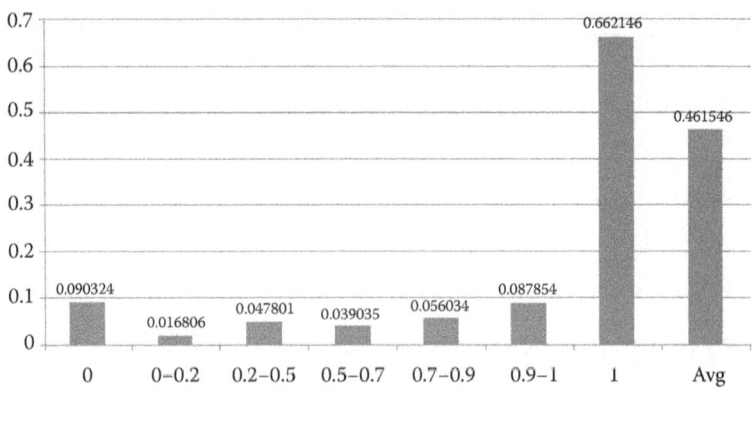

Unit: in PC

Figure 9.2 Fill rate of purchase orders.

as many as 800 kinds of parts. Of all parts, 80% are provided by around 100 suppliers. If divided by supplier purchase quantity, 20% of top suppliers contribute to more than 80% of the purchase quantity as shown in Figure 9.3.

The geographical distribution of the main suppliers is presented in Figure 9.4. The gray symbol is the location of the supplier, the symbol area is the quantity of parts provided by supplier, and the black circle is the location of YC Diesel. The distance of each supplier to Yulin from the lightest to the darkest color is 0–500 km, 500–1000 km, 1000–2000 km, 2000–3000 km, and more than 3000 km, respectively. Among the suppliers, the longest distance to Yulin is 3746 km with 68 kinds of parts supplied.

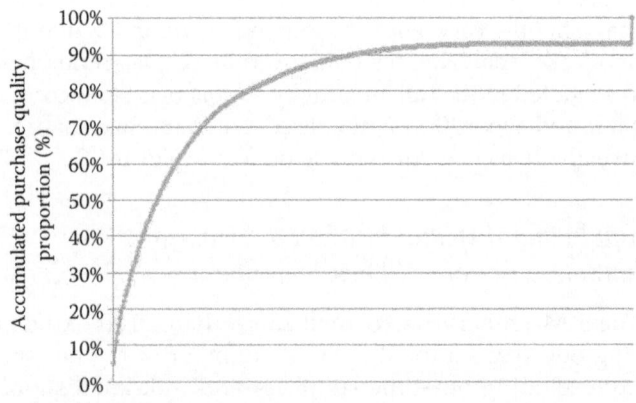

Figure 9.3 Purchase quantity Pareto chart.

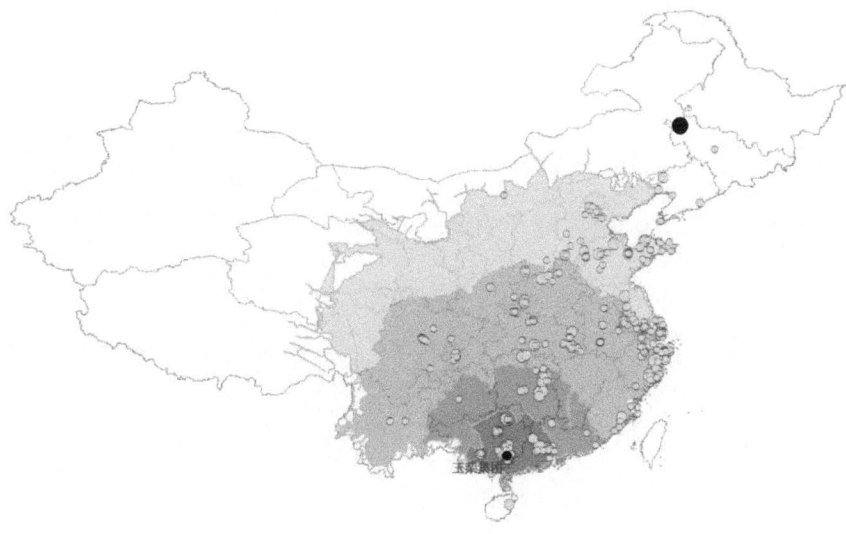

Figure 9.4 Geographical distribution of main suppliers.

9.2.3   Parts transporting assigned to Yuchai Logistics

Yuchai Logistics, which is the largest carrier of YC Diesel parts, has 20 branches, of which 12 branches are assigned to the parts transporting, namely Jieyun, Liuzhou, Hanghzou, Tianjin, Chengdu, Zibo, Zhengzhou, Xiangfan, Changsha, Xian, Wuxi, and Guangzhou. Twelve branches are assigned with the parts transporting provided by Yuchai suppliers, which are geographically distributed as shown in Figure 9.5.

Each branch serves a different number of suppliers, of which Hangzhou, Zibo, and Changsha branches are assigned with more suppliers.

9.2.4   Yuchai Logistics branch workflow of parts transporting

The workflow is divided into five phases: order receiving, truck dispatching, loading, transporting, and unloading, which covers suppliers, Yuchai Logistics branches, and YC Diesel parts warehouse (three functional departments). The workflow can be shown as in Figure 9.6.

9.2.5   Time analysis of parts transporting

Transporting time of parts is analyzed based on the ledgers of all branches. The transporting ledger is recorded with the order receiving date, loading date, dispatching date, and unloading date. These four items are closely associated with the time of parts transporting.

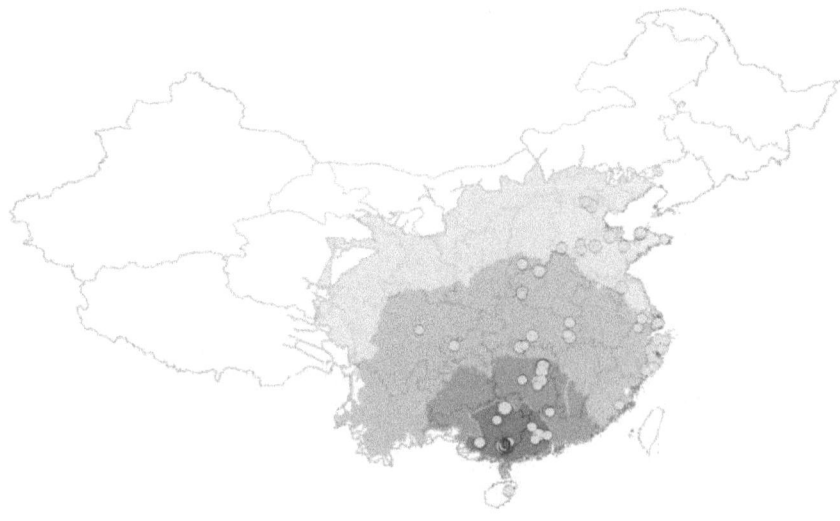

Figure 9.5 Geographical distribution of Yuchai Logistics service providers.

According to the list of suppliers submitted by each branch, we filtered out the records of parts transported from these suppliers to YC Diesel and deleted incorrect data (e.g., "unloading date" precedes the "dispatching date"), and the filtered results were analyzed as follows:

1. Order receiving date to Loading date

    In most cases, the order receiving date is the same as the loading date. If we want to know the exact time segment, the original data should be precise in hours, but it cannot be possible in reality. So we consider taking this time segment into account with the next segment from loading to dispatching.

2. Loading date to Dispatching date

    If this time segment is added to the response time before loading, we can get the time segment from order receiving to dispatching, which reflects the response speed and dispatching speed of parts transporting.

    As shown in Figure 9.7, average order receiving and loading time of all branches is ranked, and the average value of all branches is contrasted.

3. Dispatching date to Unloading date

    This time segment can reflect the actual GIT time, and GIT time averaged of all branches is 79 hours. As shown in Figure 9.8, the Zibo branch spends 132 hours as the longest average GIT time, and the Liuzhou branch spends 29 hours as the shortest average GIT time.

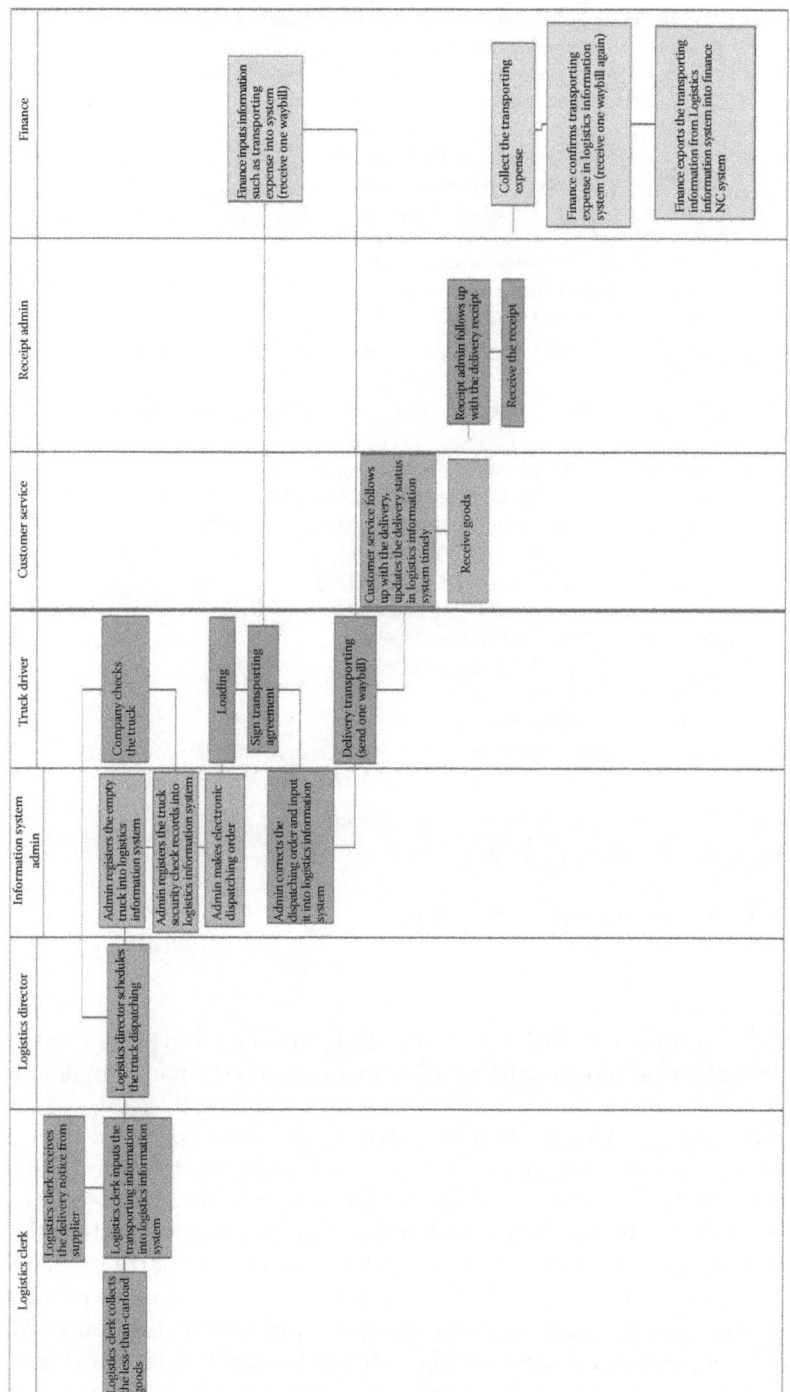

Figure 9.6 Yuchai Logistics branch workflow of parts transporting.

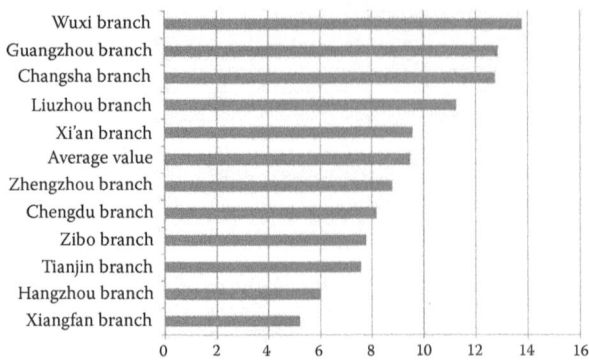

Figure 9.7 Average time from order receiving to loading of parts transporting.

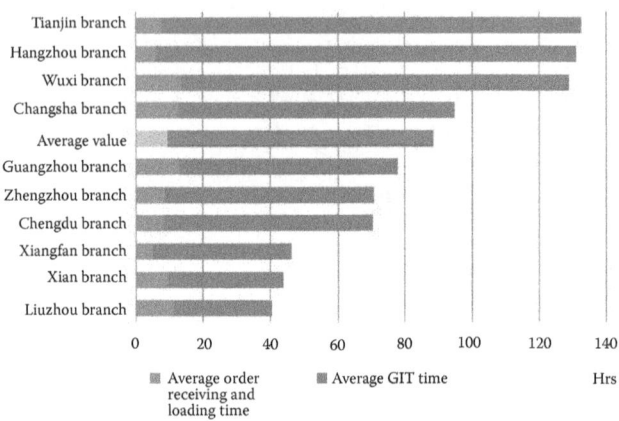

Figure 9.8 Total time of parts transporting.

1. Summary

In summary of order receiving time and GIT time, the service time of parts transporting by all branches can be provided in detail.

Analyzing the average and variance of the time spent by the 12 branches can be indicative of small variances of order receiving and loading time and that the speed of response and dispatching of all branches are almost the same. Further, most branches are able to load parts at the field of supplier within the day upon the receipt of transporting notice. The big variance of average GIT time results in the big variance of total time of parts transporting (the order receiving and loading time plus GIT time). The big variance of average GIT time can be attributed to the different distance between each branch and Yulin, which results in the different time spent in parts transporting. The longer distance from the branch to Yulin means the longer GIT average time.

9.2.6   Necessity of parts consolidation center

From the above-described analysis, a scattered supply system mounts stress on the purchasing and logistics management of Yuchai supply chain. For example, the centralized management of purchasing is very hard, quality cannot be controlled, cost and time spent in purchasing negotiation are considerable, and market fluctuation is vulnerable to high risk; purchasing logistics generate many back-and-forth transports and time spent because of multiple loadings, which results in high expense of transporting and extended lead time of purchasing. In this concern, it is therefore required to duly establish a consolidation center in the centralized supply area and integrate the scattered supply logistics.

To date, there is no available consolidation center (abbreviated as "CC") in the network of parts transporting; cargo collection at a CC can be done by the spot-to-spot picking or circulating haul (termed "milk run"), which transports parts from the supplier to the CC. Long-distance transporting from the CC to the YC Diesel parts warehouse can be done via the line haul, which runs the vehicle in big tonnage.

9.3   Design program of parts network

This program is based on the dual structure of a CC network, in which three types of nodes are arranged: supplier, CC, and factory. In the original structure of the research object, 12 branches under Yuchai Logistics have been established nationwide, but they cannot be treated as mature CCs due to the absence of warehousing or CC function. In contrast, this model aims to build a CC with such functions as loading and unloading, consolidation and short-term storage thus improving the double-level

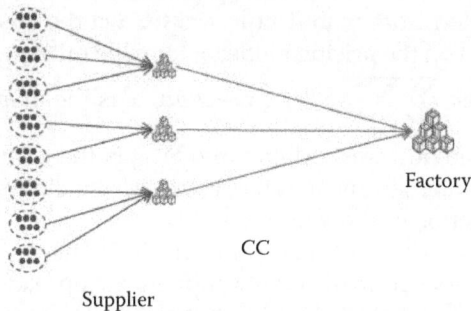

Figure 9.9 Dual distribution structure.

Dual distibution network

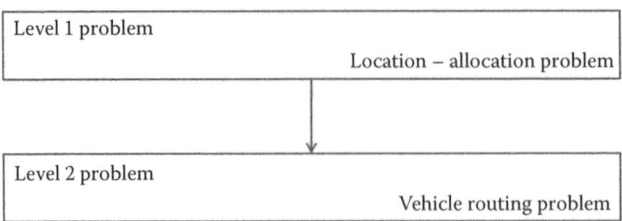

Figure 9.10 Dual subproblems of distribution model.

structure of the CC network and proposing a new mode of parts distribution (Figure 9.9).

Wherein, the supplier and factory are fixed in reality, but a CC remains to be defined for the time being. In order to establish such a network structure, three questions should be answered: (1) In which locations to set up CC; (2) which supplier should serve each CC; (3) after establishing the CC, how to arrange the CC routing.

In response to three questions, two submodels are studied in Figure 9.10: (1) the location–allocation problem and (2) the vehicle routing problem.

9.3.1   Operating cost of parts consolidation center

The subproblem of parts CC network planning mainly examines how to minimize the total cost of CC operation. CC cost includes three costs as follows:

1. Transporting costs from supplier to CC
   According to the transporting quantity per batch from the actual statistics, it is evident that parts provided from one supplier cannot fill the carload, and the rational assumption may be made that parts can be transported from manufacturer to CC via the less-than-truckload mode. Based on the pricing model of tonnage mileage, this cost can be expressed as: $\sum_{i \in I} \sum_{j \in J} \alpha_j h_i d_{ij} Y_{ij}$. Wherein, α_j is the market price rate of less-than-carload transportation mode; h_i is the annual transporting demand of supplier; and d_{ij} is the distance from the supplier to the CC.
2. CC construction and operating costs
   The fixed cost input may result from the newly established CC. Further, cargoes from different suppliers are unloaded at the CC, followed by being reloaded and transported to the factory, generating the reloading cost incurred from this unloading–reloading process.

3. Fixed CC construction costs

If a CC is constructed, some areas are required for loading and unloading cargoes or functioning somewhat as a stocking area, as the case may be, generating the rental cost. Generally, the rental price is associated with the geographical location, and the rental cost is estimated against the average CC area and the prevailing local market price. Set the fixed construction cost of alternative CC j as f_j; therefore, the total fixed construction cost is the sum of fixed construction cost at the selected CC, which can be expressed as: $\sum_{j\in J} f_j X_j$.

4. Loading and unloading costs at the CC

Loading and unloading costs at the CC are closely associated with the throughput of this CC. CC throughput is the sum of the supplier's transporting demands as covered by this CC. Suppose the loading and unloading cost of the alternative CC j is in direct proportion to the throughput of this node, λ_j is the coefficient of proportionality, h_i is the annual transporting demand of the known supplier i; therefore, the total cost of loading and unloading is the sum of loading and unloading costs at the selected CCs, which can be expressed as:

$$\sum_{i\in I}\sum_{j\in J}\lambda_j h_i Y_{ij}.$$

5. Transporting costs from the CC to the factory

Transporting cost from the CC to the factory are also determined by the length of the transporting route by following the steps: (1) Multiply the transporting quantity with this length as the weight, according to the coverage relationship between the CC and suppliers. (2) Multiply the weight with the parameter transporting rate (the carload rate is usually lower than the less-than-carload rate). (3) Sum and calculate. In the case of the research object, the cost settlement between the enterprise and vehicle owner should be calculated and paid by this method.

Suppose β_j is the market price of the vehicle transporting rate at the alternative CC j, h_i is the annual transporting demand of the known supplier i, d_{jk} is the distance from the alternative CC j to the factory; therefore, the transporting cost from the CC to the factory can be expressed as: $\sum_{i\in I}\sum_{j\in J}\beta_j h_i d_{jk} Y_{ij}$.

To sum up, the operating cost of the CC model can be estimated by the formula below:

$$\min \quad z=\sum_{i\in I}\sum_{j\in J}\alpha_j h_i d_{ij} Y_{ij}+\sum_{j\in J}f_j X_j+\sum_{i\in I}\sum_{j\in J}\lambda_j h_i Y_{ij}+\sum_{i\in I}\sum_{j\in J}\beta_j h_i d_{jk} Y_{ij}$$

9.3.2   Network planning model

As contained in this program, the location of network planning is based on the model P-Median Problems, thereby solving the concrete location of central storage at the CC in different scales and the service area of this central storage. By referring to the formula under the Classical location model, it can be expressed as

$$\min \sum_{i \in I} \sum_{j \in J} \alpha_j h_i d_{ij} Y_{ij} + \sum_{j \in J} f_j X_j + \sum_{i \in I} \sum_{j \in J} \lambda_j h_i Y_{ij} + \sum_{i \in I} \sum_{j \in J} \beta_j h_i d_{ij} Y_{ij}$$

s.t.

$$
\begin{cases}
\sum_{j \in J} Y_{ij} = 1 & \forall i \in \mathbf{I} \\[2mm]
\sum_{j \in J} X_j = P & \\[2mm]
Y_{ij} - X_j \le 0 & \forall i \in \mathbf{I}, \forall j \in \mathbf{J} \\[2mm]
X_j \in \{0,1\} & \forall j \in \mathbf{J} \\[2mm]
Y_{ij} \in \{0,1\} & \forall i \in \mathbf{I}, \forall j \in \mathbf{J}
\end{cases}
$$

With this model built, the above-described two problems under this model will be solved: (1) In which locations to set up the CC; (2) which supplier should serve each CC.

9.3.3   Selection of alternative CC

The influence of multiple factors on node layout and the available resources should be considered when selecting an alternative CC. Node layout is not only subject to some factors, including traffic convenience, material sources, local transporting price, the difference of regional economic or fixed construction cost, and other policy, economic, and cultural influence, but also considers the coordination and rational use of available human and material resources.

So far, 12 branches under Yuchai Logistics have demonstrated administrative functions, such as vehicle scheduling, material source development, and supplier management. Hence, these strengths can be tapped to expand these branches to function as the CC, and 12 branches are listed as the alternative CCs. The longitude and latitude of each branch can be queried if given with the geographical location, such as province and city. Moreover, the area that is thickly distributed with suppliers can be one

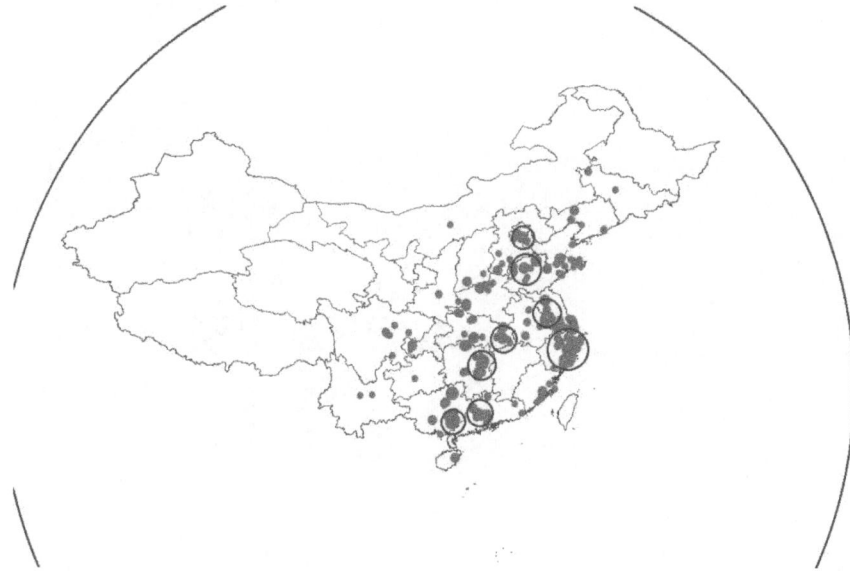

Figure 9.11 Eight areas that are thickly distributed with suppliers.

alternative, according to the distribution density of suppliers. For instance, eight areas are found with thickly distributed suppliers; an alternative CC location may be mapped on these eight areas with higher density using the centre-of-gravity method as indicated in Figure 9.11.

9.3.4   Solution algorithm of location model

To be more specific, the heuristic algorithm under the P-Median Problems Model should be performed to rationally plan the construction program of the supply network based on the 20 alternatives of central storage, according to the location and delivered quantity of YC Diesel suppliers.

9.3.5   Construction program of parts CC

When the number of CC construction at different phases (P) is given, the construction sequence of 20 alternative CCs obtained through the algorithm can turn out. For example, when P = 10, the first 10 lines of alternative CCs as contained in the table can be chosen for construction in sequence, which are Yulin, Changsha, Hangzhou, Jinan, Xiangfan, Quanzhou, Ningbo, Nanjing, Hengyang, and Langfang. Concretely, 20 CCs as marked nodes can be constructed in the sequence as given in Figure 9.12:

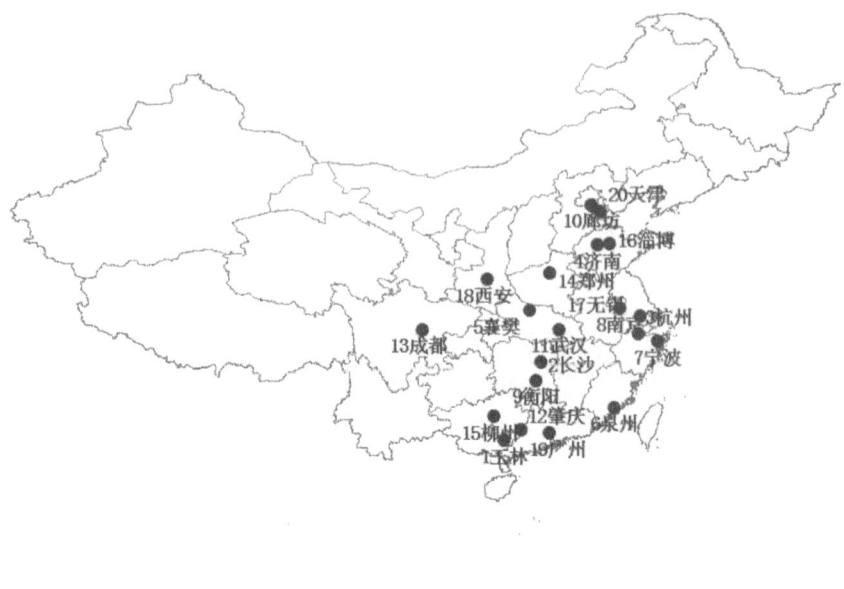

Figure 9.12 YC Diesel Parts CC.

9.4   CC vehicle scheduling problem

After finishing the CC selection of parts network and suppliers served in Phase 1, the Level 2 problem should be solved. It can be hypothesized that each CC operates independently from one another, and each CC can be treated as one subnetwork, which is composed of an independent consolidation point, all suppliers covered and the factory. As a result, the YC Diesel parts network can be divided into the combination of P subnetworks.

The suppliers as covered by different consolidation points are not intersected, which means each supplier can be only covered by one consolidation point. Therefore, P subnetworks can operate and schedule independently from one another, and these subnetworks are not intersected as well. In this phase, in which a small-sized truck uses the point to point transportation mode and a large-sized truck uses the "milk run," the self-evident rational combination and scheduling should be solved.

The CC location can be determined by taking such information into account, such as the geographical location of supplier (e.g., longitude, latitude), supplier-to-CC distance (in km), supplier's transporting demand per time (in tons), the load of different vehicle types (in tons), the freight rate of different vehicle (in RMB yuan/km, RMB yuan/ton km), etc.

9.4.1   Cost analysis of CC vehicle

The combined vehicle scheduling is mainly referred to as the consolidation model of each branch of Yuchai Logistics, and different vehicle properties are economically calculated.

As revealed in Figure 9.13, vehicle dispatching is done by taking the different vehicle properties into account, according to the load tonnage and supplier-to-CC distance classification.

9.4.2   Computing procedure of vehicle dispatching

Consolidation service consolidates the cargoes transported from the CC to Yulin, which are provided by the supplier out of the transportation route. However, if the vehicle passes by one supplier on the way from the CC to the factory, the cargoes from this supplier are not necessarily transported to the consolidation point. When the large-sized truck heads for the factory, just load the parts of this supplier by the way. The criteria: $d_{n,n+1} > d_{i,n+1} + d_{i,n} - 15$ can be used to examine if one supplier is on the route. Geometrically, if one supplier is just on the route, then $d_{n,n+1} = d_{i,n+1} + d_{i,n}$; however, if this supplier slightly deviates from the route, which results in the increased length ≤ 15 km in consideration of the reality, then this supplier is considered to be on the route.

When handling this procedure, the suppliers on the consolidation route can be ruled out, and the rest of the suppliers can be ranked from the largest to smallest according to the supply quantity. Suppose initially all

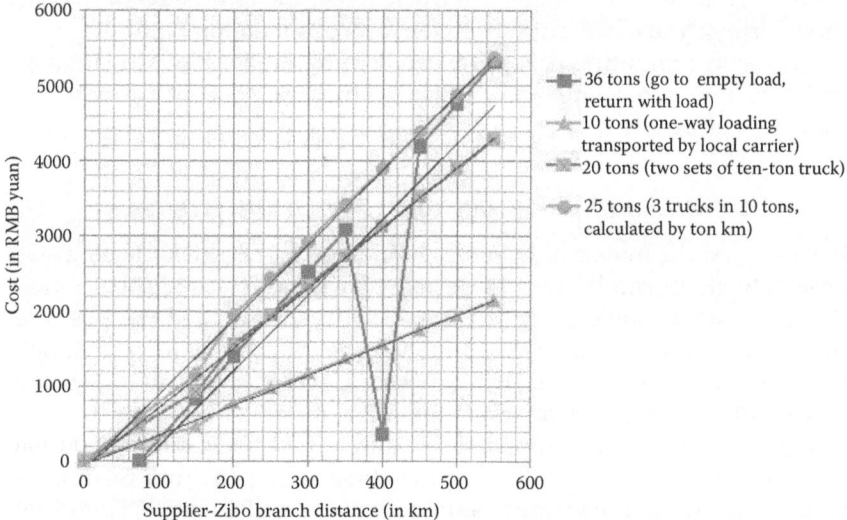

Figure 9.13 Different vehicle properties can vary with the distance and tonnage.

Figure 9.14 Interface of CC consolidation results.

suppliers transport cargoes by small-sized trucks. In this model, Z_0 is the cost incurred from consolidation using comparative computation; after it, if the supplier (s_1) with the largest transporting demand tries to use large-sized vehicle to load cargoes, compare if the transporting cost is reduced; if the cost is lower than Z_0, then this supplier should use large-sized vehicles to distribute the cargoes; if not, then this supplier should keep the consolidation model of small-sized vehicles; by following the sequence according to transporting quantity, compute if it is economical to change large-sized vehicles to transport cargoes; in the end, the vehicle selection model for suppliers and the total cost Z of final consolidation can be given.

Concerning the supplier that should use large-sized trucks as determined by the above-described procedure, the circulating consolidation route should be planned, according to the geographical location of this supplier and parts supply quantity. In the "milk run," the driving route of the large-sized truck should be computed following the shortest route algorithm. If there is any small-sized truck consolidating cargoes to the CC, then the large-sized truck should return to the CC and load the parts at the CC; if there is no small-sized truck in use, then the large-sized truck should directly drive to Yulin from the last consolidation point.

The said computing procedure can be programmed in Java language under Eclipse compiling environment, which can output the result as shown in Figure 9.14.

9.5   Summary of parts network operation model

In summary, the intersection and repetition of YC Diesel supply logistics or supply chain can be greatly decreased with CCs established nationwide. By establishing the parts CC between the supplier and the YC Diesel parts warehouse, the centralized management and operation of supply logistics can be achieved to the maximum. Another advantage of integrating the previously scattered supply logistics business is that transit storage in these CCs can function as Yulin parts warehouses, which can add the parts inventory, buffering the stress mounted from insufficient warehouse area in the vicinity of the Yulin factory (Figures 9.15 and 9.16).

By building the consolidated model of parts supply, establishing the parts CCs nationwide and tailoring the vehicle scheduling method in the

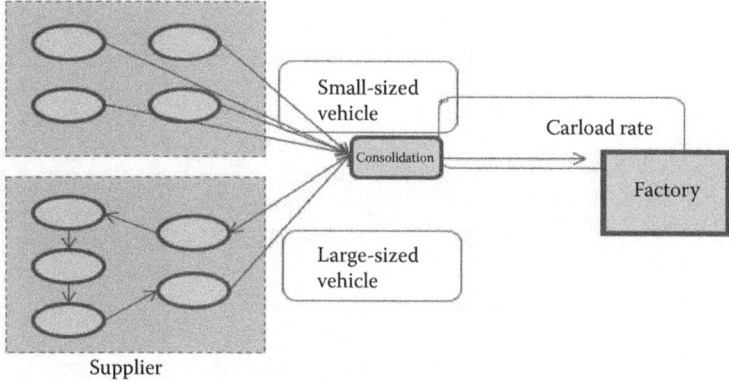

Figure 9.15 Parts supply model of one CC.

dual logistics network, the complementary characteristics of bidirectional logistics between parts and engine can be harnessed to reduce operating cost, increase the response time, and ensure the on-time supply of parts. Also, the centralized and unified system of parts supply can enhance Yuchai Machinery's in-process monitoring on the total parts supply, and a CC can be assigned with the partial function of parts QC, lessening the impact of deferred supply or QC cycle on the parts OOS.

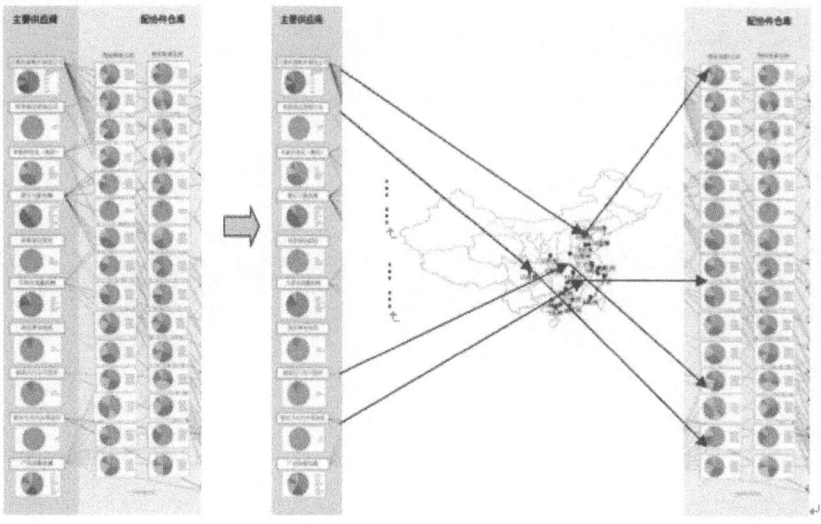

Figure 9.16 Parts supply model of nationwide CC network.

chapter ten

Indirect labor improvement project at an optical communication products company

Binfeng Li

Contents

10.1   Project overview

10.1.1   Overview of the company

10.1.1.1   Organization of the company

The company is a leading provider of optical communication products used in telecommunication systems and data communication networks. It has three branches: one each in the United States; Chengdu, China; and Taiwan. This project takes place at Chengdu.

Figure 10.1 is part of the organization structure of the company involved in the project. The blocks labeled by "QA", "warehouse", "3D-manual", and "D/T" are the objectives of the project that will be discussed later. The figure shows that quality assurance (QA) and warehouse are independent

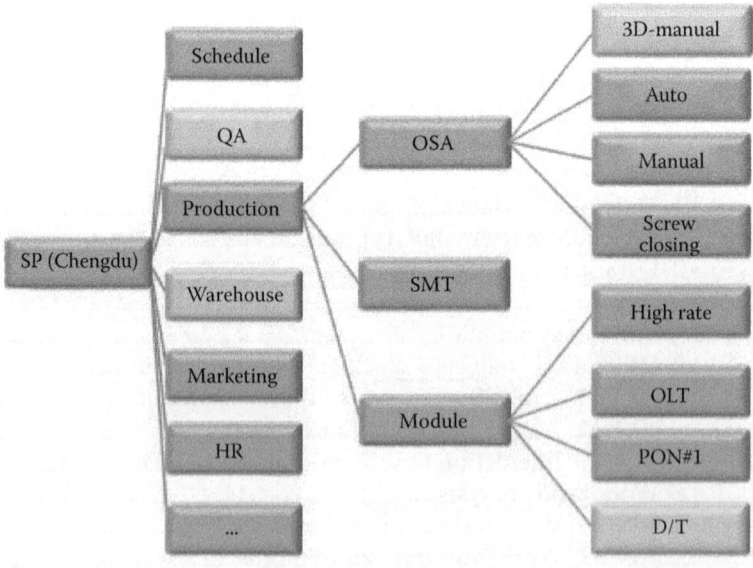

Figure 10.1 Organization structure of the company.

departments, so all the classifications of data collection, data analysis, and suggestions are based on this fact.

10.1.1.2 Product categories

The company designs, manufactures, and sells a broad portfolio of optical communication products, including passive optical networks (PONs); subsystems; optical transceivers used in the enterprise, access, and metropolitan segments of the market; as well as other optical components, modules, and subsystems. In particular, the company's products include optical subsystems used in fiber-to-the-premises, or FTTP, deployments, which many telecommunication service providers are using to deliver video, voice, and data services. Figure 10.2 shows some of the company's products.

10.1.1.3 Indirect labor at the company

At the company, indirect labor (IDL) is the employees who support production, including the production IDL, technicians, quality assurance (QA), and warehouse workers.

In 2011, production IDL, QA, production engineer (PE) on OSA line, and warehouse workers were the four largest groups of IDL, making up more than 90%. This project aims to reduce the on-line IDL, the following types of IDL were the focus:

1. Group leaders
2. Material handlers
3. Technicians (including PE, mechanical engineer [ME], New product inspection engineer [NPIE], researcher and designer [RD])
4. Trackers/statisticians
5. QA
6. Warehouse workers

Of these IDL positions, group leader, material handler, production line technician, and tracker are on the same shift of the production line, so we call them production IDL. Their workload is directly related to the yield.

For technicians, PEs and MEs work is to solve problems on the production line. If a PE cannot solve the problem within 30 minutes or if the problem is mechanical, an ME will take over. NPIEs test the procedures of new product production, and RD does research and designing.

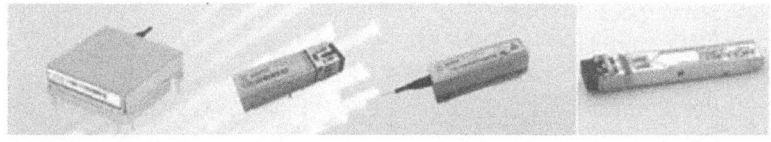

Figure 10.2 Some of the company's products.

There are five kinds of QA. Input quality controller (IQC) inspects all the materials input. Failure analyzer (FA) analyzes the failures and find the cause. Process quality controller (PQC) takes care of process quality control, and sits on the production line to test every piece of product. Output quality controller (OQC) is an output quality controller who makes random inspections when the product is ready to package. In-process quality assurance (IPQA) is a group of quality assurers who walk the production line supervising workers to ensure they are operating correctly and safely.

The warehouse department includes all workers in the warehouse. There are more than ten warehouses, but this project just treats them as two: the raw material warehouse and finished product warehouse. Thus warehouse workers are separated into these two groups.

The current ratio of direct labor (DL) over IDL is 2.89, which is relatively large and has the potential to be optimized.

10.1.2   Project description

The original reason to initiate this project was that management witnessed idle IDL on the production line and wondered whether the amount of IDL could be reduced. So the questions to be answered were:

1. What are the utilizations of all IDL types?
2. Is it possible to improve utilization without loss of production capacity?

The objective of the project was to give the utilization of different types of IDL using work sampling.

To answer the first question, there needs to be objective data collected of time utilization by work sampling. As a supplement, a questionnaire survey to provide subjective data can be conducted. To answer the second question, other data that can be utilization should be collected, a mathematical model should be built, and a logical analysis should be done.

The content of the project consists of three parts:

1. Work sampling
     Sampling
     Related information collection
     Data analysis
2. Questionnaire survey
     Questionnaire design
     Survey conducting
     Data analysis
3. Delivery
     Presentation
     Report

10.2   Study method

10.2.1   Objects of work sampling

As introduced in the "Project Overview" (Section 10.1), the IDL types at the company focused on were group leader, material handler, technician, tracker/statistician, QA, and warehouse worker. For a complete and general study, all of the IDL should be investigated. But considering the time constraints, the number of different types of IDL, and their distributions in the factory, it was decided to investigate all QA, all warehouse workers, and two production lines. The two production lines should be one of OSA and the other of FG. Even then two industrial engineer (IE) technicians were needed to help with data collection.

To choose the two lines, IDL and DL data was collected at each production line. One line was chosen from OSA and the other from the module with the lowest DL/IDL ratio, namely, the 3D-manual line and D/T lines. This is because these two lines may have more serious potential problems of IDL.

The specific IDL types of different departments are shown in Figure 10.3. Considering the characteristics of different departments, a work sampling of 3D-manual and D/T was done together, and QA and warehouse workers together.

10.2.2   Action classification

Before the work sampling, 2 days were spent observing the actions of the IDL and we talked with them about their jobs. We recorded and summarized their actions during work. After that we communicated with the IE manager and chose the actions classification of different IDL types (Figure 10.4), and the code and definition of the actions (Figure 10.5). The code is used for ease of recording when sampling. The codes were carefully selected to avoid confusion of letters when checking the records, which is why the codes are not strictly the first few letters of the names of actions.

Department	IDL type
3D-manual line	Group leader, material handler, technician, tracker
D/T lines	Group leader, material handler, technician, tracker
QA	PQC, OQA, IQC/FA, IPQA
Warehouse	Raw material warehouse worker, finished product warehouse worker

Figure 10.3 IDL types by department.

IDL type	Special work	Common work	Rest
Group leader		Count material Discuss	Chat Entertainment
Material handler	Inform scrap	Handle material Meeting Operate	Sleep Phone Wait
Tracker/ statistician		Record and cheek Sort Supervise	Walk
Technician	Adjust Repair Build station	Work with computer	
QA	Inspect		
Warehouse worker	Store material Release material		

Figure 10.4 Actions of different IDL types.

10.2.3   Sample size

10.2.3.1   Method to select sample size

To guarantee the accuracy of the result of work sampling, a proper sample size must be selected. Referring to the work of Barnes, we used

$$n = \frac{Z_{\alpha/2}^2(1-p)}{pS^2}$$

to calculate the sample size, n. We use $S = 0.05$ as the relative accuracy and $\alpha = 0.05$ as the confidence coefficient. For estimation, we use $p = 0.50$ as the utilization of work time. Then we get

$$n = \frac{1.96^2 \times (1-0.5)}{0.5 \times 0.05^2} = 1537$$

for every IDL type.

10.2.3.2   Design of sampling timetable

To design the sampling timetable, we need to calculate the cycle time of sampling first.

$$\frac{1537\,\text{observations/type}}{2\,\text{weeks} \times 5\,\text{days/week} \times 10\,\text{hours/day} \times 2\,\text{samplers} \times 4\,\text{observations/type}} = 2\,\text{times/hour}$$

Action	Code	Explanation
Adjust	A	Adjust machines
Build station	B	Build new stations
Count material	Ct	Count materials
Discuss	D	Talk about work with other colleagues
Handle material	H	Transfer materials between stations
Inform scrap	If	Inform scrap to technicians
Inspect	I	Inspect workpieces
Meeting	M	Participate in meetings
Operate	O	Help DL with work
Record and check	Ra	Make record or check orders
Release material	R	Release and store materials or products
Repair	Rp	Repair machines or product
Sort	S	Sort materials or documents
Supervise	Su	Supervise the DL
Work with computer	W	Use computer to work, e.g., check e-mail, view data, make report
Chat	C	Talk about things irrelevant to work
Entertainment	E	Use computer to do things irrelevant to work, e.g., surf Internet, view pictures
Sleep	Sp	Sleep on the production lines
Phone	P	Use or look at phones
Wait	Wa	Doing nothing on the production lines
Walk	Wk	Walk toward outside or without obvious purpose
Absent	Ab	Be absent from the production lines

Figure 10.5 Code and definition of actions.

The estimated cycle time is 0.5 hour. Then we use the random number generator in Excel to generate the start time of every cycle to guarantee the randomness of observations. Then we get the sampling timetable (Figure 10.6). The sampling timetable includes a week for night shift data collection for comparison between day shift and night shift.

10.2.3.3   Presampling

To adjust the estimated utilization of work time, p, 2 days were used for presampling of production lines, and one day for presampling of QA and warehouse workers.

Daytime	7/18/2011	7/19/2011	7/20/2011	7/21/2011	7/22/2011
8:30	8:32	8:44	8:30	8:46	8:31
9:00	9:14	9:02	9:04	9:08	9:15
9:30	9:30	9:41	9:37	9:31	9:48
10:00	10:14	10:01	10:15	10:18	10:03
10:30	10:42	10:33	10:39	10:42	10:44
11:00	11:11	11:08	11:04	11:07	11:17
Lunch	–	–	–	–	–
13:10	13:21	13:10	13:18	13:12	13:27
13:40	13:43	13:41	13:58	13:55	13:53

Figure 10.6 Timetable of work sampling (partially displayed).

Since lower utilization leads to larger sample size, we need to select the sample size by the IDL type with minimum utilization. The minimum utilization of IDL on production lines is 0.4968 and the cycle time is still 0.5 hour. The minimum utilization of QA and warehouse workers is 0.4118 and the cycle time adjusted to 0.75 hour is 45 minutes.

10.3   Data summary

10.3.1   Overview of data

During 3 weeks of sampling, we tracked 24 shifts (5 night shifts and 19 day shifts), made 2746 sampling, and got 25,141 independent observations. The rate of progress is the ratio of the number of observations we made over the number of observations we should do with 5% relative error at a confidence level of 95%, which indicates the progress of our work. The rates of progress of all types of IDL are larger than 110%, which includes necessary redundancy for data elimination by the control chart and validity test.

10.3.2   Data summary by IDL type

To answer the first question of the project, the basic information is the time utilization of different types of IDL. The utilization of group leader, material handler, technician, and tracker on both the 3D-manual line and D/T lines were summarized. It was found that the time utilization of all the types of IDL on the D/T lines is higher than that on the 3D-manual line, especially for group leaders and material handlers, whose difference of utilization between the two lines reaches 20%. Because the basic functions of IDL are similar, it is possible to make the IDL on the 3D-manual line achieve the same level of utilization of IDL on the D/T lines.

The utilization of trackers on both lines is almost the same and at a relatively high level, which means a small potential of development is possible under the current situation. The reason for this is that a tracker's work is repeated with low complexity, which is more like a direct labor task than an indirect labor task. So the development of a tracker's utilization should focus on action reduction and optimization.

The lowest utilization on both the 3D-manual line and D/T lines occurs with technicians. Though the difference of utilization between the two lines is significant, it does not matter. However, it should be noted that the compositions of technicians on the two lines are totally different. The technicians on D/T lines are all production-line technicians, whereas those on the 3D-manual line are PEs, MEs, NPIEs, and RDs. So the difference between the two lines can be partially explained by composition differences. However, the problem is the low utilization, which is related to the content of technicians' work. Technicians' work is to keep the production line running properly; there is a trade-off between utilization of technicians and utilization of machines. To achieve high utilization of machines, which can produce great profits, the utilization of technicians is partially sacrificed. The question we should think about is how to make the trade-off, and we will discuss it later.

The utilization of PQCs, OQAs, IPQAs, IQC/FAs, raw material warehouse workers, and finished product warehouse workers was also summarized. As these types of IDL do not have fixed workstations, we cannot record status of "absent," because we do not know where they should be. So the only statuses of the six types of IDL are "work" and "rest."

The first impression of this summary is that the utilization of the six types of IDL is lower than production IDL. IQC/FA gets the highest utilization of 62.6% among QA. PQC's time utilization is only 47%. The intuitive explanation is that a QA's work is discrete rather than continuous, and a great amount of time is squandered on the waiting time at the intervals of work. What is more, especially for PQC and OQA, the task on a workstation is overload for one but too little for two, which will cause low utilization. For IPQA, which has a utilization of 52%, a lot of rest time is spent walking back to their rest area and it does not seem reasonable to require them to supervise the production line at all times.

In the warehouse, warehouse workers in the finished product warehouse have higher utilization times than those in the raw material warehouse. Of course, the quantities of input and output have a great influence on utilization, and we have to take that under consideration in the analysis process.

10.3.3   Data summary by shift

Because the factory runs 24 hours per day, data collection should also be conducted during the night shift for comparison to a day shift. So we spent a week on night shift with production IDL and technicians.

According to the collected data, on average, day shift utilization is significantly higher than that at night. However, in the case of trackers, the night shift utilization almost remains the same as that during the day shift and it is even converse on the 3D-manual line. Recall the fact that a tracker's work is more like direct labor rather than indirect labor, and the small difference between day shift and night shift utilization indicates a small potential for development.

The utilization of technicians on the 3D-manual line also varies within a small interval, which supports the theory of trade-off, telling us the analysis and suggestion for technicians on the 3D-manual line should be treated independently.

Other types of IDL had a significant falling in utilization from day to night. To make it clear, action analysis should be conducted, and then we can explain why the falling occurs and know how to increase the utilization.

10.4   Data analysis

10.4.1   Validation of data

Before data analysis, we must validate the collected data to ensure it reflects the normal working conditions observed. To validate the collected data, we use a P control chart analysis. Here p is the rate of busy time during work time. The upper control limit is

$$\text{UCL}_p = \bar{p} + 3\sqrt{\frac{\bar{p}(1-\bar{p})}{n}}$$

The lower control limit is

$$\text{LCL}_p = \bar{p} - 3\sqrt{\frac{\bar{p}(1-\bar{p})}{n}}$$

$\bar{p}$ is the average busy rate of the specific IDL type. Since the data are about humans rather than machines or products, we cannot be too strict. The control charts show that 3D-manual technicians (mainly PE technicians) and IPQA have large fluctuations of utilization. They have lots of outliners without specific reasons. This is because the work content of PE technicians is to handle exceptions and adjust machines, which is very uncertain. Thus they will be treated differently from the others in the following analyses. The main task of IPQA is periodic supervising, which means the random starting time of work sampling can cause large fluctuations. Besides, we find that the two warehouses may have periodicity and correlation of utilization, and how to confirm this is shown in the following steps.

10.4.2   Utilization analysis

10.4.2.1   Time effect on utilization
Time effect analysis is concerned with how utilization changes with time in a day. The collected data are graphically presented and they show that there is an obvious drop of utilization at every break, which is to be expected. This phenomenon causes utilization fluctuations, thus increasing the number of IDL needed to overcome the peak time of workload and decrease the whole utilization. And, the periodicity and correlation of utilization of the two warehouses are confirmed. The two warehouses are both busy at the beginning of work time.

These data state that there is room to enhance utilization of IDL by easing their workload. For example, if we can ease the work of raw material warehouse workers by extending the production schedule to provide more time for them to prepare materials, the number of warehouse workers needed to handle peak time can be reduced, and the whole utilization will increase.

10.4.2.2   Autocorrelation of utilization
Autocorrelation analysis is to find whether the utilization of a specific IDL type of one observation has correlation with the next few observations. After matching observations with a time series, we can classify IDL types into two kinds:

1. Autocorrelated, which means the utilization at this moment has a significant effect on the utilization at the next observation point. It indicates that the general working time for one task of this job is larger than the interval between adjacent observations.
2. Non-autocorrelated, which means the utilization observed at different time points are independent.

The result is summarized in Figure 10.7. Autocorrelated IDL types have fewer transitions between work and rest, which means they have less setup costs. If the arrival of work is relative stable, their utilization should have a higher expected value.

Autocorrelated	Non-autocorrelated
Technician on 3D-manual line Group leader on 3D-manual line Tracker on 3D-manual line	Material handler on 3D-manual line
Tracker on D/T lines	Technician on D/T lines Group leader on DT/lines Material handler on D/T lines
OQA	PQC, IPQA, IQC/FA Finished product warehouse worker Raw material warehouse worker

Figure 10.7 Summary of autocorrelation of utilization.

10.4.2.3   Interaction of utilization between IDL

If we want to reduce IDL by combining two or more different jobs, the necessary condition is that the utilization of the two jobs have a negative correlation, which means when one IDL type is busy and the other is idle. So we tested every pair of jobs in the same workshop. We could not find a negative correlation within any pair.

10.4.3   Work ratio analysis

We use word *utilization* to describe the overall efficiency of each IDL type, and *work ratio* to describe the record of efficiency of each IDL type. This means utilization is the summary of work ratio records.

10.4.3.1   Work ratio distribution

The distributions of work ratio of different IDL types represent the number of observations of the specific IDL type at different work ratio levels.

The distributions of work ratio of different IDL types are classified into two types:

1. Centralized—The variance of work ratio is relatively small. The observations are centralized around one utilization level. This distribution means the working condition of the specific IDL type is relatively stable. If the current utilization of the IDL type is low, the workload can be increased to improve utilization. Thus the expectation of utilization of this distribution type could be higher.
2. Nearly uniform—The variance of work ratio is relatively large. This distribution means the working condition of the IDL type varies largely from idle to busy. Increased workload may lead to a shortage of manpower at some time. Thus the expectation of utilization of this distribution type should not be too high.

Based on the analysis, we can set high utilization for OQAs, PQCs, IQC/FAs, D/T material handlers, raw material warehouse workers, D/T group leaders, IPQAs, and 3D-manual trackers. OQAs, PQCs, IQC/FAs, raw material warehouse workers, and IPQAs have low utilization according to the work sampling results.

What is more, as mentioned in Section 10.4.2.1, the workload of some IDL types may be able to be smoothed. Then the distribution may be transferred from the nearly uniform to the centralized.

10.4.3.2 Work ratio trapezium model

For DL, there should be an obvious correlation between utilization of work time and output because their work is to manufacture product directly. Thus it should be a monotone increasing fitting curve passing (0, 0) and the overlap line utilization equals 1 after a certain point on utilization versus earned hours (number of finished workpieces times standard labor time) chart. But for IDL, the correlation cannot be so significant and thus it is hard to determine the theoretical optimal number of IDL. So we create a work ratio trapezium model (Figure 10.8) to estimate the current and optimal situation. Here we use earned hours to stand for output.

The work ratio trapezium model consists of three parts (Figure 10.9): a fully redundant area, partially redundant area, and full loaded point. If the current work condition of the IDL type is in the fully redundant area, it means there are always lots of IDL types being idle no matter how much work needs to be done. If the current work condition of the IDL types is in a partially redundant area, it means the work ratio increases with output although there is someone being idle. If the current work condition of the IDL types exceeds the full loaded point, it means the IDL types are sometimes fully loaded and need more manpower.

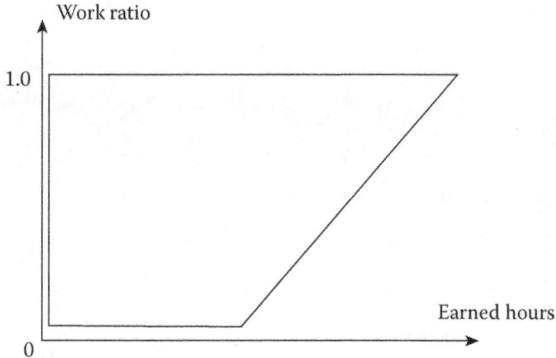

Figure 10.8 Work ratio trapezium model.

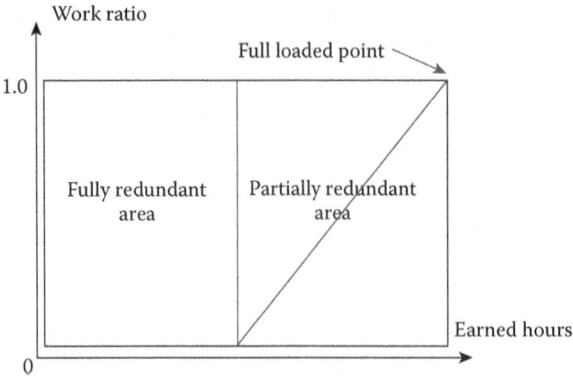

Figure 10.9 Explanation of work ratio trapezium model.

To take group leaders on the 3D-manual line as an example (Figure 10.10), it crosses the fully redundant area and the partially redundant area, which means it has a lot to be improved. The steps of the algorithm to find the estimated optimal solution are as follows (also see Figure 10.11):

1. Find the point with the largest output O and with the work ratio $R < 1$.
2. Find the straight line through the point that rejects $\alpha/2 = 2.5\%$ of all the points, and the rejected points must be with a lower work ratio than the chosen point.

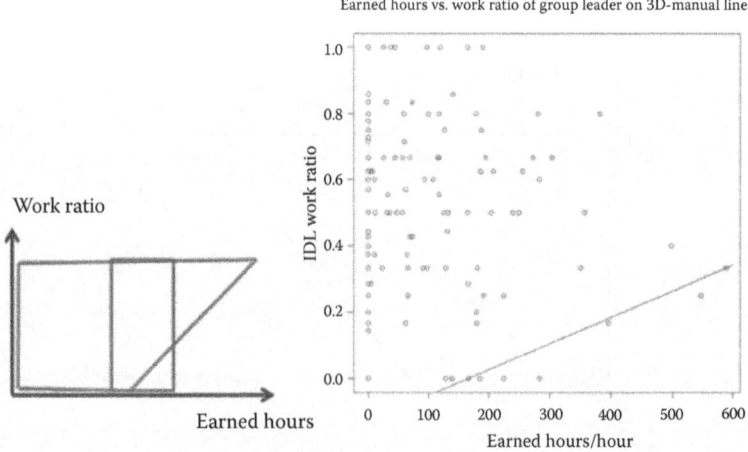

Figure 10.10 Work ratio trapezium model of group leader on 3D-manual line.

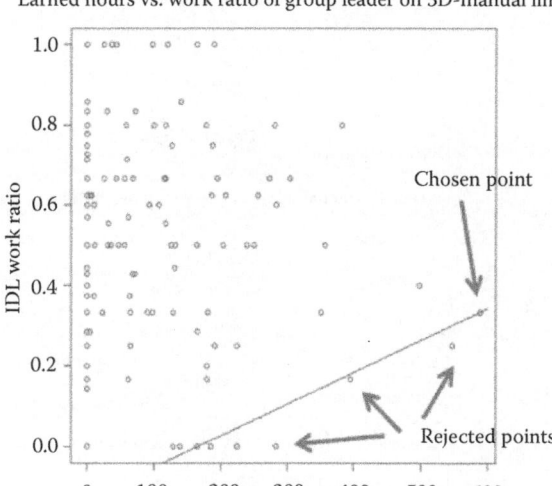

Figure 10.11 Algorithm of work ratio trapezium model.

3. Use the straight-line function to find the output of the full loaded point, namely, *FO*.
4. Reduction percentage (15% allowance) = *O*/(*FO* * 0.85).

The calculated results of group leaders on the 3D-manual line are

O = 588.48 (hour)
R = 0.333 < 1
Number of rejected points = 205 * 2.5% = 5
FO = 1435.16 (hour)
Reduction percentage = *O*/*FO* = 51.8%

The result shows that to reach the full loaded point, the number of group leaders on the 3D-manual line should be reduced 51.8%.

In summary, for technicians on the 3D-manual line, we will use queuing theory for analysis due to the special characteristics stated earlier. For QA, we do not have the proper data to describe their output, so we cannot use this model. For warehouse workers, we need more data to use this model since 2.5% of all the points are less than 1. The negative reduction percentage means the IDL sometimes exceeds the full loaded point and needs more manpower. But if the delay of work is allowable, we can remain with the current situation, namely, reduction percentage is 0.

Type	Action	Type	Action
Type I work	Adjust	Type II work	Discuss
	Build station		Meeting
	Count material		Sort
	Handle material		Supervise
	Inform scrap	Type I rest	Wait
	Inspect		Walk
	Operate	Type II rest	Chat
	Record and check		Entertainment
	Release material		Sleep
	Repair		Phone
	Store material	Ab	Absent
	Work with computer		

Figure 10.12 Action types.

Since the algorithm only gives estimated theoretical reduction percentage, the results should be used as a reference or trial point rather than the final decision.

10.4.4   Action analysis

Up to now, we have only focused on the utilization of IDL types, namely, their efficient use of work time. We also need to analyze the effectiveness of work time of different IDL types. The action is classified into five types (Figure 10.12):

1. Type I work—Work within purview and must be done in time
2. Type II work—Unessential work or work that is not strictly controlled
3. Type I rest—Ordinary rest
4. Type II rest—Forbidden kinds of rest or entertainment
5. Ab—Absent (only applies to production IDL)

10.5   Questionnaire survey

10.5.1   Overview of questionnaire survey

Work sampling gives us objective data about the time utilization. After analysis, we get the information about how the utilization changes and the possibility of optimization. However, after all, this data is based on our observation that may lose some other essential factors. The most obvious one is workload. Work sampling can tell us how much time a worker spends on work actions, but it tells nothing about the workload of every type of IDL, which is one of the major factors that should be considered in target setting and labor reduction. A questionnaire is a good method to reveal subjective evaluation of the workload and to validate the results of work sampling.

In sum, the purpose of conducting a questionnaire survey is to get subjective data as a supplementation to work sampling, to get information about workload rather than utilization, and to listen to workers' opinions.

In order to make a reasonable comparison between work sampling and the questionnaire survey, all the subjects of the work sampling are invited to complete the questionnaire survey. Supervisors are also invited. Because IQC and FA work in the same workshop, which is relatively separated from other IDL, to get a reliable evaluation the type of IQC/FA in work sampling is separated into IQC and FA.

The subjects are listed next:

- Supervisors on both D/T lines and the 3D-manual line
- Group leaders on both D/T lines and the 3D-manual line
- Material handlers on both D/T lines and the 3D-manual line
- Technicians (PE, ME, NPIE, RD, production line) on both D/T lines and the 3D-manual line
- Trackers on both D/T lines and the 3D-manual line
- PQC
- OQA
- IQC
- FA
- IPQA
- Raw material warehouse workers
- Finished product warehouse workers

It was a web-based questionnaire survey. We used a professional questionnaire survey website, http://www.sojump.com, to conduct the survey. Figure 10.13 is the first page of the questionnaire. To distinguish the questionnaires that are answered unceremoniously, we designed a test question that asks what company you are working for. If one's answer to this question was wrong, the questionnaire would not be accepted by the system.

10.5.2 Validation of data

To get reliable information about the real situation from the questionnaire survey, the validity of data is quite important. A three-step filter was used to eliminate unreliable data.

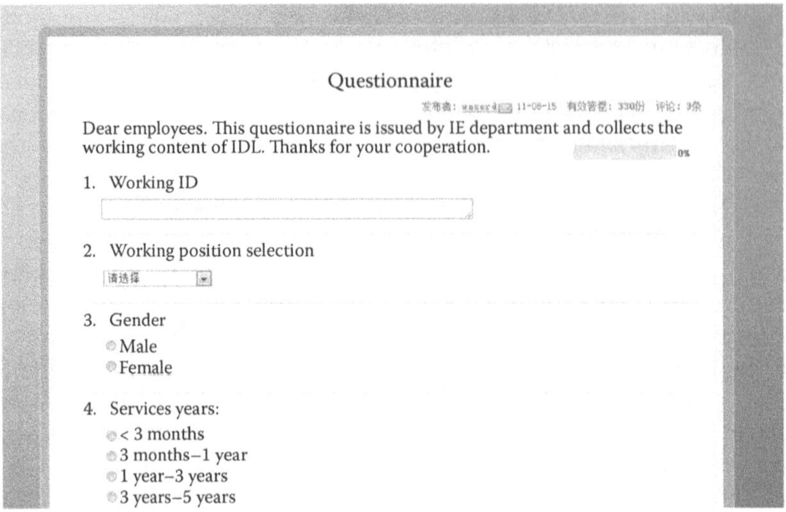

Figure 10.13 First page of questionnaire.

Step 1—Test question. As mentioned earlier, a test question of "What company are you working for?" is inserted in the questionnaire to refuse the unceremonious questionnaires. After this step, 312 questionnaires remain.

Step 2—Skip freshmen. Perhaps a questionnaire is answered seriously, but it still has little value if the respondent does not know the company or other's work well. So we designed a question about working years and deleted all questionnaires completed by respondents who had been with the company 3 months or less. As most IDL at the company are promoted from operators and few IDL are freshmen, only 23 questionnaires were deleted and 289 remain.

Step 3—Paradoxical mistake between grading question and ranking question. In the questionnaire, question 5 is a matrix question asking the 7-level grade of workload of different types of IDL. Question 7 is a ranking question to ask the rank of workload of different types of IDL. It is logical to require the answers to these two questions to coincide.

The transformation is easy to understand. We give the value of 1 to 7 to the 7-level grade; the higher grade one type of IDL gets, the higher workload it has. Transmit the value of grade to the ranking question, and we can get an array of numbers that indicate the workload of different types of IDL. If the two answers totally coincide, the array should be a monotone decreasing array. If a number behind is larger than the number before it, we call it a paradoxical mistake. And if there are more than two paradoxical mistakes in the array, the questionnaire is deleted.

10.6   Conclusion

To figure out the work conditions of IDL and put forward improvement suggestions, we used utilization analysis, work ratio analysis, and action analysis for the data from work sampling, and conducted a questionnaire survey.

1. Utilization analysis—Time effect analysis of utilization analysis tells us that the work conditions of IDL vary within a day; some IDL types are especially obvious. Thus there is room to smooth workload to reduce IDL.

   Autocorrelation analysis shows the continuity of the work of each IDL type and helps to decide the expected value of utilization of different IDL types (Figure 10.7). The 3D-manual group leader and tracker, D/T tracker, and OQA should have higher expected utilization according to this analysis. The 3D-manual technician is not mentioned because of its special work content.

   Interactions between utilization of different IDL state that there are no negative correlations between IDL types and thus no possibility to combine work of different IDL types.

2. Work ratio analysis—Distribution of work ratio gives us qualitative references for target utilization of different IDL types. IDL types with centralized distribution of work ratio and low current utilization have more potential to be improved, namely, OQA, PQC, IQC/FA, raw material warehouse worker, and IPQA.

   The work ratio trapezium model shows the relation between work ratio and output, and provides quantitative references for reduction of IDL (Figure 10.10).

3. Action analysis—Action analysis compares the action component of different IDL types and shifts, and gives qualitative evaluations, and weighting analysis gives quantitative evaluations of actions of IDL types and shifts. IDL types with a weight score below 65 and utilization below 65% means they may have more potential improvement.

4. Questionnaire survey—We received 312 questionnaires in total and 272 of them were valid. The questionnaire survey provides subjective evaluation of IDL's work conditions. The results from the work

sampling were compared to that from the questionnaire survey. The comparison shows no big differences except for PQC, and this situation will be discussed in the next section.

10.7   Discussion

10.7.1   Standard of utilization

We have the utilization of all types of IDL by work sampling. With data analysis, we know why the utilization of some types of IDL is high or low. Now, the problem arises as to how to set the standard of utilization of different types of IDL. All the factors that have effects on utilization should be taken into consideration. Besides current utilization, three other factors are considered.

1. The ratio of error rate over workload—This ratio represents the marginal cost of workload. With the increment of workload, the utilization will increase and the error rate will increase. Only when the benefit of utilization increment meets the loss of additional failure can we increase workload.
2. Variance of workload—The larger variance of workload means we have to prepare for the peak and loss efficiency at trough. So types of IDL with small variance may have a higher expected utilization level. There are two factors effecting variance: one is the change of workload, the other is duration of each task. If every task lasts for a very short time, frequent setup costs will be an obstacle to high utilization.
3. Influence on the whole line utilization—If the work of one type of IDL is the bottleneck of the whole production line, the increment of its utilization will cause lots of loss at other steps. If the influence on other steps is smaller, it is easier to make changes.

If the levels of them are low, it means the target of this type of IDL can be expected to have a high utilization.

10.7.2   IDL reduction

After setting standards of utilization, it is time to start discussing how to reduce IDL.
1. The first thought of IDL reduction is to combine tasks of different types of IDL. Unfortunately, in Section 10.4.2.3, we proved that job combination of two or more IDL types is impossible.

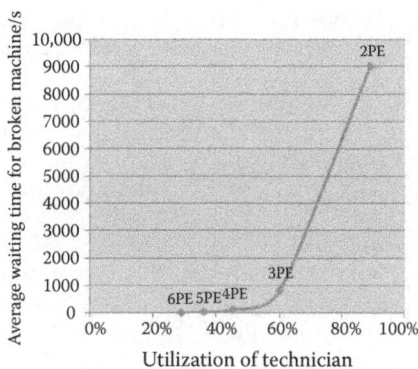

Figure 10.14 Result of simulation example.

2. Work ratio trapezium model's suggestions—For production IDL, assume that the product capacity is stable in the long term. Using the work ratio trapezium model we can get the largest reduction ratio.
3. For PEs, MEs, and other technicians, their work is to solve problems in production, so the information about frequency of problem occurrence and time spent on solving problems is extremely important. If we have all the time information, according to queuing theory, we can treat it as a G/G/s server system. By simulation, we can make a trade-off between cost of labor and cost of waiting for a broken machine.

Figure 10.14 is the result of an example of simulation. When the number of PE technicians reduces from six to two, the utilization increases from 30% to 90%, but the average waiting time for a broken machine increases rapidly. If we have the information about marginal cost of labor and machine waiting times, we can make the best choice to maximize profit.

Bibliography

Barnes, Ralph M., *Work Sampling*, 2nd ed., John Wiley & Sons, New York, 1957.
Gouett, Michael C. et al., "Activity Analysis for Direct-Work Rate Improvement in Construction," *Journal of Construction Engineering and Management.* Submitted September 3, 2010.
Gurumurthy, Anand, Kodali, Rambabu, "A Multi-Criteria Decision-Making Model for the Justification of Lean Manufacturing Systems," *International Journal of Management Science and Engineering Management*, vol. 3, no. 2, pp. 100–118, 2008.
Niebel, Benjamin, Freivalds, Andris, *Methods, Standards, and Work Design*, 11th ed., McGraw-Hill, 2003.

Thomas, H. Randolph, "Labor Productivity and Work Sampling: The Bottom Line," *Journal of Construction Engineering and Management*, vol. 117, no. 3, September 1991.

Thomas, H. Randolph, Daily, Jeffrey, "Crew Performance Measurement via Activity Sampling," *Journal of Construction Engineering and Management*, vol. 109, no. 3, September 1983.

Thomas, H. Randolph, Guevara, Jose M., Gustenhoven, Carl T., "Improving Productivity Estimates by Work Sampling," *Journal of Construction Engineering and Management*, vol. 110, no. 2, June 1984.

Tzeng, Gwo-Hshiung, Chang, Chun-Yen, Lo, Mei-Chen, "The Simulation and Forecast Model for Human Resources of Semiconductor Wafer Fab Operation," *IEMS*, vol. 4, no. 1, pp. 47–53, June 2005.

chapter eleven

Groundwork for industrial engineering inside enterprises

Jiarao Zhu

Contents

The implementation of industrial engineering (IE) tools typical of Lean Production (LP) in enterprises in China has proven to have a crucial effect on production operation management. For more than 20 years, information technology (IT) has been rapidly applied in the manufacturing industry. During this period, IT system-based IE management ideas and tools (e.g., enterprise resource planning [ERP], material requirements planning [MRP], supply chain management [SCM], manufacturing execution system [MES], and advanced planning and scheduling [APS]) have been intensively used. Only with the support of these IT systems could the LP-based and cross-functional IE management concepts and tools be harnessed.

However, not all IE can be successfully practiced in enterprises and those management ideas packaged in an IT system can never replace a manager's expertise. With a view to LP, sometimes the user may have a mixed feelings: Managers hold different opinions about LP, though they are always interested in discussing the "good" LP potential, while the arrival of the mysterious LP would often give rise to doubt or opposition. Some IT systems that should have enhanced LP management are construed as shackles instead of facilitators in the opinions of some enterprise managers. Moreover, the behavior or conduct of employees is not changed with the rapid application of these management tools. That is why some practitioners of LP call these stylish management words buzz-words, meaning words that are only popular for a short time and soon disappear from usage.

What impedes IE from being successfully applied in enterprises? How can an enterprise root an IE tool into its culture rather than making it become a passing trend? In this study, I introduce the essential factors for successful IE application in enterprises by sharing my experiences.

11.1   Industrial engineering (IE) and enterprise strategy

An enterprise strategy provides a guideline for the survival and development of an enterprise, by which all managers need to define and design the vision of development and make the overall roadmap for this development vision. Briefly, it clearly sets out what the enterprise wants to become and how to make it happen for the next 5 or more years. Guided by this development roadmap, enterprises have to face different challenges during different development stages, and as such management teams should focus on different key points, solve different problems, and use different methods and tactics. Furthermore, due to limited resources (human resources, investment, time, etc.), professional managers should choose the right management tool to be applied and ensure that the most cost-effective tool is used at each stage.

As a management tool, IE tools should serve the enterprise strategy and fit the current development stage. To be healthy and well rounded, a child needs to eat different food, do various exercises, and receive different kinds of education at different ages. An unbalanced diet or education may be unfavorable to his or her growth. The difference between raising a child and managing a company is that guidelines have been (relatively) established to help organize the accommodation and education of a child, while there are no fixed rules for managers to follow. They have to consider the overall situation of their business and make the choices they believe to be appropriate and reasonable. A simple copy of good practices from industry benchmarks will not necessarily bring success. The idea of implementing IE tools to show off will put the enterprise even further away from real leanness. Inside one industry, different companies are at different stages of development. Some companies can start the implementation of a relatively complex management tool (such as flexible manufacturing), whereas some others still need to make efforts to put the most basic management tools into daily practice (such as a production plan). Successful implementation of complicated management tools often means that the enterprise's management has reached a high level. However, a successful manager will not use complicated tools only as elegant decoration. Particularly for those enterprises that are still suffering from weak management, mangers should focus on the basics. To implement IE tools, good managers should be determined, clear minded, and down to earth. They should set goals for each phase of company development and take solid steps to transform management ideas into reality. They believe that there is no silver bullet to solve all problems. And because of the importance of details, a small error of implementation of management tools may quite often result in big deviation from the initial intention.

Major problems faced by manufacturing enterprises at early development stages can often be summarized as unable to sell or unable to produce. Enterprises at this stage usually need to struggle to find effective channels to sell its products and services. And they may also have difficulties delivering the products and services to fulfill customers' demands. In such cases, research and development (R&D) and sales at both ends of the value chain should be in the focus of management, because the initial presence of an enterprise is decided by these two ends and its overall efficiency largely depends on these two ends. Yet when the enterprise has strengthened its presence in the industry, overall operation efficiency will significantly improve the enterprise's competitiveness. Managers should then introduce different types of management tools, including IE tools, according to the company's needs. Continuous improvement of management system effectiveness and operation efficiency is necessary.

For example, I was once in charge of the production management in a manufacturing enterprise. A few years ago, the parent group to which this enterprise is affiliated was implementing IE and LP, and the subsidiary as well began to exercise some of the LP strategies, such as inventory reduction, total productive maintenance (TPM), and flexible production. Yet many concerns were hidden in the trend, or at least we thought so. It quickly occurred to us that the state of the subsidiary was quite different from the parent group. First, the subsidiary was a new industry for the parent group, and the technology of new production lines was not sufficiently mature. In addition, product design covered the intended models of main products, but the products were found with unexpected design defects after production and sales, due to underdeveloped R&D processes and the insufficient experience of technical personnel. So we had no choice but change the product design. This resulted in a frequent change of design after products were launched and further triggered the continuous change of technology, raw material supply, and inventory allocation. The change of technology made it hard to fix the standard unit time and the change of raw material volatized the material supply. Because of these factors, it was hard to balance production workload and we often saw workers waiting for materials and low work efficiency from personnel. In addition, the marketing channel had a loose grip on market information and insufficient knowledge of customer demand. In peak season, marketing was inclined to overrate the sales plan for competing products;

and in the down season, marketing underrated the sales demand with the concern of overstocked products, which led to the ultralow accuracy of sales forecast, enlarged volatility, and more difficulty executing the production–sales–inventory plan. Last, most of the managers did not have systematic training and did not understand well management tools including IE tools. After applying a tool, low- and middle-level managers might easily focus on the form of the management tool instead of the actual effect. For example, unit time of different work stations, equipment operation data, and other basic data required for capacity calculations were far from being reliable, and though capacity models of every production process were built for capacity calculations and capacity bottlenecks identified, these have little instruction for actual work. And the production plan still greatly depends on the available experience.

Likewise, there were many other problems and the management team believed that it was not good timing for rapid implementation of LP. In consideration of the overall strategy, it was crucial to improve the product design, improve the management of marketing channels, standardize the sales plan, establish the internal production management system, and address material problems that trouble the day-to-day operation. No IE or LP would take effect in the company if these problems were not addressed. After a review of the company operation with the management team, we decided to decrease the extent of business and concentrate the management resources on the basic corrections of operation and management. For the introduction of IE, the basics is an important part. For our company at that stage, more important task should be make preparations for introduction of LP/IE tools rather than implementation of these tools.

After confirming the overall strategy of improvement, the improvement of R&D and sales management was given the highest priority. With respect to production operation, we cooperatively standardized the purchase and production flow, and improved the S&OP (sales and operations plan) based on the modified product design. Also, some basic data such as output, quality, and personnel from the daily production and operation was recorded. Thanks to these efforts, internal operation was readjusted to the standard-running path.

At that time, most IE tools applied in our company were just basics, such as work measurement and 5S implementation. These most basic functions were done to gradually develop the systematic thinking of managers and supervisors, quantify management habits, focus on details, and train those industrial engineers to be qualified. In the previous 5S implementation, for example, workshop mangers emphasized regular checks and supervised workers to clear and clean the site as per the criterion, mainly depending on personal feelings. However, no one tracked or reviewed the work progress concerning solutions to on-site problems after each check, which led to slow correction. We immediately corrected those detailed problems, which were previously neglected, after we concentrated on the basic IE. We made the criterion of quantitative evaluation of on-site 5S. Thus, the organizer was able measure the overall progress of on-site correction and employees could also know the result of their own efforts. The tracking of on-site problems not only accelerated the on-site correction but also developed the habit of lower-level managers of following the·plan–do–check–act (PDCA) duty-cycle operation. And in previous measurement of unit time, the measurer did not divide the measured unit time into takt time, equipment time, or labor time (the definition of unit time and the problem of work measurement are also described in Section 11.3.1), because he did not focus on the purpose of unit time for production capacity planning, cost accounting, or personnel planning. Further, in the process of work measurement, the job division used by the measurer was discrepant with the activities of actual production, meaning the measured data of unit time could not be directly applied in workshop management. Moreover, engineers were unable to contrast those measured data due to low degree of standardized documentation of work measurement, worsening the work inheritance. In the process of reinforcing the basic management, we also solved these problems.

In the process of executing these basic functions, not all common staff could quickly adapt to this kind of Lean management; someone even contended that this Lean management was too harsh. We often noticed the feeling of low-level managers that there was too much know-how with a tool. It was these ideas that we hoped to impart on those low-level managers. These tasks seemed to be easy, but the intended effect could be hardly reached if we did not have it applied strictly and carefully. Yet without a solid management base, the rapid implementation of more advanced and complex management tools would be double the work.

These preparatory works appeared to be common enough, so after about half a year, we began to import other IE tools into production operation, such as kanban, quick production exchange (similar to Toyota single-minute exchange of die [SMED]), inventory optimization, and labor productivity management. For most low-level mangers, IE remained a novel management tool, though they had some experience. It was hard for them to immediately take an active part in these reforms only depending on limited experience of IE practices. Therefore, we piloted a tool before extensive application, thereby significantly improving the progress and effect.

Through this practice, I am convinced that for IE tools implementation, even if we are forwarding on the right direction, the sequence of tools implemented and the right timing are also prerequisites for success.

11.2 *IE function and organizational structure*

The discussion on the identity of management, either in technology or art, is often heard. IE as a management tool can be best described as the technology with the content of art. Its common feature with other technologies is that this management tool is a kind of science that can be either attested or falsified. In IE practice, the expected target will not be fulfilled if it goes against the objective law. Different from any other technology, the practical application of an IE tool greatly depends on the human factor, which reveals the strong artistic feature of IE. Organizational structure or human resources (HR) management methods and rules can determine the success of IE.

First, the position of the IE function in the enterprise organizational structure should be focused upon. Small-sized enterprises might no employ a full-time industrial engineer or solicit external IE advice from a consulting company or individual consultant. However, if the enterprise grows to some scale, it is a must to enable a specific IE function. In most cases, the IE function is unlikely to become one of the major functions, except for some special industries (e.g., the management consulting industry). Despite this, it can function as the multiplier of business efficiency. This multiplier can be installed in different positions of an organization. Its position in organizational structure will always determine its functionality.

The industrial engineer can assist the decision of main business to improve the advisory role of operation efficiency. Some decisions are associated with production management instead of daily life. For instance, is the current capacity insufficient or excessive? Should we adjust the number of production operators? Should we invest more or increase the amount of equipment? Should we change the layout of the workshop? If so, how should we change it? Although these tasks do not create direct output, they have an essential effect on improving the efficiency of production operations. The industrial engineer can play a major advisory role in making such decisions when a direct report to the manager in charge of production operation regarding the organizational structure will be an appropriate choice (see Figure 11.1).

Under the organizational structure in Figure 11.1, the IE function stands in parallel with the production operation unit (i.e., production workshop) and a direct report is given to the chief operating officer (COO), compared to the relationship between headquarters and frontline troops. This organizational structure will highlight the IE function of assistant decision making, and allow industrial engineers to identify the problems and possible improvements in connection with daily production operation due to the close relationship between the industrial engineer and daily production operation. Improvement action is easily done as they belong to the operations department. In actual work, the drawback is that the industrial engineer may act in deviation of the intended IE course and

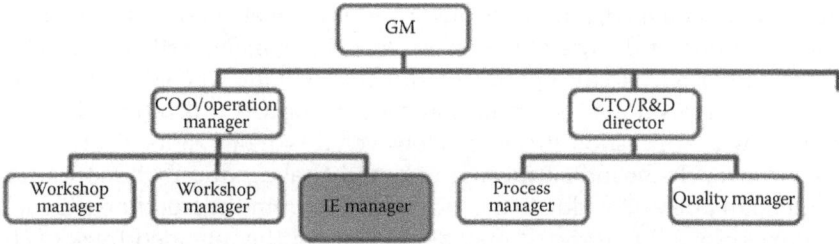

Figure 11.1 Advisory IE function.

would do some firefighting work or daily management work instead of improvement work if the COO could not under. In contrast with the following supervisory IE function, industrial engineers under this organizational structure have a weaker monitoring effect on operation efficiency.

Another setup emphasizes the monitoring effect of the IE function on the efficiency of production operations. When the enterprise grows to some scale and operation efficiency has led to enterprise competitiveness, it is necessary to monitor the operation efficiency of the operations department with respect to the organizational structure. For instance, the manufacturing enterprise monitors the financial efficiency of the overall production operation. This is usually done by the finance department, which does the cost accounting of production operations. Financial efficiency is the comprehensive measurement of the efficiency of all resources used in production operations. In addition to this evaluation of overall efficiency, some enterprises also need to monitor the efficiency of other special resources, such as personnel efficiency (labor productivity) and equipment efficiency (equipment comprehensive utilization). In such cases, managers can assign the user or allocator of resources to different departments or functions under this organizational structure in order to vigorously monitor the utilization of resources. This effect can be enhanced provided that the IE function and production operations department can be separated in this way (see Figure 11.2).

While under this organizational structure, any demand of resources (e.g., personnel, equipment, site) from the production department will be

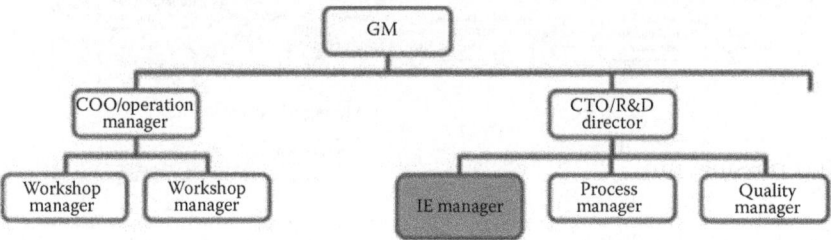

Figure 11.2 Supervisory IE function.

measured, calculated, and confirmed by the relatively neutral IE function (subject to little influence of the operations department), rather than it be decided internally. In the opinion of the general manager (GM), this monitoring mechanism makes for the control of resources utilization, thereby improving the operation efficiency. However, this organizational structure has a drawback: the industrial engineer might have reduced knowledge of production operations due to the long distance from the frontline production operations. Moreover, it may deviate out of the intended track of IE work. In a word, the improvement project being carried out by industrial engineer is not necessarily the urgent problem of production operation.

In reality, these two methods cannot be simply assessed, but managers should rationally set up an organizational structure according to the degree of tacit cooperation and company culture between the members of the management team.

Besides the setup of the internal organizational structure, another problem influencing the IE function is the career path of IE personnel. An industrial engineer who can serve for a long time should be treated as vital for any enterprise, but few such engineers exist. Most industrial engineers are facing a realistic problem: What will I do after 3 to 5 years of IE work? What is my career path? Supervisors or managers of IE departments may also face a problem: how to maintain the team operated by a rational personnel structure. The career path of an industrial engineer that offers a potential solution to this problem is shown in Figure 11.3.

Junior industrial engineers can be recruited from polytechnic graduates or internally recommended from other departments. In general, an engineer is able to deal with all IE functions in the responsible area after about 3 years of relevant training and practice. In such cases, they have to make the first choice in career path. There are definitions given to different development paths, such as the technician path, expert path, and

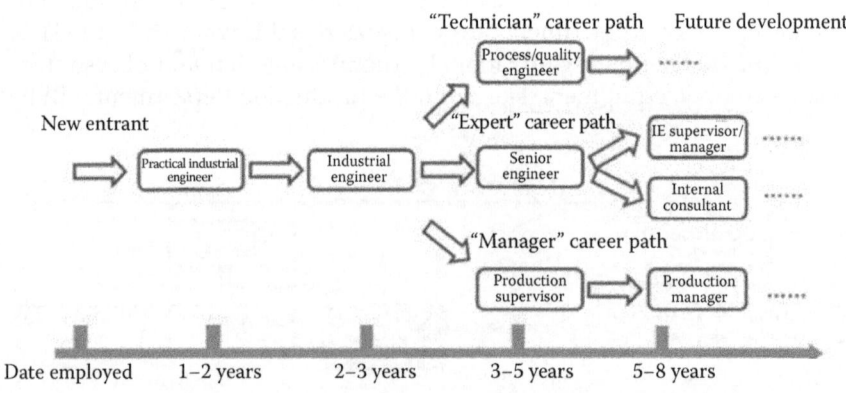

Figure 11.3 Career path of an industrial engineer.

manager path. If an engineer decides to go deep into the IE field, he or she will be given an opportunity to become an IE expert who provides support for internal improvement and consulting, or the leader of IE team. In the technician path, an industrial engineer can move to any work emphasizing product technology like process and quality. Although this development path appears to not be associated with IE, these technical departments often cooperate with IE departments in actual production. To that regard, the cooperation between the technical team and IE team can be enhanced with some IE knowledge. The third career path indicates that an industrial engineer can also be the production manager. Due to systematic training received and practicing of operation improvement, industrial engineers normally fit in the role of operation management within short time. By doing so, another benefit is production improvements can be easily supported and smoothly initiated when the production, operation, or management manager has a well-founded knowledge of IE. If the enterprise gives support for this career path with respect to the personnel system, the IE manager will not restrain the internal flow of industrial engineers. Instead, personnel flow can make for the internal transmission and popularization of IE-related knowledge to some extent.

I once worked in a subsidiary under a parent group. When I was hired, its IE was in the early phase of implementation. Some basic functions such as work measurement had been carried out, but these functions were not mature and standard. The majority of IE tasks were finished by part-time industrial engineers; only a full-time engineer was responsible for work measurement and management, while a large amount of standard unit time was set by process personnel and material consumption in the designing phase. As to salary, the engineering technician was well-acknowledged with a high salary, whereas industrial work was treated as management or assistant work. As a result, many engineers in charge of IE functions were less motivated. On the other hand, the workshop was still loosely managed because IE technicians were not notified of the workshop status, though a lot of basic IE tasks needed to be carried out. It was known to all that many corrections should be done in the entire production, but we could not figure out a way. For instance, the workshop manager and production planner often calculated the capacity and personnel demand at a regular time, due to capacity and personnel problems. Process personnel also often defined the unit time, but they could not figure out how to apply the unit time data to do capacity calculations and personnel allocations. The workshop manager lacked the systematic management skills and basic IE expertise. This was due to the physical long distance of the IE technician from actual production tasks. Industrial engineers should continuously push for reform, but not all industrial engineers can do so. They could study the process technology-related problems, but they were not good at pushing for organizational reform and process improvement.

A consensus was reached after several discussions with the production supervisor and R&D director. To be exact, it was decided to reinforce the instructive (advisory) function of IE for the current phase of production operation. To this end, we organized an IE team led by the production supervisor. Work measurement, as previously carried out by process engineers, was assigned to IE team. Some process engineers were transferred into IE team and became industrial engineers. And personnel calculation, capacity calculation, and other functions, as previously carried out by the production workshop, were reallocated to the IE team. Within this IE team, several industrial engineers were in charge of production workshops. They were held responsible for calculating and planning the personnel and capacity in each workshop, and carried out the improvement project in each workshop as per the annual plan from the production supervisor. This organization not only coordinates the resources (mainly personnel and site) required by several workshops under the unified command of the production supervisor but also enables the IE team to monitor workshop efficiency. In the end, we adjusted the salary payable to industrial engineers and kept it basically in the same level with that of product R&D personnel.

In the initial phase of reorganization, some young engineers showed concerns when transferred from the function of process technology function team to the IE team. They feared that they would not enjoy the social status as internal technicians and doubted their future career development. In this case, the first-time support from the management team played an important role. The IE team first worked with the process department and workshop to confirm unit time, capacity, and personnel based on the product structural adjustment. The IE team also went about some improvement projects, among which were two projects led by the production supervisor and GM. Industrial engineers had a sense of achievement for the progressing basic functions and improvement projects. Two years later, we thought that the IE team was fully encouraged when for the first time an industrial engineer was promoted to workshop manager.

A few years later, when I flashed back on the reform with my former colleagues, it was agreed that a concentrated focus on IE work and a reduced command chain were the right decisions in alignment with the company management strategy. Of course, any kind of change will be in vain if the manager is not clear about the aim of reorganization. The most appropriate organizational structure may vary from enterprise to enterprise. Managers should be aware of the effect of the IE position on work progress, and they must take all factors into account before making a final decision.

In the process of IE implementation, human factors include the requirements of a person responsible for IE (generally the IE manager). The IE manager is considered as the lead of IE implementation and is required to do the following. First, the IE managers should expertly operate various IE tools, which means that they should not only know the exact procedure for operating each tool but also be fully aware of the effects and application conditions of each tool on the company's operations. Second, IE managers should be familiar with every detail at the microlevel and understand the macrosituation, because IE is a transitional function between enterprise strategy and operational improvement. At the macrolevel of enterprise strategy, IE managers should know the features of the industry and the executed strategy. At the microlevel of enterprise operation, IE managers should know the product and service, the processes applied in production operations, and the internal standards and systems. Last, as the executive of reform, IE mangers should be effective communicators and continuously do cross-functional communication. In the process of IE implementation, IE managers should know the demand and concern of each function. IE managers should help employees at different levels to reach a consensus. For first-line operators, IE managers should describe the company strategy in plain words to help those operators understood their role; whereas for the enterprise manager, IE managers should be able to summarize the essential points of detailed problems in concise and simple words, allowing these managers to rapidly grasp all kinds of strategic problems to be solved in the company's operations. It is commonplace to get resistance during the reform, but the IE manager needs good communication skill and a strong will. All these requirements can lead to an excellent IE manager who is vital to the enterprise. The model for a competent IE manager is shown in Figure 11.4. Technological factors

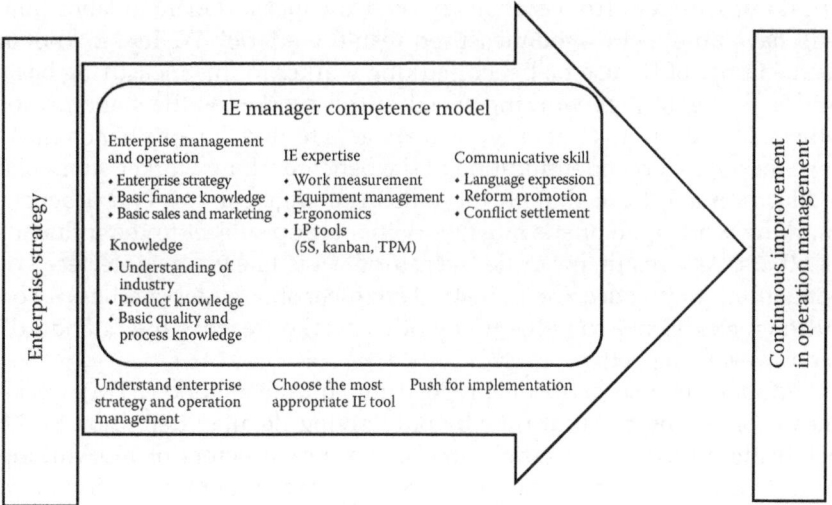

Figure 11.4 The competence model of an IE manager as determined by the work mode of the IE manager.

can only succeed when combined with the human factor, which is also the charm of IE.

11.3 IE tool combined with enterprise operation system

As described earlier, IE is a technology depending on human operation (staff members), while the behavior of staff members is standardized and guided as per the operation system of the enterprise. To that regard, IE application should be colocated with the operation system. If properly colocated, they can promote one another.

11.3.1 Case study 1: Work measurement and personnel system

As well accepted, work measurement is the prelude to most IE tasks. Capacity calculation, cost accounting, personnel management, equipment management, and other related work are based on reliable data of unit time. Unit time can be divided into the components of two different properties, namely, equipment time and labor time. To be exact, equipment time is determined by the process parameters of production equipment, and its error will be controlled within a small range, provided by normal operation of equipment and fixed process parameters. But the measurement of labor time is special and operators are specially measured. A rational person will consider the possible influence of this measurement and change his own behavior with justified reason. Briefly speaking, different from equipment time, the measurement of labor time will have an effect on the measured result. Frederick W. Taylor, known as the father of IE, used a benchmarking worker for his measuring basis and took the average working speed of this worker as the standard for others. However, it is not easy for managers to find Taylor's benchmarking worker, and the identification of the benchmarking worker should be well grounded. Even though the industrial engineer can find a benchmarking worker, he or she must work in a group subject to the influence of others. As a result, his or her measured act is likely to be interfered or influenced. In practice, the industrial engineer should depend heavily on his own experience to judge if any other worker reaches the "standard" labor working speed.

Modern IE also performs predetermined time method (PTM), which means obtaining the unit time by only giving detailed definition to the act of the measured operator without any measurement of his working time. However, in some cases, it is hard to give the standard definition to the act of the operator as per PTM (the operation needs to be broken down into motion elements or motion element mixes), especially when there is any repeated sequence and low frequency. And for some cases, unit time cannot be accurately obtained by either direct measurement or PTM. With manual operation of inspection, for instance, the inspection speed may sometimes conflict with the quality of inspection. Further, manual time also contains too many considerations in addition to physical action, and the speed of consideration is hard to measure. Although some PTM tools like the methods-time measurement (MTM) and the Maynard operation sequence technique (MOST) give definition to some unit time with consideration, there are nothing else but several kinds of simple check actions. PTM is often questioned in that PTM comes from an operator's measurement of motion element, which does not classify gender. Plus, PTM imposes a heavy workload on the industrial engineer.

When I was a young industrial engineer, a lot of time was spent considering the aforementioned problems and it was almost impossible to determine the most perfect, closest to theoretical value and most equitable manual unit time. When I was promoted to a management position, I came to realize that the measurement of manual unit time is not only a technical problem but also an overall management problem. It touches the management of front-line operators and the company's human resource policies. The purpose of unit time measurement was not to find out the unit time value closest to absolutely correct, but to provide data support for production and operation management. If work measurement serves this purpose well, it is successful. A company's management system will influence work measurement. What is important is that a good mechanism should be constructed that ensures appropriate work measurement, and allows both employees and the employer to benefit from the improved efficiency.

Many enterprises use partial or full piece rate wage for first-line operators. Quite often, these kinds of incentive payment systems will increase operators' output. This wage system can be described in different incentive curves as shown in Figure 11.5. Different modes are reflected in different incentive curves. In mode 1, a worker's bonus increases with the output. In mode 2, the bonus increases gradually slow down with the increase of output. In mode 3, the bonus gradually increases with the output. The incentive curve as shown in mode 1 is the most frequently used due to the wage accounting and is the most rational curve as directly felt by most viewers. As a matter of fact, these different incentive curves will directly influence the performance of operators, causing an effect on the measured result, as shown in Figure 11.6. In mode 1, the operator's performance level (i.e., output level) is evenly distributed due to the bonus increasing with the output. In mode 2, the growing rate of the bonus is gradually decreasing and the operator's performance level is roughly distributed around the inflection point. In mode 3, the growing rate of the bonus is gradually increasing with the output, which may easily result in the polarization of an operator's performance.

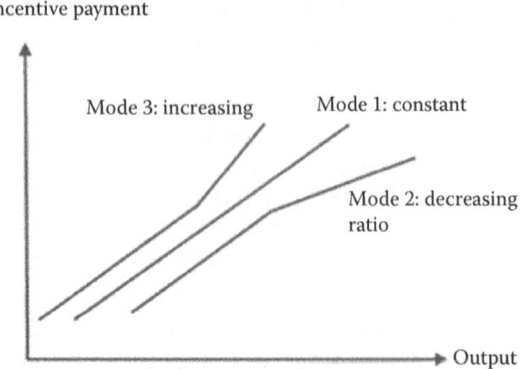

Figure 11.5 Different modes of incentive payment system.

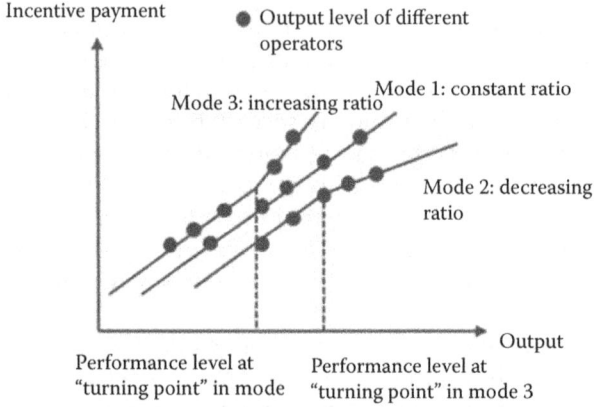

Figure 11.6 The influence of different patterns of incentive payment on the distribution of operators' performance.

Previously an enterprise used mode 3 to give high premium to an operator who was able to reach the high level of output. Although some employees are able to reach the higher level of output, it gives less incentive until the inflection point, which encourages operators to keep the performance away from the turning point as much as possible. As a result, this incentive curve will widen the output gap between employees. The incentive curve of mode 2 and mode 3 will have a direct effect on the average performance level of the operators. The measured result of any worker will be susceptible to this salary system.

As a matter of fact, the mode 2 incentive curve makes for decreasing the performance level gap between the operators and makes it convenient for production scheduling (production managers usually expect that the performance level between different operators can be as high as possible with little gap), but the inflection point in the mode 2 incentive curve should be properly set. It needs reliable basic data. In other words, data produced from work measurement will have an effect on the setup of the incentive curve, and in turn the set incentive curve will have an effect on the subsequent work measurement.

Moreover, the types of activity covered in work measurement should also take the design of the salary system into account. Many enterprise managers believe that the first-line operator should be wholly paid by the piece rate wage method in order to get them motivated to increase the labor productivity. However the designer of such a system usually overlooks the abnormal unit time. Besides the normal unit time, operators often deal with some anomalies resulting from actual work, thereby giving rise to an abnormal unit time. For example, one needs to polish a part with burr, or a part is erroneously processed in the upstream or requires an additional rework in the downstream. All similar anomalies or uncertainties are found in standard operations. If such anomalies frequently take place, one must measure these anomalies to keep the labor and equipment efficiency correctly measured. However, if the percentage of anomalies is negligible, neither the unit time used for wage accounting or capacity will be of referential value. In such a case, it is important to remove the uncertainties of production operations, and the salary system can be based partially on fixed salary and partially on merit pay. By doing so, we only need to measure the main work actions, while estimating the total time of assistant actions and correction actions on anomalies.

Last, the personnel management system is not connected with the work measurement, but it affects the work measurement. A growing number of enterprises have begun to focus on the effect of labor productivity. Shojinka, as mentioned in Toyota Production System (TPS), for instance, is a typical idea of increasing labor efficiency. Similarly, the enterprise can increase production efficiency by improving technology or management, and one of the key components of such improvement is to do work measurement on the operation in need of improvement. Gradually, some low-level managers and employees may hold that once an industrial engineer starts the work measurement, it means either downsizing the number of first-line operators or increasing the workload of each operator. However, many of them have trouble reallocating the downsized personnel. In reality, some low-level managers and employees are often fed up with the work measurement and efficiency improvement. The root cause is that there is no mechanism for sharing the result of efficiency improvement among employees. I am impressed by a management principle that 1/3 of improvement effect is used for cutting the cost of products and services, enhancing the overall competitiveness; 1/3 is used for increasing salaries; and the remaining 1/3 is acquired by the enterprise.

A few years ago, I exchanged ideas with managers from domestic enterprises on the management method of work measurement and personnel efficiency. With respect to the importance of these IE tools, many of them were indifferent and believed that a labor shortage will not occur in the years to come. Therefore, precise work measurement and personnel management may bring certain benefits for enterprise. But these management tools are sure not worth managers' great efforts. Yet with the growth of the social economy, technicians are paid higher salaries, especially in economically developed areas, and the income gap between blue-collar workers and laborers has gradually narrowed. Many enterprises, particularly SMEs (small and medium-sized enterprises), are vulnerable to some difficulties such as expensive and insufficient manpower. In the short run, this difficulty might make manufacturers suffer during the transformation and upgrading phases. Whereas those enterprises emphasizing efficiency management and establishing a win-win mechanism between the enterprise and employees during the implementation of IE tools will gradually have the competitive edge over their peers.

11.3.2 Case study 2: Work in process (WIP) inventory optimization and supply chain strategy

Inventory reduction is a major tactic used to improve enterprise operations. Also, due to the emphasis on capital efficiency, the level of enterprise operations and management is marked somewhat high if it can operate with less inventory. Hence, through monitoring, each section of value chain is doing its best to reduce inventory. However, from the angle of the entire supply chain, local optimization cannot evolve into global optimization of the entire supply chain, because negative effects may arise from it. In such cases, company managers should stand from the angle of the entire supply chain to check if the optimization strategy is effective in each section.

As illustrated by inventory control of production, inventory not only has an effect on capital efficiency (average inventory means the capital usage) but also determines the inventory area required for production facilities (the maximum inventory determines the required inventory area), exerting an influence on the layout of production facilities. With this in mind, production managers will try to avoid any fluctuation of work in process (WIP) inventory in addition to the reduction of average inventory. As shown in Figure 11.7, the change of inventory depends on the difference of upstream and downstream production speed. Therefore, levering

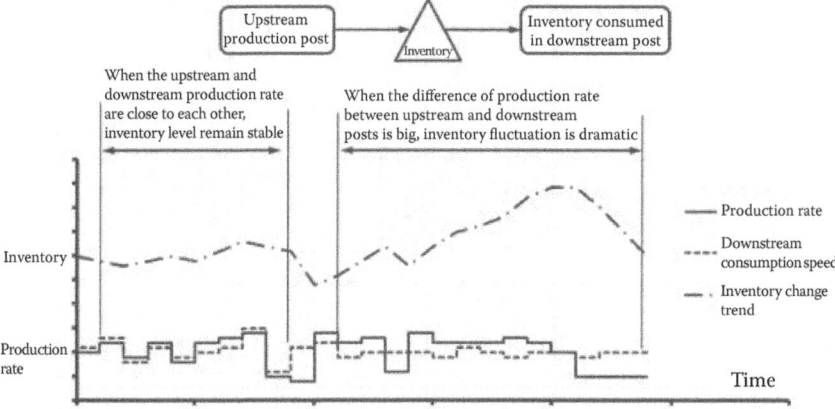

Figure 11.7 Inventory change in the supply chain as determined by the difference of production speed and consumption speed.

the upstream and downstream production rate will reduce inventory fluctuation and control the inventory level within a lower range.

For each section, inventory can be compared to a reservoir and the change of inventory is determined by the difference of inflow and outflow. To reduce the inventory, output speed should match the consumption speed as much as possible. When they are close, WIP inventory will stay unchanged; when they are greatly different, inventory will exist. For the control of inventory, section managers will try to level the production and consumption speed between the upstream and downstream inventory as much as possible. This measure is very rational. However, I once experienced an opposite example. One company used a dumbbell-shaped internal supply chain system as shown in Figure 11.8. After entry into the raw materials warehouse, different raw materials were processed in the same main production line. Then these materials were transferred to different storage areas of semifinished products, processed to finished products through workstations, and then entered into the finished products warehouse until they were launched for sales. In each section of the supply chain, the speed of inventory turnover was chosen as one of the key performance indicators (KPIs) of managers. For the purchase director, the turnover speed of raw materials or the annual turnover of raw materials (e.g., four turnovers or more a year means the inventory of raw materials shall be no more than 3 months of consumption) is the KPI. For production operation supervisors and marketing directors, the flow rate of WIP and product are their KPIs.

The sales quantity of each model has changed little with the time, allowing production planners to accurately plan the replenishment time

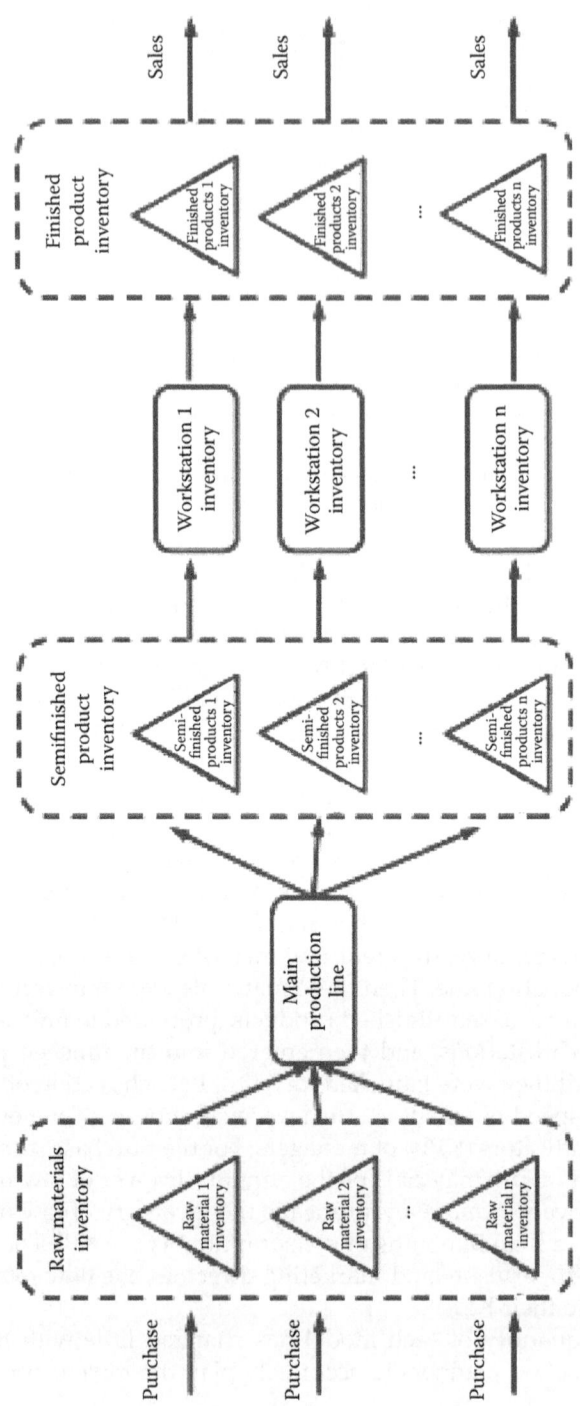

Figure 11.8 A supply chain in a dumbbell shape.

of each type of finished products. And due to the production exchange of the main production line, planners try hard to decrease the frequency of production exchange and ensure the semifinished products are produced in large volume from the main production line before the production exchange in order to reduce the capacity loss resulting from this process. To immediately consume the inventory of semifinished products, workstations are also set up with large capacity. For production managers, this production plan can be obviously advantageous: the main production line appears to be the bottleneck of the entire production system. Its total utilization can be increased by reducing the frequency of production exchange, which seems to improve the overall efficiency. Moreover, fitting the production speed of workstations to the main production line reduces the inventory fluctuation and average inventory of semifinished products.

Yet with respect of the entire supply chain, the optimization of production does not lead to the improvement of overall efficiency, as determined by two features of this supply chain. First, the workstations used in the production are a kind of special equipment, but each unit of equipment is only used for the production of one product. The larger the workstation capacity required for the production of each product means larger investment density and production area for the production utilities. Second, the inventory of finished products, either inventory value or inventory cost, will be much higher than that of semifinished products and raw materials. Therefore, the higher inventory of finished products means lower efficiency of the overall inventory. It is quite worthwhile to increase the inventory of semifinished products and reduce the inventory of finished products.

For each product, the production speed and selling speed, and the inventory change of semifinished products and finished products are as shown in Figure 11.9. To reduce the inventory of semifinished products, the workstations of finished products are equipped with larger capacity to eliminate the difference of production speed between the main production lines. However, this causes a bigger difference between workstations and the sales of actual inventory, which will result in a bigger fluctuation of finished products and lower utilization of workstations. In such a system, the local optimization of production gives rise to the worst performance of the entire supply chain. And this local optimization seems to be justified in that "inventory is the root of all evil."

As shown in Figure 11.10, the improvement strategy is quite clear: keep the production speed of workstations in alignment with the selling speed to reduce the inventory of finished products. Also cut the demand for the capacity of workstations and save the investment on equipment; further, increase the production exchange frequency of the main production line to avoid the inventory of semifinished products being too large. In other words, decrease the efficiency of partial inventory of semifinished

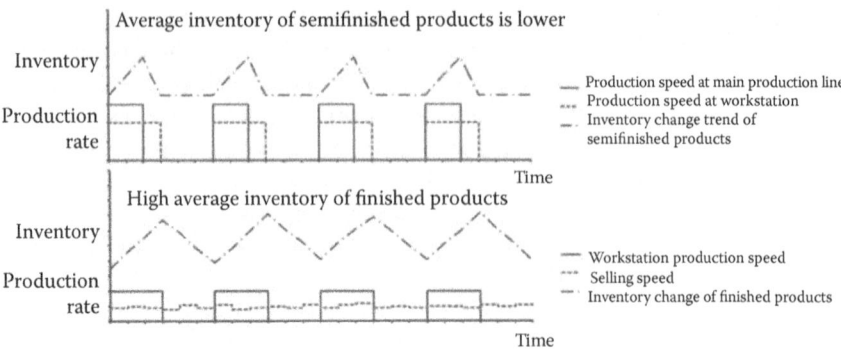

Figure 11.9 Preoptimization production model.

products and the production efficiency of the main production line in exchange for high efficiency of finished products inventory and utilization of workstations.

Although the theory of optimizing this supply chain does not appear to be complicated, it is really difficult to implement. First, it is not easy to identify the problem of this local optimization resulting in the decrease of overall performance. The reader might be surprised at this conclusion, but it is the hard truth. Managers at each section always focus on their own performance indexes. Only those professionals with a good knowledge of supply chain features can identify the problem from the global view. As a matter of fact, this enterprise did not figure out this problem until the monthly meeting on supply chain analysis. In this meeting, the GM and other persons in charge of purchasing, production, and marketing reviewed the change of supply chain performance. It could be imagined that this problem would have likely remained hidden under a pile of datasheets if not for the cross-departmental meeting.

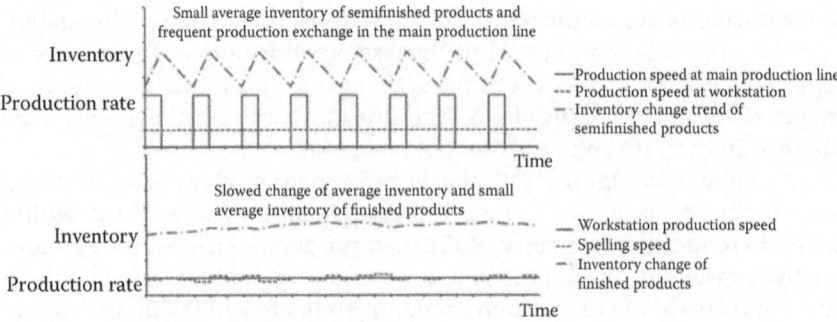

Figure 11.10 Postoptimization production model.

Second, each decision maker and operator should be fully aware of the background knowledge of each project. Although the required knowledge is not complex, each team member should spend some time learning it. The enterprise makes the entire management team fully knowledgeable of operation law of the supply chain. This is particularly true of production planners; only knowing the effect of these methods on the entire supply chain could they support this project, though some correction measures appear to contradict the common sense of production management. Through training and communication the person in charge of each section is fully aware of the effect on the inventory on the entire supply chain. Each of them could consider the supply chain strategy from the angle of the enterprise operation.

Last, after adjusting the entire inventory strategy, the production and operations departments should tailor the management measure. Besides the management of WIP inventory, they amended the inventory index accordingly and added the special index-measured production system to fit the degree of the new supply chain control mode. For instance, they encouraged the production department to increase the frequency of production exchange and set up the index of "flexible production" accordingly. This index and inventory turnover have evolved to be key components of the production plan.

Through this adjustment the enterprise saw the inventory of raw materials and semifinished products increase by about 70% and 30%, respectively, and a 100% and 40% increase in inventory area. The entire supply chain has been paid off with the cut of average total inventory value of around 20%, also with some decreases in inventory area.

With the case studies, I hoped to explain how IE tools constitute the organic entity of enterprise management system. Few separate IE tools are used and any improvement should be executed with the supporting management measure. Only by focusing on the full system and measured can IE tools be successfully executed.

11.4   Conclusion

Every enterprise where IE can be implemented has several features in common. First, the enterprise has reached a certain development stage, where operation efficiency is strategically important for enterprise competitiveness. Second, concerning the structural organization, IE should be rationally positioned, so that expert-level IE know-how can flow into the enterprise and help the management team understand the effect and limitations of the application of IE tools. The person who is in charge of IE functions should be able to push for the implementation and development of IE tools. The management infrastructure, such as the internal culture or work habits of employees, can support Lean management. Last, the key elements of the enterprise's management systems, such as the human resources system, supply chain management, sales management system, and other related functions can continuously evolve with these IE tools to create an organic management system. Only in this way can those IE tools function as intended. These features are not obtained inherently but can only be acquired through continuous efforts by all employees, especially if they require sufficient time and resources be contributed by the enterprise management team.

Competitive companies will not be underestimated by their peers provided that advanced product technology, market influence, and a highly efficient operation management are kept in alignment.

chapter twelve

Implementation of cell manufacturing mode of aerospace enterprise

Yin Liu

Contents

12.1 Enterprise status

The Beijing Satellite Manufacturer is the main factory that develops and produces the main structures of spacecraft in China, such as satellites and spaceships. Before 2000, it produced less than one whole satellite annually

or produced one whole satellite over several years. Its main work was feasibility analysis, technology exploration, technique research, and testing,
etc., which belonged to research-oriented production, and it was focused
on breaking through the technical route; confirming technology status;
and solving problems, such as feasibility of technique and production
conditions, etc. It did not emphasize the schedule, quality, and urgency
of production. After entering the 21st century, its production tasks were
increased dramatically, even exponentially. Figure 12.1a shows the task
quantity of each stage with lead time continuously shortened. As shown
in Figure 12.1b, the manufacturer needs to meet the quality requirements of spacecraft with a small sample that must be right the first time
and to complete growing production tasks in the same time. It posed a

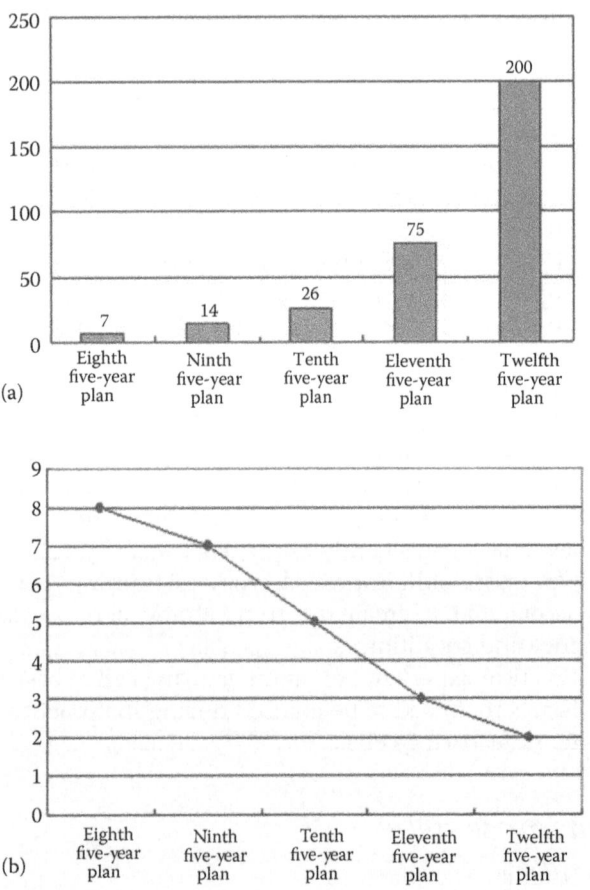

Figure 12.1 Model task quantity and lead time changes. (a) Model task quantity.
(b) Lead time changes of new large satellite.

severe challenge to the manufacturer in terms of production capacity and organizations.

To meet the growing production tasks, with the help of relevant research, Beijing Satellite Manufacturer first analyzed the characteristics of aerospace products and compared typical domestic production systems, such as the automated assembly line, Toyota production system, cell production, and Boeing parallel production, etc. Then they put forward methods to improve the spacecraft production system by flexible application of cell production and took the capacity, flexibility, and reliability improvement as basic principles for production system optimization. This work helped to accomplish the main production cells logic partition and cell construction and met the practical requirement for model production. To meet the new task requirements and enterprise further development, Beijing Satellite Manufacturer optimized the resource allocation and operating mechanism of cells by simulation and then established the production system, which is mainly featured in cell manufacturing and supports the research and production requirements of future satellites or spaceships.

Composite material (hereinafter referred as composite) products are an important part of satellites' main structure, including the bearing cylinder, honeycomb sandwich products, and connectors, etc., and accounts for a great proportion of the production task in the factory. This chapter starts with an analysis of the overall production situation of composite products, takes the production mode change of the solar array base plate of the honeycomb sandwich products as its object, and introduces how to implement the process in typical aerospace products' cell manufacturing.

12.2   Problems and contradictions

Composite products mainly include the central bearing cylinder, honeycomb sandwich plate, solar array base plate, and solar array connector. Molding technology of composite structures determines the production characteristics of composite structures: a high ratio of manual work in production, low standardization degree, and difficulty in controlling the stability and reliability of product quality. This may lead to low-level quality problems, low production efficiency, and high production costs. Before 2000, composite components were basically in the stage of exploratory application research and product development, which was produced by a "small workshop," namely, a piece of the component, from production arrangements to product delivery, was completed by the same group of operators. The workshop consisted of three work teams: Bonding Team 1, Bonding Team 2, and Benching Team.

Such organization has a strong adaptability to the change of product types, and one position or worker can conduct operation upon various products. It is able to meet production requirements with great product variety and small task size and also has lower comprehensive costs. However, there are obvious problems with such organization:

- Workers' operation skills are not targeted, with poor operation extension, and their proficiency in operation and potential skills haven't been fully utilized.
- One worker needs to conduct operation upon different products and to re-prepare the production when switching product types.
- Product equipment and resources are scattered, which will increase the time to circulate during production.
- Shared equipment is hard to coordinate. Resource shortage easily occurs due to task conflicts.
- Dispatching management is complicated, and any production bottleneck is hard to identify and reposition. Production capacity improvement for a specific product is less targeted.

The above problems greatly affected production efficiency, stability, and reliability of product quality, tending to become the bottleneck after the growth of tasks. To respond to the growing tasks, the manufacturer put forward the general idea of adjusting the production system of the Composite Workshop. The idea was proposed after comprehensive analysis and refining of task requirements for various composite products, technological characteristics, and work characteristics and combining the specific conditions of the composite plant, sites, equipment, and implementation. It is to be dominated by cell production, to center on general characteristics of technology, to form flexible production capacity according to different products, to establish product units if conditions permit, and to meet stable and efficient production requirements of special products so as to form a composite aerospace product production system with greater flexibility and quick responsiveness, and it is able to meet a certain task size.

According to the plant condition and equipment characteristics, we confirmed the objects of the product cell as the following:

For the bearing cylinder, structural plate, and solar array base plate, these three kinds of products occupy more than 90% of the total production of composite products. A bearing cylinder is a kind of large-scale product, which needs a bigger space, so it is hard to collect all the resources and equipment required in one place. The size of the operation desk used by the structural plates is smaller, and in order to fully use the space, it is placed in a smaller room, but this leads to a separation between the slab and the other resources of process. In addition, both the bearing cylinder and structural plate need to use an autoclave, but installation of the autoclave requires a specific environment and security conditions, and an autoclave is hard to allocate into a specific product cell. Therefore, limited by the plant conditions in the old plant, these two kinds of products haven't been formed into physical product cells, and their operations are driven by the production process of the products, and the concept of product cells is reflected by operation management.

There are three kinds of base plate products: large, medium, and small. The size of the operation desk during production is larger than the size of operation desk of the structural plates, so the operation desk needs a certain amount of space and cannot be placed in a small room. Environmental and security requirements for the equipment, such as the oven, are not as harsh as the autoclave's and can be met by the ordinary plant. Thus, improvement costs of centralized deployment of the main equipment and resources of the base plate are relatively lower. Then the divided sites were reallocated according to the task trend of the base plate and have been first to realize the construction and application of the manufacturing cell of the base plates. To deal with the further substantial increased task requirements after 2010, based on new plant construction and relocation of the composite production area, the firm planed manufacturing cells for eight kinds of products in the form of compartments and realized fully cell production of various composite products in the new plant.

12.3 Implementation and technique methods: Implementation of cell manufacturing mode of solar array base plate

12.3.1 Products

The solar array base plate is one type of honeycomb sandwich product and is a main structure of the mechanical solar array. It is mainly composed of panels, embedded parts, and honeycomb cores.

The basic production process of the solar array base plate is divided into six stages: preparation for compositing, compositing, solidification, processing, subsequent processing, and testing.

Before establishing the manufacturing cell, specific problems are as follows:

1. Targetless Operation
   Influenced by past production, before establishing the base plate cell, there were many problems, such as scattered production sites; dispersed resources; long logistics routes; operators having to participate in production of various products, such as the base plate, structural plate, and bearing cylinder, etc.; large operation variance; targetless operation; slow improvement of proficiency; and difficulty in cultivating operational skills.

Because of sharing plant and production resources with other composite products, the production layout is relatively scattered, and production resources cannot be evenly used. Winding machines, operation desks, and oven equipment are frequently used during the production of the base plates, but this equipment is also used for production of other composite components when it is used for molding the base plate: Most composite components need operation desks during molding; during the production of the base plate, panel solidification, honeycomb core coating, and solidification after assembling the ovens are used, but the ovens are also frequently used in molding of other composite components. Therefore, shared resources, such as the operation desk and ovens, are key factors that restrict production improvement of the base plates.

The duration of manual work on processes such as compositing and assembling and placement is long. Except that some productions are able to use equipment such as the oven, there is lots of manual work, and lots of workers are needed in different sites and to move products between equipment rooms; thus, staff shortage has become the most obvious bottleneck. How to save staff occupation or to meet more production requirements with fewer workers is the most urgent problem that needs to be solved during the production process of the basic plate.

To deal with the growing production requirements of the solar array base plate, the main idea of establishing a base plate manufacturing cell with the cell concept is according to the specific conditions of the plant and process requirements of various production factors of the solar array base plate, to conduct centralized allocation so as to meet the production requirements of various processing steps, and to form a base plate manufacturing cell that can meet the requirements of the production task.

12.3.2   *Elements and implementation solutions*

1. Elements

According to the requirements of cell manufacturing mode, and considering specific production characteristics of the solar array base plate, the elements of solar array manufacturing cells can be divided into two parts: physical elements and rule elements. On one hand, physical elements include the product classification and technical processes, the equipment and facilities in production, the logistics forms and storage condition of base plate production, the documents and data for production, the organization of production staff and team, the status of cell production environment and security control, and the information-based equipment for cellular manufacturing.

On the other hand, rule elements include the production management and quality management as shown in Figure 12.2.

2. Introduction to the Implementation Solution

Primarily, the implementation solution for the base plate cell include, first, analyzing components of the base plate manufacturing cell, then collecting and sorting relevant data and information on the basis of this; second, allocating physical elements to form the physical foundation of the manufacturing cell; finally, designing and promoting rule elements, so as to achieve stable operation of the cellular manufacturing cell (Figure 12.3).

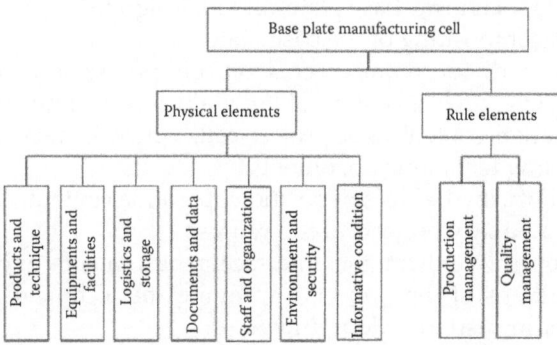

Figure 12.2 Element diagram of solar array cell.

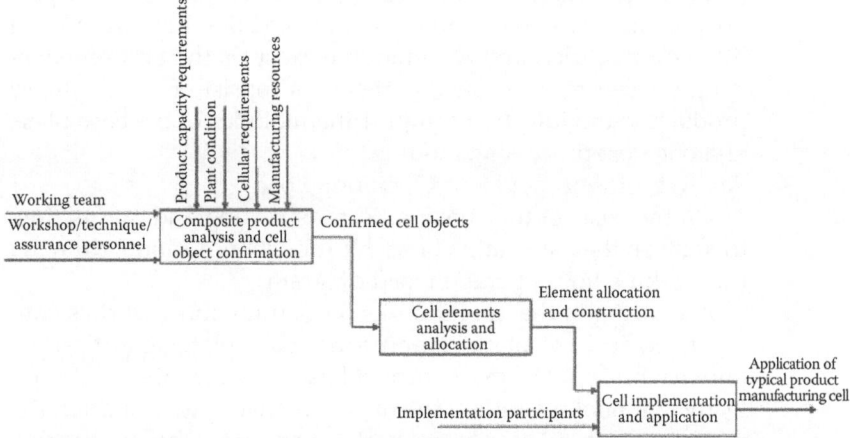

Figure 12.3 Implementation diagram of solar array base plate cell.

12.3.3 Implementation and completion

Sorting and Analysis of Products and Techniques

1. Specifications of Cell Product

   For the manufacturing cells that are used for base plate production, the main target products are three specifications of solar array base plate to cover different satellite platforms.

2. Composition of Cell Product

   Fully to sort the production composition of the base plate manufacturing cell and to form a detailed components (units) list.

3. Technique Combining and Optimizing

   At first, to figure out the main developing line, which focuses on the solar array base plate, and to design the technique process from compositing of the base plate.

   Second, for requirements of cell production organization, refine the project content of the main process line and take each operation content as a project cell, which is also the basis for dividing technique documents.

   Third, on the basis of position operation content, refine important auxiliary projects of the process.

   Fourth, confirm the interface requirements and guarantee conditions of the project, improving the technical guidance of the administrative department.

   Conduct solidification on the basis of optimization of the technique process, guiding technique documentation, and organization of production cells.

   According to the production process of the solar array base plate, the production of the base plate is divided into four production modules with product and logistics moving among different modules, and each module executes the task independently. These four production modules consist of one auxiliary production module, two compositing modules of the base plate, and one post-processing module.

4. Analysis of Working Hour Operation Load

   On the basis of the above process of sorting and optimizing, to analyze the load status of each production position, drawing the working hour operation load diagram.

   According to the operation load diagram, most work is concentrated in base plate compositing, base plate solidification, and machining. The main form of base plate compositing is collective manual operation; the most effective way to shorten the production circulation is properly to organize the production, to organize production according to manufacturing cell, and to make products moving in each cell. Two composite cells are in parallel for continuous production of base plates, solidification in a sequence, and most steps after solidification are serial steps.

Equipment and Facility Allocation

1. Cell Layout

According to the area and structure of the plant, the layout of equipment and facilities of base plate manufacturing cell are shown as Figure 12.4, which is basically a U-shaped layout:

2. Fixture Allocation of Base Plate Cell

Combining the fixture used in production, the specific list is shown as Table 12.1.

3. Allocation of Raw Materials and Standard Components

List all raw materials of base plate production, and the allocation list is shown as Table 12.2.

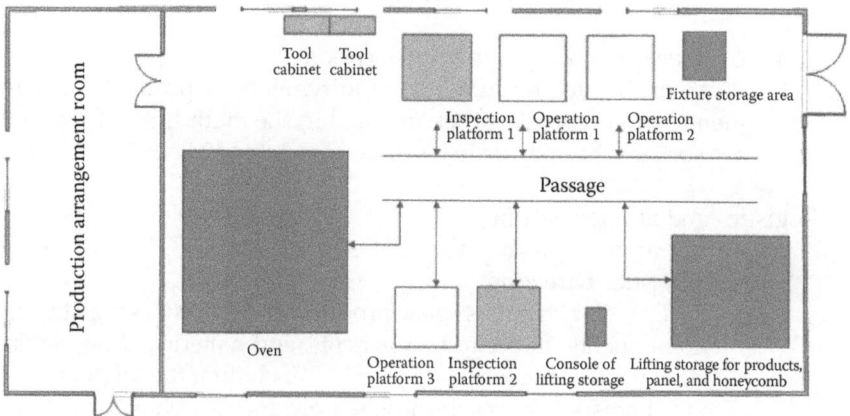

Figure 12.4 Layout of base plate cell.

Table 12.1 Fixture list for each platform of solar array base plate

Platform	Basic fixture of platform 1	Basic fixture of platform 2	Basic fixture of platform 3
1	Molding mold 1	Molding mold 3	Molding mold 5
2	Winding mold 2	Molding mold 4	Molding mold 6
3	Molding mold 3	Embedded fixture 2	Molding mold 7
4	Embedded fixture 3	Panel sample 1	Panel sample 2
5	......	......	......

Table 12.2 Material list of solar array base plate

Platform	Platform 1	Platform 2	Platform 3
1	Adhesive 1	Adhesive 1	Adhesive 2
2	Film	Film	Adhesive 1
3	Adhesive 3	Adhesive 3	Film
4	......	......	......
5	......	......	......

Table 12.3 Instrument and equipment list of main lines

No.	Name	Qty
1	Oven	1
2	Freezer	1
3	Multi-meter	2
4	Operation desk	3
5	......	......

4. Instrument and Equipment Allocation

According to the molding requirements of product, instruments, and equipment that are used in the main lines of the cell are sorted. The specific list is shown as Table 12.3.

Logistics and Storage Setting

1. Specification of Product Circulation
   a. Logistics Direction

   The solar array base plate production cell includes logistics of components, standard components, and materials. Main logic directions are shown as Figure 12.5, including the following:
   i. Logistics from semifinished product warehouse or standard component warehouse to compositing position
   ii. Logistics from material warehouse to production arrangement cell
   iii. Logistics from compositing position to machining position
   iv. Logistics from machining position to subsequent position of base plate
   v. Logistics from compositing position to finished product warehouse
   b. Transportation Matters

   The transportation of the solar array base plate between plants uses vehicles driven by battery. Plates should be protected by sponges and at least two people and be protected from bumping. Other logistics circulation uses a traveling bogie.

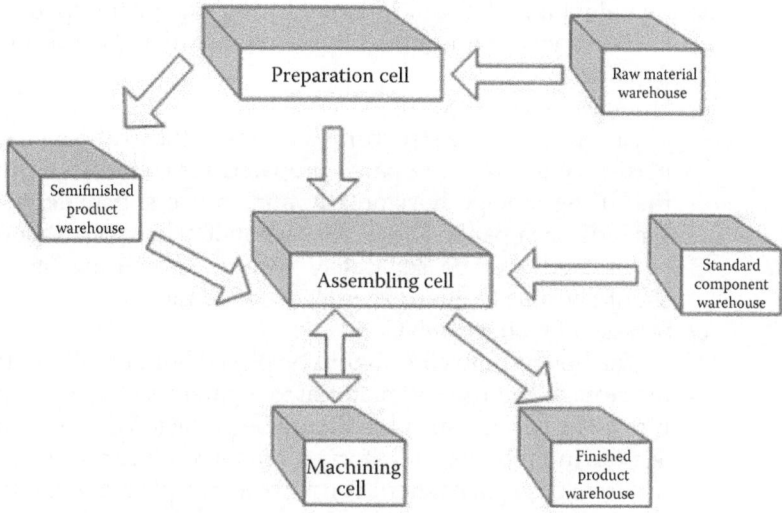

Figure 12.5 Logistics diagram of solar base plate production.

2. Product Storage

To set the storage area in the base plate production process, the position and function of each storage area are shown as Table 12.4.

In practical production, except for the above storage area to store corresponding panels or base plates, some components are still placed on the operation desk, which becomes a temporary storage site.

a. Storage of Base Plates

Use automatic lifting storage: This lifting storage is designed for storing base plates with a total of 18 layers of pallets. Each pallet can store two horizontal medium base plates, four small base plates, making full use of the space and greatly reducing the storage yard. The lifting storage is managed digitally, and plates can be placed by models, facilitating free access to plates. At the same time, this storage has quite good tightness, which can effectively protect the base plates against pollution of dust or

Table 12.4 Statistics of storage area

Name	Position	Function
Raw material storage area	Plant 6	Store raw materials
Storage area of plant 1	High-rise warehouse	Store high-rise plate and panel
Storage area of plant 2	Layer room, hall	Finished panel

accidental injury. The application of lifting storage has made the production spot become more tidy and standard and contributed to 6S management.

b. Storage of Grid Panel

According to the structural features of the grid panel, it is necessary to place the panel prepared for compositing into the lifting storage horizontally and to use soft material to separate each panel. Panels for each model base plate should be marked and be separately placed. The lifting storage should be able to guard against dust and moisture.

c. Storage of Honeycomb Cores

The honeycomb core should be placed horizontally in lifting storage and use soft material to separate each honeycomb piece. The honeycomb core should be protected against damage during moving and storage. Honeycomb cores used for different base plates should be marked and placed separately. The lifting storage should be able to guard against dust and moisture.

d. Storage of Components of Base Plates

Components, such as pre-embedded components and corner reinforcement parts, should be stored on shelves and be packaged separately with proper environmental requirements.

3. Product Marking and Traceability

According to the requirements of the program file in the company, the solar array base plate produced by the cell should be build-marked exclusive so as to confirm the products meet technique requirements and maintain traceability.

a. Before production, the research and production department should confirm the batch marking being accordant with the operation plan, diagram, and technique documents.

b. The technique department should stipulate marking requirements and method of batch marking in the design and technique documents. The production department should make the batch marking according to the requirements and fill in or record batch marking in a quality control recording card.

c. During production, products with similar shape or requiring multiple batches of heat treatment and surface treatment should be tagged or placed separately and strictly distinguished.

d. Products purchased should be reviewed by batches, set up in an account, carded, and stored or allocated by batches.

e. The warehouse keeper should store the products (semi-finished products and finished products) from different batches separately and mark them.

Documents and Data
1. Management of Design Documents

   Design documents consist of diagrams, technical notes, and technique requirements, etc. They should be countersigned and be managed and issued according to related requirements, charging by the supervisor.
2. Combing and Standardization of Technique Documents
   a. Standard Requirement of Technique Documents

      Technological procedure should meet relative standardization requirements. Technological documents should be detailed and quantized with strong operability. Technological documents should fully cover the production process, and all technological documents should be coordinated, accordant, clear, complete, and meet production needs. The making, proofreading, checking, inspection marking, approval, and signature of technological documents should be complete and meet related approval and signature requirements for technique documents. Modification of technological documents should be submitted with a corresponding modification form with clear modification items and a complete description of modification contents.
   b. Technological Documents Sorting

      The solar array base plate of each platform has complete product molding; during the molding process of the product, review and sorting has to be conducted according to completeness, correctness, and guidance. Based on the solar array base plate technology molding, optimizing and solidification has been conducted according to the technique process of the solar array base plate, formed relevant documents, and allocated molding documents to manufacturing cells according to cellular organization. The cell staff completes the production work according to the molding documents and current processing and assembling equipment, which has effectively guaranteed the manufacturing cells' operation.
3. Product Guarantee Documents

   When establishing the cell, according to production characteristics and product guarantee requirements for the solar array, a product guarantee outline of the base plate cell has been made. Meanwhile, related program documents have also been equipped to ensure the quality control of the production process.
4. Other Regulations and Rules and Documents

   Except for the general enterprise's regulations and rules, the regulations and rules of the production site also include inspection specification, safety production administrative regulations, and position operation instruction for the product manufacturing cell and these documents have been equipped and guide production when the cells are established.

5. Data Packet Management

    During the development of the solar array base plate, the product assurance engineer is responsible for collecting, summarizing, analyzing, and reporting product assurance information, such as design deviation, manufacturing overproof, substitution material, modification of technique status, defective goods handling, technique problem solving, and clearance of quality problems, etc. When the above information occurs during product manufacturing, the responsible department should be able to report the product assurance information in time after review, and the data package should be standard, complete, and available with traceability.

Staff and Organization

1. Staff Organization Mode

    According to the molding characteristics of composite materials and key points of the manufacturing process, combined with the workers' ability and specialties, the company changed the teams' classification from based on the type of work to teams based on production units. Teams consist of many positions and a set team leader and position principal (booth principal). The team leader has to be responsible to the workshop leader, and the booth principal only has to be responsible to the team leader. This fulfills the functions of the team leader and core personnel and totally changes the situation from one task completed by one worker or one team to cell production. It can not only utilize the potential capabilities of workshop workers, but also shorten the production cycle and greatly improve product quality, which can solve stabilization and reliability problems of product quality to some degree.

2. Position and Staff Setting

    Integrate work types of a composite workshop and cancel the differences between bench workers and bonding workers, combining them into the bonding of composite materials to meet the requirements of cell production. Set three kinds of positions for the base plate cell: technique position, operation position, and inspection position.

    Cell position and staff setting are as follows (Table 12.5):

3. Idea Transformation of Staff

    Within the cell-manufacturing mode, position functions of operators have been greatly adjusted. Because the molding process

Table 12.5 Cell staff settings

Position	Operators	Technologists	Inspectors
Quantity	4~6 people	2~3 people	1~2 people

of the composite materials is dominated by collective work; thus, teamwork is a necessary part of the manufacturing organization. The advantages of cell-manufacturing mode cannot be fully appreciated until the team's collaboration ability and the executive force have been improved by changing the ideas of the staff, advocating cooperation and coordination and cultivating team spirit.

During cell manufacturing, operators are basically fixed in a production cell and repeat similar work every day. They are bound to be bored with their work. Thus, the motivating methods for them must be changed as well, advocating operators to improve their proficiency and skills.

4. Skill Training
    a. Certification and Training for All Staff

        According to the requirements of manufacturing and position instruction, all staff of manufacturing cells should obtain related operating licenses so as to realize employment with certificates. Except for the basic requirements of the position, the capabilities of the professional team should also be paid attention according to characteristics of the solar array base plate. Conduct targeted professional training so as to ensure staff capabilities and quality to meet requirements of cell construction.

    b. Composite Molding Training System Improvement

        With cell production, operators are basically fixed in each production cell, which is beneficial for improving their proficiency for a certain product, perfecting their operational skills, and making their skill turn into technical and professional type. Operational proficiency and technique are key points to ensure stability and reliability of product quality and building a foundation to improve production efficiency. Therefore, the evaluation measurement of the skills of operators should also be changed correspondingly so as to meet the needs of cellular production and to truly promote operators to learn and practice their skills.

Environment and Safety
    1. Environment Control of Producing Spot

        According to relevant requirements, a workshop should include supervision of temperature, humidity, and cleanliness so as to ensure the workshop meets the environmental requirements of the solar array. According to 6S management rules, it is necessary to clean the plant in time and conduct regular examination of the production spot.

2. Technique Safety Analysis and Measures

   To ensure the manufacture of the solar array base plate is smooth; to effectively control the risk factors that exist in the production process; and to ensure the safety of the staff, production, instruments, and equipment during manufacturing, stipulate corresponding administrative regulations for safety production, analyzing risks during production and finding existing risks, and stipulate corresponding targeted safety measures.

3. Disintegration and Implementation of Safety Production Responsibility System

   Specific staff safety responsibility, strict product safety, and staff safety and health have been stipulated in the safety production management of manufacturing cells of solar array base plates; safety control measures upon risks identified during manufacturing of solar array base plates has been proper implemented; no product quality problem caused by production environmental problems occurred, nor did security incidents.

4. Product Protection

   Moving, storage, and protection measures for products of the solar array base plate should be executed according to the program document of the enterprise. Details include the following:

   a. Operators should wear work clothes. All kinds of workers (including outsourced workers) should wear clean gloves when they touch components and products to avoid mechanical damage and pollution to component surfaces or products during operation.

   b. Use special handling fixtures for transportation, using special packaging boxes for storage and transportation so as to protect products from bumping and damage.

   c. During manufacturing, operators should operate according to related technique documents, managing protective work for products, and protecting base plate products from mechanical damage and oil contamination.

5. Personnel Protection

   a. Generally, use hoisting to move heavy equipment. If it is manually operated, workers must wear canvas gloves and work footwear and avoid equipment sliding and workers being injured.

   b. Moving is forbidden when the metal mold is at a high temperature.

    c. Hoisting and transportation of the base plates should be strictly consistent with the "packaging and transportation requirements" and "administrative regulations of safety production" of products to avoid personnel injuries.

6. Emergency Response Plan and Personnel Security Training

A production security incident response plan has been stipulated, and the related department should conduct safety education for the staff of the cells each month. Equipment, such as cranes and lifting appliances, should be within validity and in good condition, and the operation spot should be fully equipped with fire hydrants and extinguishers within validity.

Information Conditions

In terms of manufacturing cells, the information system that is directly relevant is the quality system and manufacturing execution system (MES).

1. Quality System

The quality information system has realized electronic computerization for five kinds of sheets, such as the technique handling sheet, defective good handling sheet, waste product notification, materials alternation sheet, and the unqualified purchased product sheet. In the cell, inspectors, operators, and workshop leaders are able to handle the sheets on-site, which obviously improves handling efficiency for quality problems. The quality information system also has realized functions, such as the production process, task dealing, qualification documents, statistical analysis, standardizing information, and comprehensive inquiry, as well as functions, such as process management, task box management, and customization of sheet specification and process, to further facilitate quality information dealing on-site.

2. Manufacturing Execution System (MES)

MES is the information system to conduct management, such as task management and progress control of the workshop. It has realized applications, such as reception of plant-level production tasks, planning of workshop-level production tasks, assigning teams and operators, getting feedback from operators upon tasks, and displaying cellular and position production information, which makes the workshop plan and dispatching management more convenient and improves the efficiency and accuracy of production management.

Production Management

1. Combining and Optimizing the Production Process

   Process optimization starts from working contents, first, to separate serial working from concurrent working and to conduct concurrent working so as to shorten the whole production process, and then, according to the whole task and deliver node, to arrange production uniformly and try to reduce the production proportion of undersupplied products as much as possible. During the whole production process, each production node should be completed according to schedule and make the whole process be processed smoothly. Certainly, the manufacturing process should be continuously improved and perfected. As to the working ahead or delay of production nodes, the reason must be cleared, analyzed, and solved as well as adding the improvement into the production process.

2. Confirmation of Production Operation Management Regulations
   Make a Specific Production Plan

   a. During the operation process of the manufacturing cell, make detailed production plans, which should cover whole components and subassembly; then subdivide the plan into steps, confirming the undertaker and completion time of each step, and conduct the track management by monthly or weekly plan. In addition, arrange the production style according to product features and requirements: batch production or simplex production.

   b. As to batch production, it is necessary to confirm production quantity and completion time of each batch, arrange order of propriety according to the user's requirements, and make a production plan separately so as to meet the requirements of a whole spacecraft.

   c. Be in accordance with requirements strictly related to the third-level document of the enterprise, to conduct batch management and quality control upon all levels of products with immutable technique status and above two sets, such as single aircraft and components or subassembly, so as to make sure products meet the technique requirements with traceability.

   d. As to simplex production, to produce and assemble production for each component, strictly follow the schedule with closed-loop management.

3. Using and Operating Rules of Production Equipment

   a. Use of Operation Desks

      The operation desk is the facility that is used most frequently. Its specification and size determines the type and quantity of products the can be placed. Its specification and the quantity of each specification are important factors that affect the production capacity of cells. There are three medium operation desks, two medium inspection platforms. The operation rule is that one medium operation desk can only process one medium base plate or two small base plates once.

b. Operation Rules of the Oven

The oven is used to heat the honeycomb coating and compositing solidification of the base plate and can be used to heat one medium base plate or two small base plates once.

c. Personal Dispatching Rules

There are four permanent workers in the base plate cell, and they are responsible in the whole process of compositing, heating, and post-processing. If there is a backlog of production tasks, dynamically call other workers to help in production.

4. Fixture Management

Fixture status analysis: fixtures of the manufacturing cell should be inherited from the model product. All fixtures used in component and base plate production should be reviewed according to stipulations and issued qualified certification for qualified marking after the assay is approved to mark validity according to requirements of technique documents.

Fixtures for positioning of the base plate should be uniformly managed by a certain person and the management system of positioning fixtures stipulated, which can effectively control the use of positioning fixtures, and is able to report quantity and service conditions to the technical department and production department in time to provide a powerful guarantee to effectively implement the production plan of the base plate. The uniform management of positioning fixtures is able to avoid misapplication of positioning during the compositing of the base plate; meanwhile, it shortens the preparation time of the fixtures of the base plate, which has greatly improved the development quality and production efficiency of base plate.

5. Raw Material Management

a. All the raw materials have qualified certification and are reviewed; materials exceeding the storage period should have an extended review conducted according to stipulations.

All the raw materials for production of the solar array base plate are directly purchased by the manufacturer and have qualified certification issued by the manufacturer. They should have conducted an admission review in strict accordance with related stipulations after they enter the factory and applied for model production after passing inspections by inspectors if they are qualified. The materials exceeding the storage period should have an extended review conducted according to stipulations and be applied to model production after passing the review. Materials that are unable to pass the review should be subject to a scrap process.

b. Standard fasteners should have qualified certification and be reviewed according to stipulations.

Standard fasteners for the model should have qualified certification issued by the manufacturer and be reviewed according to the review stipulations, then applied to the model products after passing the review.

c. Administrative Regulations and Recording of Unqualified Products

According to relevant administrative regulations and procedures, unqualified goods and materials should be conducted with return and exchange handling or accepted conditionally according to the review report and "unqualified products handling sheet" issued by inspectors. And the unqualified products should be stored separately. As to goods and materials that have exceeded the storage period, they should be handled with the "degradation (scrap) application form" filled out by keepers and degraded, scraped, and stored separately.

6. Instruments and Equipment Management

Monitoring and measuring devices, such as calipers, micrometer, feeler gauge, flexible rule, straight edge ruler, and multimeter, used for production should be used within validity of verification and with proper marking.

Quality Control

1. Formulate Guarantee Outline of Cell Products

Incorporate production management requirements into the quality management system and product quality assurance system and put forward specific requirements for condition control of workers, machines, materials, rules, and environments so as to ensure production quality of products, to effectively control the supplied materials, production process, tests, detection, and acceptance check with correct and completed records and traceable quality information. The product quality and consistency are able to meet design requirements.

2. Product Inspection Procedure

The development process and product delivery inspection procedures of the solar array base plate have been made, and general requirements for inspecting the solar array base plate, the examination requirement for the main performance index of product delivery inspection, and requirements for inspection items, inspection methods, and selection of measuring equipment during the manufacturing process have been stipulated. Corresponding inspection procedures have been stipulated.

3. Execution Rules of Production Inspection

    Completed third level management documents have been stipulated to conduct correct, complete record and signature during product inspection, according to the requirements of the rules and regulations of the quality manual, program document, and third level documents, etc. Inspectors of the cell should be able to strictly fulfill functions, such as controlling, prevention, monitoring, and reporting, and to clearly show the inspection state of the product and whether the inspection has been passed or not, and to ensure the traceability of inspection records.

4. Redundancy Control

    a. During the assembling process of the base plate, operators should carefully examine the quality and quantity of fixture and product components according to the diagram and technique documents and conduct assembly after verification; then conduct checks and examinations in time after assembly so as to avoid redundancy.

    b. Clear redundancy of products after each process, before assembly, order transfer, and entering the warehouse, and strengthen redundancy control during storage and delivery.

    c. The assurance engineer for the model products should conduct site supervision and inspection according to the standards and technique documents so as to ensure the redundancy is effectively controlled.

    d. Use a vacuum cleaner to clean products or around redundancy during assembling, filing, and machining. According to the definite requirements of the technique documents, technologists should clear the installation hole of instruments so as to ensure the product is delivered without redundancy.

    The redundancy control of the production field is effective, and no quality problems have ever occurred due to unqualified redundancy control during the manufacturing process.

5. Quality Information Management

    Product assurance information produced during the manufacturing process of the base plates includes manufacturing variances, materials replaced, modification of the technique state, unqualified product handling, technique problem handling, and quality problem clearance, etc. The product assurance engineer is responsible for collecting, summarizing, analyzing, and reporting product assurance information. When the above information is produced during the production process, responsible departments should report relevant information to the product assurance engineer in time by telephone, recording, emails, quality information systems, etc. The product assurance engineer summarizes the information and submits it to superior leaders for review. According to analysis of the results of the product assurance information, the product assurance department should organize related departments to make correct measures, and conduct supervision and inspection on implementating the resulting measures.

12.3.4 *Continuous improvement of the manufacturing cell of base plates*

1. Application of System Modeling and Simulation in Construction of Base Plate Cell

   With the planning and construction of the composite plant in the new plant district, taking advantage of the plant moving, based on construction of the base plate cell and practice, to build a new base plate cell in the composite plant. To accurately analyze the production bottleneck of future tasks of the base plate cell, to correctly identify resource conflict, and to scientifically conduct resource allocation for the manufacturing cell, the plant introduces a production system simulation tool, implementing modeling system simulation, analyzing and optimization measures, discovers problems during base plate production by analysis on production capacity, and provides solutions. Based on this, combining task trends for the base plates in the plant in the new district, to provide specific suggestions about resource allocation for base plate production and guide cell construction of the plant in the new district, specific approaches are used:

   a. First, establish an operation simulation model of base plate production, including a composite module, post-processing module, and auxiliary production module.

   b. Implement the simulation model and take targeted task size as objects, analyzing each resource occupation after completing the task, identifying and positioning the production bottleneck, and conducting countermeasure analysis.

   c. According to problems reflected in the operation results of the simulation modules, to improve corresponding resource allocation, conduct the same task so as to check whether the improved allocation can meet the production requirements or not; if not,

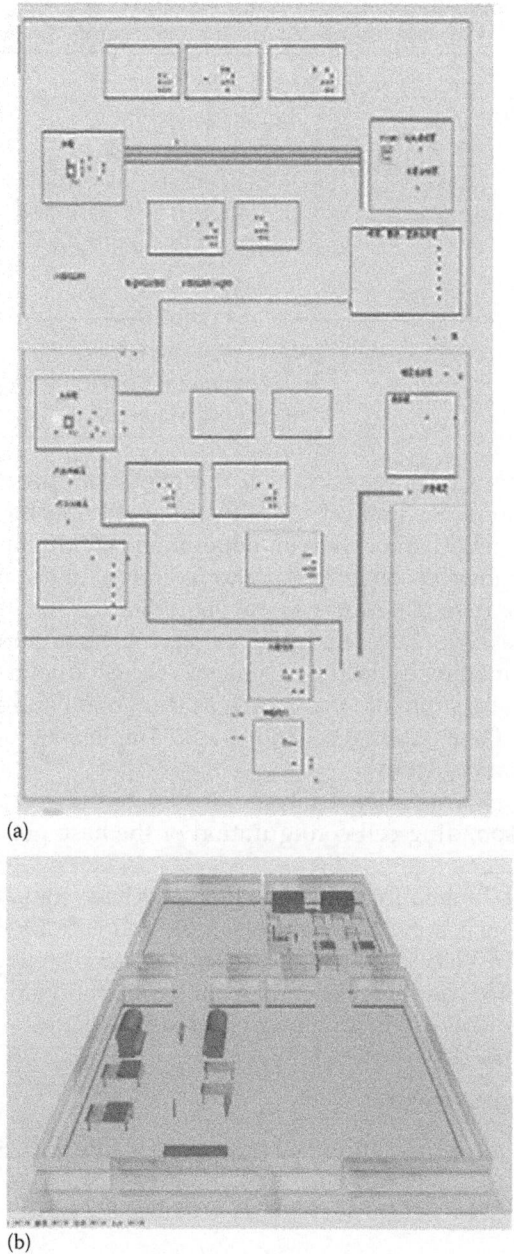

(a)

(b)

Figure 12.6 Base plate cell simulation model based on Witness. (a) Two-dimensional simulation model of base plate cell. (b) Three-dimension presentation model.

Table 12.6 Targeted allocation for winding cell and base plate compositing and assembling cell of composite plant

Cell	Name	Basic allocation
Auxiliary cell (District A)	Winding machine	2
	Large operation desk 1	2
	Medium operation desk 1	3
	Oven 1	2 layers
	Machining operation desk 1	1
Compositing and assembling cell (District B)	Oven 2	2
	Large operation desk 2	2
	Medium operation desk 2	3
	Small operation desk 2	3
	Inspection platform	2

    adjust the resource allocation and operation rule and keep doing the simulation analysis until the requirements have been met.

d. Based on cell allocation and operation rule by simulation analysis, guide the construction and operation of the manufacturing cell for base plates in the new plant district.

    The base plate manufacturing cell simulation model was implemented by Witness, which is the production system modeling and simulation software, as shown in Figure 12.6.

e. Base Plate Cell Allocation and Implementation Based on Simulation Analysis

    As to simulation analysis for targeted task size, provide the corresponding cell configuration of the base plate, as shown in Table 12.6.

    The Tangjialin Composite Plant has been completed on schedule. The compositing and assembly cell and auxiliary cell have been in place according to the provided resource allocation in Table 15.6, and this has built a good foundation for completion of subsequent production tasks for the base plates.

12.4   Improvement

The results of implementating the base plate manufacturing cell are as follows:

12.4.1   *Obvious improvement of production capacity: Successful completion of model production task*

Before establishing the manufacturing cell of the base plates, preparation of technique documents and worker organization according to cell manufacturing are totally charged by one team with lots of serial process and long preparation time; phenomena such as device conflicts and staff shortages, etc., frequently occurred. In 2005, according to the production site condition and equipment-specific conditions and facilities, such as the operation desk, a new manufacturing cell for the base plates was established, which met capacity pressure brought by the great rise in annual average tasks, and it completed two times as many tasks as in 2004. The annual average production capacity has been improved to more than three times the former capacity. The production capacity has been obviously improved, and the model production task has been successfully completed without an obvious increase in manufacturing resources.

12.4.2   *Obvious improvement of production efficiency*

Improvement of production efficiency is mainly indicated as the following two aspects:

- Operation time of some processes has been shortened: After establishment of base plate cells, production time per piece has been shortened more than 15% on average. The reasons for less manufacturing time are as follows:
    - As to the procedure for manual work: For manual operation by workers on the operation desk before establishment of cells, workers needed to conduct production operation as well as to get materials and to move products by themselves. After establishment of the cells, workers in this process are able to concentrate on manufacturing activity; thus working time has been shortened; meanwhile, due to concentration of similar operations and workers' proficiency in operating, the working time has also been shortened.
    - As to the procedure for mechanical processing: Mechanical processing is conducted in the machining workshop, and products to be processed need to be delivered or fetched. Before establishment of cells, deliveries were not in batches and took lots of time. After establishment of cells, it is easier to deliver by batches, thus time has been shortened; in addition, batch production has also improved operation proficiency and efficiency.
- Optimizing the whole production process includes the following:
    - Changing some serial operations into concurrent jobs by refining the operation: After detailed analysis of the technological processes, they separated the subordinate line operation and main line operation of production; thus, the subordinate line operation is able to be parallel with the main line operation instead of occupying the time of the main line operation.

- Separating the production operation and production preparation to improve the professionals, further improving the efficiency: For example, the management of fixtures has been changed into centralized management by a specially assigned person from decentralized management by operators, and fixtures preparation time has been saved. Also, materials delivery has been changed into delivery by a specially assigned person from being fetched by operators, and related preparation time has been saved.
- Strengthening field management, emphasizing the standard operation: To conduct layout in strict accordance with the 6S concept, the spot should be clean and tidy with clear marking. The standards are that a replacement worker can find auxiliary materials or fixtures within one minute when the specially assigned person is not on-site. Meanwhile, a simple and practical set of rules and regulations has been stipulated so as to ensure reception and dispatch of auxiliary materials or fixtures will be correct and efficient.

12.4.3 *Products meet all quality requirements, and quality level has been steadily improved*

Since implementation of the manufacturing cell for the solar array base plates, the operation is smooth, and the process route is feasible with steady technique. All solar array base plate products delivered have met the designer and user's requirements. The controlled and quality conditions are good during the product development. Major quality problems have not occurred during the product development or after the delivery, and products have met all quality goals and requirements.

Measures that greatly improved the quality level of cell products include the following:

- Standard management: With the increasing of development tasks by small batches, typical collective operation has been standardized and technique content has been detailed and fixed by the corresponding system. It is able ask workers to keep discipline as specified during operation and ensures that production of every product is produced as specified. Summarizing, fixing, and forming operation specification are effective ways to ensure product quality, and these measures have become long-term mechanisms of cells now.
- Refining operation: After refining of cells, a high repeatability of work and more efficient operation of operators has improved their skills in repeat exercises, which has ensured the product quality and promoted the steady improvement of product quality.

12.4.4   Operators' skill improved, workers' training accelerated, to create the group atmosphere

Before establishment of cells, the same operators may have needed to conduct operation upon different products. After establishment of cells, the work of operators has become more concentrated, and the corresponding training has become more specific, which has made them familiar with operational content in a shorter time and kept them improving their operation skills.

In addition, turning the traditional production mode into refined management is able to make miscellaneous procedures become clearer. The work content has also become more scientific and refined. Effective communication and management mechanisms have been established among staff, a group atmosphere has been created, and a sense of ownership has been improved. It has provided power for further improvement of productive forces.

12.5   Experience and cognition

12.5.1   Practical experience of manufacturing cell of base plates

By implementing the cell-manufacturing mode and establishing the manufacturing cell for the base plates, the production capacity has been improved without an obvious increase in manufacturing resource investment. The main experience is as follows:

- By establishing product cells to do the same production tasks intensively, to assemble the same resources intensively, to arrange layout reasonably, and to concentrate production process in a certain place, the logistics route is shorter and production efficiency is improved.
- As to production processes with a high proportion of manual work like the production of base plates, changing manufacturing mode, refining the operation process and concentrating the operation object can not only improve skill proficiency and operation maturity, but it can also accelerate workers' training, which is good for improving their skills.
- During the establishment and implementation of cell-manufacturing modes, ensuring stable and efficient operation of the cell is done by optimizing the production process, changing serial operation into parallel operation, separating the main line and auxiliary lines, improving the professionalism of each operation, and stipulating strict and standard systems and regulations.
- Cell-manufacturing mode is an effective way for aerospace enterprises to solve production bottlenecks and to respond to growing

production tasks, and it is also an important way for them to become industrialization-oriented from research-oriented enterprises. It requires a complete and systematic cellular concept, taking advantage of promoting deep application of cell-manufacturing mode, and making continuous improvement to the manufacturing capacity of aerospace products.

12.5.2   Issues that need to be focused during the practice of manufacturing cells

During application of the manufacturing cell of base plates, there are some aspects that require more attention:

- Establishing special product cells so as to meet the production requirements of corresponding products exclusively from corresponding resources. As to the fluctuant size of astronautic tasks, the plant should pay more attention to allocating equipment and workers scientifically, so as to improve the production capacity of typical products as well as to have flexibility in manufacturing.

Establishment of cells is conducted for certain task quantities. With growing production tasks, new bottlenecks and problems will appear, and scientific methods and tools need to be implemented, to analyze the influence of changes of task quantity on cell production capacity, and to conduct continuous improvement and optimization of cell configuration so as to meet new production requirements.

chapter thirteen

User-centered design

Patrick Rau

Contents

With the development of science and technology, simply providing products, functions, or technology cannot meet the needs of human beings. People pay more attention to feelings over functions of products. "User-centered" means designing according to users' requirements through the deep understanding of the users and verifying the design of the using. The success of the product relies on the consideration of user demand in multiple aspects to assist in the design of products of a high level rather than merely raising the concepts by designers. Numerous cases indicate that to meet the demand of users, functional technologies must be provided and the user experience should be satisfied, which leads to commercial success.

Take Google, for example. The first of the 10 principles of the company says, "Focus on the user and all else will follow." Google took its best user experience as its tenet upon which to build its foundation. The design and development of its Internet browser as well as the minor modifications to the appearance of the home page is done with great care, ensuring the satisfaction of user demands. Supported by a dedicated team, the strong research and development capability is user-centered and focuses on the using experience. Researchers with different professional backgrounds gather in the User Experience Research Department, work together with the involvement and feedback of users in different stages of products designed by different methods, adjust the design through the methods of collecting user advice, and survey and observe the use of products by users to make the products easier to use.

Such research teams exist in modern enterprises universally. More and more companies are regarding product design with research of the user experience as important as the research and development of the technology. The following cases in this chapter will explain how to use different research methods to design user-centered products.

13.1   Will fragrance affect mood?

The internationally renowned enterprises of bath products started the research in this field. In order to make their products (including cleanser, shower gel, shampoo, etc., that we use commonly) widely accepted by customers, their teams hope to design bath products that meet the demand of customers' psychology through research on the change of mood when smelling different fragrances. If the relationship between fragrance and emotional response is understood thoroughly, fragrance formulas that can trigger the positive emotional reaction of target users would be developed, improving the bath experience of customers through which the acceptance of the products is increased among customers.

The method of the mood test, in the past, has been filling out questionnaires by participants after the experiments. Although easy to implement, its results do not always represent mass customers due to different measuring standards of individuals. Therefore, in addition to a questionnaire, the mood reaction should be explained through measuring instruments that collect objective physiological data, which includes the myoelectric response and skin resistance. By collecting data, the error due to the different standards of participants when filling in the measuring forms can be avoided. Moreover, people in different countries react differently to the same fragrance stimulation. Therefore, the company expects to design fragrance of bath products that is in accordance with the demand and preference of Chinese customers by researching their relationship between fragrance and emotional reactions.

13.1.1   How to measure physiological reaction

The research team needs to know the purpose and the procedure of the research first. The problem they are faced with is how to measure a change of emotion. Several suitable methods for measuring emotion have been found:

1. Subjective self-report is referred to as questionnaires filled out by people to indicate the type and intensity of mood. However, such a method is of strong subjectivity as different individuals have different evaluation standards for the same mood.
2. The common methods to measure certain physiological indexes that reflect the mood changes by devices are myoelectric response and skin resistance. Myoelectric response makes use of electrical stimulation to examine nerve excitation and conduction function by which the muscle activates in response to the fragrance. Skin resistance describes the conductivity of the skin. The myoelectric response and skin resistance under different emotional states is different, so it can be measured by special instruments. Although it is more objective, it does not easily explain the change of mood by the measured physiological indexes.

After determining a measuring method, the standard experimental procedure should be designed, including the sequence of filling self-measuring forms and the measuring of myoelectric response and skin resistance to avoid data inconsistence due to different sequences of procedures. Environmental temperature and humidity of the experiment should be controlled consistently and should be operated by two fixed individuals.

After confirming the procedure, the researchers will recruit participants for the experiments. Researchers will set some standards of recruiting participants to reduce the influence of irrelevant factors on the results by screening background information. The experimental procedure is as shown in Figure 13.1. The subjects fill in the questionnaires each time after they smell a kind of fragrance, and the physiological values are recorded by devices, and finally they are asked about their overall feelings.

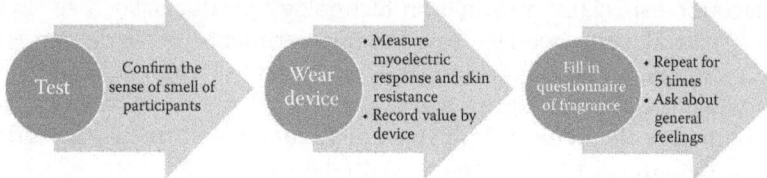

Figure 13.1 Fragrance experiment procedure for mood response.

Staging can be considered when designing the experimental procedure. Pre-experiments can be conducted before initiating formal experiments, or the experiment can be conducted directly according to the kinds and demands of the experiments. In pre-experiments, which are carried out before mass recruitment of subjects, a few subjects are invited. The current research methods and procedures are observed to see whether there is any defect or room for improvement. By observing the effect of pre-experiments, the researchers could adjust the experimental scheme in order to ensure the accuracy of data of the formal experiment.

13.1.2   What is found?

In such an experiment, the gender and number of subjects will usually be considered based on the demand of the products. More than 100 or even 200–300 subjects may be needed to obtain effective data. For the choice of gender, female subjects usually tend to be chosen for some female-oriented products. But sometimes, women will take part in an experiment of general public consumer goods or male-oriented consumer goods because of the high sensitivity to fragrance of women.

At the end of the experiment, the research results concluded from the statistical analysis indicate that most fragrance is related to the change of mood and the fragrance mainly triggers a positive mood. According to the results above, the international well-known company of bath products could understand the different mood reactions of customers to different fragrances, which would be the reference for making the aroma of different bath products, thus making the bath products more popular among customers.

In addition, there are similar experiments in research and development teams of wine, cigarettes, or perfume, through which products that are more conforming to the requirements of customers can be developed.

13.2   How to design IT products suitable for senior citizens

Design of some products seldom takes the demand of special groups into account, especially information technology products, such as smart phones. Special teams need to design user experiments for the demands of specific groups. In developed countries, such as the European Union and Japan where population aging has advanced, related basic research has been studied in the fields of industry and academia to consider design to meet the demand of special groups. With the social trend of aging, China will be faced with a lot of needs of aging society in the future.

The background of the next project in Europe, where the aging society has emerged already. As aging society is a serious problem, most people may think that the infrastructure construction for an aging society is to build nursing homes. In fact, it is not the case. The nursing home is a part of the infrastructure, but the infrastructure and other information technologies also consider the issue of aging. Because of the change and influence of the whole society, more retired elders want to work at home and look for a healthy life, for which more infrastructures are needed.

The project is a service platform that realizes their desire to live independently and improves the quality of life in all aspects, introducing a user-centered design concept. The platform integrated 12 kinds of services for the elderly, including the demands of the elderly in aspects of independent living and autonomous and intelligent work space, etc.

The global delaying of the retirement age means that elder people will continue to work, and among them more will work at home or part time, which involves relevant auxiliary information technology. China is now faced with the same problem.

The actions taken by countries entering aging society, such as European countries, the United States, and Japan are expressed in this section. Here we take traffic information and route guidance service designed for the elderly, for instance, to introduce how to design IT products suitable for the elderly (Rau, 2009).

13.2.1   How to design the user experience of the elderly

This research uses the method of focus group. The so-called focus group refers to a series of discussions on a certain topic by a small group guided by a host, from which researchers obtain useful information. Considering the characteristics of the elderly, such a research method for designing products for the elderly should not be too tedious, and the last period of the experiment should not be too long if the elderly are invited.

The design of the research procedure starts after determining the research method. The process framework is shown in Figure 13.2. First, the host carries out a warm-up activity to inform everyone of the purpose of the discussion in an easy way. The second part is case description. Design and function detail of a few representative apps for the mobile phone and route guidance service apps are introduced. The third part is personal assessment. The researchers ask the elderly people to fill in

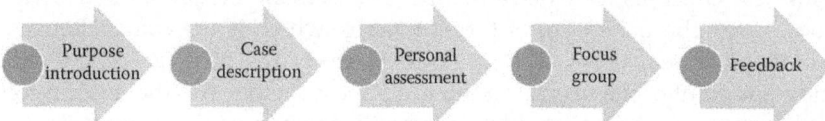

Figure 13.2 Experiment procedure for focus group.

a function assessment questionnaire about the case introduced before including the quality and necessity of the current traffic information and route guidance service functions. The results of the survey are used to distinguish the significance of functions. The fourth part, which is the most important, is to discuss the traffic information and route guidance service introduced before by the elderly; the fifth part is the subjective feedback.

13.2.2 Traffic information and route guidance service in the opinion of the elderly should be...

After the experiment, the researchers summarized the results of the questionnaires and discussions, learning factors about traffic information and route guidance that attracted the most attention of the elders.

The study summarized the demand of traffic information and route guidance service for the elderly, for whom using the guidance service in the real world or in the virtual environment are both difficult because as long as the obvious buildings change a little, they will get lost. In addition, in the aspect of using the guidance service, the elderly tend to use it on short trips as most elders in China will travel accompanied for long-distances when they are not greatly relying on the guidance service. In addition, traffic information and the route guidance service designed for the elderly will take factors, such as their eyesight and physical factors, into consideration and provide more thoughtful service and tips on health.

13.3 How to design products with a new concept for the elderly

People are drawing more attention. The study is a project carried out by Taiwan Industrial Technology Research Institute in order to ensure good health conditions for the elderly over the age of 65 years old and is aimed at exploring the life demands of the elderly and exploring the method of developing relevant technologies and producing relevant products (Liang et al., 2012).

The research team hopes to use the representative information and design four persona models representing four different life types of elderly people, which can be the tool for designers in designing products with new concepts for elderly people. The persona model is the description of real characteristics of target groups, which is the comprehensive prototype of real users. The researchers study the objectives, behavior, and viewpoint of the users and integrate these factors into the description of the users of typical products abstractly to support the determination and design of products.

13.3.1 How to develop persona

The design of a persona model needs qualitative research whose procedure is shown in Figure 13.3. First, conduct a user survey, including the three stages of questionnaire, observation, and interview. Second, establish a persona model according to the results of the user survey, and finally, design products for the representative group with the tool of rule model.

The researchers recruited about 20 elderly subjects with chronic disease at first. Then the researchers seleced half of the subjects aged from 65 to 82 according to gender, age, and life behaviors, who were suffering from chronic diseases and needed to take medication regularly. Next, the researchers sent the self-report forms to the subjects to record the use of drugs and health care instruments. The subjects were asked to fill in the self-report form every day for one week. In the following month, the research team visited each subject, during which the results of the interviews were recorded by photo and audio.

The persona models established as a result of the research are derived from the subjects of the interviews and observation. The researchers selected different representative subjects as persona models after classifying the data based on a card classification method.

When the researchers classified the data with the card classification method, they picked out groups representing different types as persona models. The so-called card classification method is to record data on a card according to certain rules and then organize ideas by people through card classification. The card classification method can be applied in fields including training, website design, visual search, and other fields. In this study, each card stands for the data collected from each subject. The researchers classified the cards of similar subjects into one group and named it and then repeated the classification procedure until no new classification method could be discovered, and finally the persona model is determined by the discussion of classification by researchers.

In order to determine the final persona model, a seminar on design was held by four designers who have master's degrees in design-related fields and working experience of more than 8 years. In order to make the designers more familiar with the topic, description, and situation

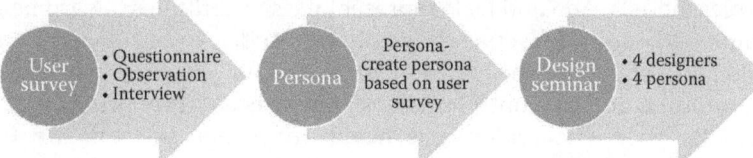

Figure 13.3 Research procedure for new concept product design.

of persona models, the research team sent the user survey data to the designers one week in advance so that they could understand the content in advance and begin to think about their design concept.

13.3.2   How to design products with new concepts for the elderly

According to the data from the self-report of the subjects and the observation of their behaviors, the subjects usually need to take drugs punctually on schedule. However, they are likely to forget to take their drugs on time due to reasons such as being busy, being out, or visits from friends. The elderly will often put their drugs in a convenient and fixed place, especially in the place associated with the use time of drugs, which acts as a reminder. For example, some subjects will put the drugs that are to be taken before or after dinner on the dinner table or in a cupboard in the kitchen; some subjects put the drugs to be taken before sleeping or after getting up in a cabinet beside the bed. Some subjects put drugs beside the kettle or the water cooler because most drugs need to be taken along with water.

It was discovered that the subjects use the medical health care instrument in three modes: occasionally without constant record, every day without constant record, and every day while recording. The children of the elderly people usually purchase these health care instruments, but they seldom are involved in the use of the instrument. The elderly people using the instrument complete the operation of measuring and recording all by themselves. If they have any doubt in reading the data indicated by the instrument, they will explore by themselves many times, and they usually will measure several times and take an average.

The forms used for recording by the elders can be divided into three formats: brochure provided by the hospital, self-made form, and blank paper to record data simply.

The result of interviews indicates that the elderly people are concerned with the following aspects of drug reminders: (1) health management, (2) help, (3) ease of preparation and use, and (4) taking their drugs on time.

The designers created four persona models after weighing gender and balance according to the result of self-report, observation, and interview, and they are named Ada, Betty, Cliff, and Dale, respectively. In the four roles models, Ada and Dale represent passive attitudes toward health care, and Betty and Cliff represent positive attitudes. In addition, Ada represents those who obey the instructions of doctors; Betty represents those who exercise regularly, control their diet or use health care products; Dale represents those who seldom use health inspection instruments; Cliff represents those who often monitor physical measurements. In terms of other qualities, Ada seldom stays at home; Betty often collects health care information; Cliff makes his own special health care tools; Dale wants to change his living habits.

In the seminar, each designer demonstrates about 10 ideas, some of which are designed for specific persona models, and some are designed for all persona models.

Sometimes it is difficult for designers to develop a concept from the data of user research, especially when the designers were not involved in the user research. In this study, the research team created the persona model derived from user research. When designers are designing conceptual products, the persona model, as an effective tool, allows designers to concentrate on user needs. The design ideas and design seminar not only provide some conceptual products to meet the needs of the elderly people after the recession of their capability but also meet their demands in multiple aspects. For example, the design of the drug reminder for Ada and the recording device for blood pressure for Cliff provide a sense of accomplishment. Applying the persona model to the design of conceptual products helps in the design of products that meet the demands in multiple aspects and levels, providing guidance for the design of actual products.

13.4 Conclusion

In the era when science and technology cannot satisfy human needs any more, user demand should be taken into consideration in more aspects; only by doing this can the high-level products that satisfy customers be designed. Through the cases mentioned above, user-centered design is not as easy and simple as imagined. It needs to undergo experiment design: instrument measurement, data analysis, meeting, and discussion. Many companies are now developing in this direction, taking the genuine needs of customers into account when developing products. Experiments and surveys can be conducted to see whether the actual needs of users and researchers and developers are consistent. When the user experience is increasingly diverse, designs that are less people-oriented will be eliminated. In the future, user-centered design will become an important indicator of product research and development.

References

Liang, S. F. M., Rau, P. L. P., Zhou, J. et al. A qualitative design approach for exploring the use of medication and health care devices among elderly persons. *Human Factors and Ergonomics in Manufacturing & Service Industries*, 2012.

Rau, P. L. P. Requirements of Transport Information Service and Route Guidance Service for Older Adults. Proceedings of 17th World Congress on Ergonomics, 2009.

Index

A

Acrylic plant, 141
Aerospace enterprise, *see* Cell
 manufacturing mode,
 implementation of (aerospace
 enterprise)
Agile manufacturing (AM), 248
AM, *see* Agile manufacturing
Auto continuously variable transmission, 92

B

Base plate products, 323
Benchmarking management (cross-
 disciplinary), 243
Bottleneck
 constraint, 185–187
 "moving around," 40
 workstation, 39, 40
Brand awareness, 225
Break-even price, 145, 176
Butadiene, 161

C

Card classification method, 353
Cause–effect map analysis, 204
Cell manufacturing mode, implementation
 of (aerospace enterprise), 319–346
 enterprise status, 319–321
 experience and cognition, 345–346
  issues that need to be focused
   during the practice of
   manufacturing cells, 346
  practical experience of
   manufacturing cell of base
   plates, 345–346

implementation and technique
 methods, 323–342
 continuous improvement of
  manufacturing cell of base
  plates, 340–342
 documents and data, 331–332
 elements and implementation
  solutions, 324–325
 environment and safety, 333–335
 equipment and facility allocation,
  327–328
 implementation and completion,
  326–340
 information conditions, 335
 logistics and storage setting,
  328–330
 production management, 336–338
 products, 323–324
 quality control, 338–340
 rule elements, 325
 skill training, 333
 sorting and analysis of products and
  techniques, 326–327
 staff and organization, 332–333
 targetless operation, 323
improvement, 342–345
 group atmosphere, 345
 production capacity, 342–343
 production efficiency, 343–344
 quality level, 344
 refining operation, 344
 standard management, 344
problems and contradictions, 321–323
 base plate products, 323
 increased task requirements, 323
 operation desk, size of, 323
China South Industries Group Corporation
 (CSGC), 228

357